THE WILSON W. CLARK
MEMORIAL LIBRARY
UNIVERSITY OF PORTLAND

Errors
in
Organizations

The Organizational Frontiers Series

The Organizational Frontiers Series is sponsored by The Society for Industrial and Organizational Psychology (SIOP). Launched in 1983 to make scientific contributions to the field, the series has attempted to publish books on cutting-edge theory, research, and theory-driven practice in Industrial/Organizational psychology and related organizational science disciplines.

Our overall objective is to inform and to stimulate research for SIOP members (students, practitioners, and researchers) and people in related disciplines including the other subdisciplines of psychology, organizational behavior, human resource management, and labor and industrial relations. The volumes in the Organizational Frontiers Series have the following goals:

- Focus on research and theory in organizational science, and the implications for practice
- Inform readers of significant advances in theory and research in psychology and related disciplines that are relevant to our research and practice
- Challenge the research and practice community to develop and adapt new ideas and to conduct research on these developments
- Promote the use of scientific knowledge in the solution of public policy issues and increased organizational effectiveness

The volumes originated in the hope that they would facilitate continuous learning and a continuing research curiosity about organizational phenomena on the part of both scientists and practitioners.

The Organizational Frontiers Series

SERIES EDITOR

Eduardo Salas
University of Central Florida

EDITORIAL BOARD

Tammy Allen
University of South Florida

Neal M. Ashkanasy
University of Queensland

Adrienne Colella
Tulane University

Jose Cortina
George Mason University

Lisa Finkelstein
Northern Illinois University

Gary Johns
Concordia University

Joan R. Rentsch
University of Tennessee

John Scott
APT Inc.

SIOP Organizational Frontiers Series
Series Editor
Eduardo Salas
University of Central Florida

Hofmann/Frese: (2011) *Errors in Organizations.*
Outtz: (2009) *Adverse Impact: Implications for Organizational Staffing and High Stakes Selection.*
Kozlowski/Salas: (2009) *Learning, Training, and Development in Organizations.*
Klein/Becker/Meyer: (2009) *Commitment in Organizations: Accumulated Wisdom and New Directions.*
Salas/Goodwin/Burke: (2009) *Team Effectiveness in Complex Organizations.*
Kanfer/Chen/Pritchard: (2008) *Work Motivation: Past, Present and Future.*
De Dreu/Gelfand: (2008) *The Psychology of Conflict and Conflict Management in Organizations.*
Ostroff/Judge: (2007) *Perspectives on Organizational Fit.*
Baum/Frese/Baron: (2007) *The Psychology of Entrepreneurship.*
Weekley/Ployhart: (2006) *Situational Judgment Tests: Theory, Measurement and Application.*
Dipboye/Colella: (2005) *Discrimination at Work: The Psychological and Organizational Bases.*
Griffin/O'Leary-Kelly: (2004) *The Dark Side of Organizational Behavior.*
Hofmann/Tetrick: (2003) *Health and Safety in Organizations.*
Jackson/Hitt/DeNisi: (2003) *Managing Knowledge for Sustained Competitive Knowledge.*
Barrick/Ryan: (2003) *Personality and Work.*
Lord/Klimoski/Kanfer: (2002) *Emotions in the Workplace.*
Drasgow/Schmitt: (2002) *Measuring and Analyzing Behavior in Organizations.*
Feldman: (2002) *Work Careers.*
Zaccaro/Klimoski: (2001) *The Nature of Organizational Leadership.*
Rynes/Gerhart: (2000) *Compensation in Organizations.*
Klein/Kozlowski: (2000) *Multilevel Theory, Research and Methods in Organizations.*
Ilgen/Pulakos: (1999) *The Changing Nature of Performance.*
Earley/Erez: (1997) *New Perspectives on International I-O Psychology.*
Murphy: (1996) *Individual Differences and Behavior in Organizations.*
Guzzo/Salas: (1995) *Team Effectiveness and Decision Making.*
Howard: (1995) *The Changing Nature of Work.*
Schmitt/Borman: (1993) *Personnel Selection in Organizations.*
Zedeck: (1991) *Work, Families and Organizations.*
Schneider: (1990) *Organizational Culture and Climate.*
Goldstein: (1989) *Training and Development in Organizations.*
Campbell/Campbell: (1988) *Productivity in Organizations.*
Hall: (1987) *Career Development in Organizations.*

Errors in Organizations

Edited by
David A. Hofmann • Michael Frese

Routledge
Taylor & Francis Group
New York   London

Routledge
Taylor & Francis Group
270 Madison Avenue
New York, NY 10016

Routledge
Taylor & Francis Group
27 Church Road
Hove, East Sussex BN3 2FA

© 2011 by Taylor and Francis Group, LLC
Routledge is an imprint of Taylor & Francis Group, an Informa business

Printed in the United States of America on acid-free paper
10 9 8 7 6 5 4 3 2 1

International Standard Book Number: 978-0-8058-6291-1 (Hardback)

For permission to photocopy or use material electronically from this work, please access www.copyright.com (http://www.copyright.com/) or contact the Copyright Clearance Center, Inc. (CCC), 222 Rosewood Drive, Danvers, MA 01923, 978-750-8400. CCC is a not-for-profit organization that provides licenses and registration for a variety of users. For organizations that have been granted a photocopy license by the CCC, a separate system of payment has been arranged.

Trademark Notice: Product or corporate names may be trademarks or registered trademarks, and are used only for identification and explanation without intent to infringe.

Library of Congress Cataloging-in-Publication Data

Errors in organizations / edited by David A. Hofmann, Michael Frese.
    p. cm.
    Summary: "This volume is dedicated to creating a single source that both summarizes what we know regarding errors in organizations and provide a focused effort toward identifying future directions for research. The goal is to provide a forum for researchers who have conducted a considerable amount of research in the error domain to discuss how to extend this research, and provide researchers who have not considered the implications of errors for their domain of organizational research an outlet to do so"-- Provided by publisher.
    Includes bibliographical references and index.
    ISBN 978-0-8058-6291-1 (alk. paper)
    1. Organizational learning. 2. Organizational behavior. 3. Errors. 4. Errors--Research. I. Hofmann, David A. II. Frese, Michael, 1949- III. Title.

HD58.82.E77 2011
302.3'5--dc22   2010049337

Visit the Taylor & Francis Web site at
http://www.taylorandfrancis.com

and the Psychology Press Web site at
http://www.psypress.com

Contents

Series Foreword ix
Preface xi
About the Editors xv
Contributors xvii

1. Errors, Error Taxonomies, Error Prevention, and Error Management: Laying the Groundwork for Discussing Errors in Organizations 1
   David A. Hofmann and Michael Frese

2. Learning Through Errors in Training 45
   Nina Keith

3. The Role of Errors in the Creative and Innovative Process 67
   Michelle M. Hammond and James L. Farr

4. Revisiting the "Error" in Studies of Cognitive Errors 97
   Shabnam Mousavi and Gerd Gigerenzer

5. Collective Failure: The Emergence, Consequences, and Management of Errors in Teams 113
   Bradford S. Bell and Steve W. J. Kozlowski

6. Team Training as an Instructional Mechanism to Enhance Reliability and Manage Errors 143
   Sallie J. Weaver, Wendy L. Bedwell, and Eduardo Salas

7. Learning Domains: The Importance of Work Context in Organizational Learning From Error 177
   Lucy H. MacPhail and Amy C. Edmondson

8. Errors at the Top of the Hierarchy 199
   Katsuhiko Shimizu and Michael A. Hitt

9. When Things Go Wrong: Failures as the Flip Side of Successes 225
   Erik Hollnagel

10. The Link Between Organizational Errors and Adverse Consequences: The Role of Error-Correcting and Error-Amplifying Feedback Processes 245
Rangaraj Ramanujam and Paul S. Goodman

11. Cultural Influences on Errors: Prevention, Detection, and Management 273
Michele J. Gelfand, Michael Frese, and Elizabeth Salmon

12. A New Look at Errors: On Errors, Error Prevention, and Error Management in Organizations 317
Michael Frese and David A. Hofmann

Author Index 327

Subject Index 341

Series Foreword

Our science is rich with theory, constructs, metrics, and methodologies that help us better understand human performance at work. While much has been accomplished in previous decades on this subject, much remains to be studied and understood, especially when the stakes for effective performance are high. Preventing and managing human errors is a goal of our science. Dealing with errors and learning from them is an imperative in our science and, of course, our practice, particularly where it matters—because lives are at risk or the economic well-being of organizations is at risk. We must provide the aviation, mining, nuclear, space, financial, oil, military, health care, and manufacturing industries (and others) with scientifically rooted solutions that prevent and manage human errors and learn from them as much as possible. That is the motivation of this volume—to develop a science that can provide a foundation for designing, developing, delivering, and evaluating interventions aimed at understanding and preventing human errors. So, in our field, this volume is a first of a kind, a much-needed volume on a topic not discussed that much in our journals, meetings, and conferences. It should be, given recent world events. This is a welcome contribution to our science and our field.

Michael Frese and Dave Hofmann have done a tremendous service to our science. They have created a well-balanced, thought-provoking, and engaging volume on human error. They have assembled a multidisciplinary set of experts with different theoretical perspectives and approaches to deal with this complex issue. There is much to learn in this book. We learn in this volume that the organizational climate matters in predicting and eliminating human error. Well-designed and -aligned human resources systems matter. The better the human-system integration is, the better the human performance will be. Leadership, of course, matters. Teams help. Training and team training help as well. The list goes on. At the end, we have a great collection of multilevel theoretical perspectives—a science—that can influence practice.

On behalf of the Editorial Board of the Organizational Frontier Book Series for the Society for Industrial and Organizational Psychology (SIOP), thanks are extended to Michael and Dave. We hope this volume motivates those in science and those in practice as well to generate evidenced-based solutions to prevent and manage human errors—the field, industries, and our society need them.

Eduardo Salas
University of Central Florida
Organizational Frontiers Book Series Editor
November 2010

Preface

Since we first met and became friends in State College, Pennsylvania, in the early 1990s, we have frequently met at conferences and exchanged research ideas—each time leaving with the same parting comment, "We should work on something together." Yet, busy schedules, overcommitment, and research that was complementary but not necessarily overlapping proved to be hindrances to any such collaboration. As my (Dave Hofmann) research on leadership and safety climate continued to grow and develop, Michael Frese's work on human error started to move from a focus on software training to the error orientations of individuals and organizational cultures. Thus, over time, our research started to move closer together. The result of this gradual merging of our research interests is this edited volume. Now, we have finally had the opportunity to "work on" something together.

The starting premise of this volume is that individuals will make errors (Reason, 1990). There is simply no getting around this fact. Yet, much of the research literature and practical advice around errors seems to indicate that if one puts in place the correct systems and designs it is possible to prevent errors from happening. Certainly, putting in place well-designed systems and human–technology interfaces can prevent some (and perhaps many) errors. Our position is that they will not be able to prevent all human error. It is like one of those party candles that no matter how many ways you blow it out, after a few seconds it rekindles. If this is the case, then organizations need to adopt not only an error prevention mindset, but also an error management approach.

Error prevention approaches are obvious—they are designed to prevent humans from committing a particular error. An error management approach, however, is perhaps not so clear. This approach assumes that individuals will make errors and builds systems to help manage errors after they occur so that they do not lead to disastrous consequences. Subtle error management systems are all around us, even though we might not recognize them as such. When your computer asks if you really are sure you want to delete a file, it is an error prevention system (i.e., "Are you sure that you want to delete this file, I mean are you really, really sure you want to delete this file?"). Forcing individuals to answer yes or no to this question is error prevention. An error management approach suggests that the software designers will assume that, despite this prevention mechanism, individuals will occasionally make a mistake and want to retrieve a file that was deleted. Hence, the invention of a "trash can" that holds deleted items where they are relatively easy to recover. This is a balance between error prevention and error management. Our approach, and the approach that we encouraged our authors to take, is to look at both sides of this coin.

There is something else about errors. Many people assume that errors are always bad: Error = Bad. Yet, there are many examples of significant discoveries occurring through "happy accidents." And, Michael's work has shown convincingly that encouraging individuals to make errors during software training sessions can lead to better learning and reduced stress when errors do occur. So, there are also positive things that come about from errors. We have encouraged our authors to think about this alternative view of errors as well.

Overview of the Volume

Michael and I begin the volume (Chapter 1) with an overview of the human error literature, looking at how we define errors, detect errors, and prevent and manage them. We also briefly discuss the notion of collective, or organizational, errors. This would be a good place to start for readers who are largely unfamiliar with this literature. In Chapter 2, Keith provides a review of the research investigating one positive aspect of errors; namely, how they can aid individual learning during training and lead to less stress when errors do occur after the training. Another area in which errors can play a central, and often positive role, is in creative and innovative efforts. Hammond and Farr (Chapter 3) discuss not only how errors can lead to creative efforts but also how errors can occur during innovation. Mousavi and Gigerenzer (Chapter 4) describe how the traditional, experimental approach to documenting various "cognitive biases" has not sufficiently taken into account the practical goals of human judgment (ecological validity). They forward the idea of ecological rationality as the matching of decision-making strategies to the structure of information in the environment. This perspective is then applied to efforts to "debias" decision-making efforts with specific application to the interpretation of medical statistics.

Bell and Kozlowski (Chapter 5) start a series of chapters focused more on the implications of errors for groups and teams in organizations. They describe how errors in teams develop, the consequences they may have, and how they should be managed. In Chapter 6, Weaver, Bedwell, and Salas describe how team training can increase reliability and improve error management. They review a number of different team-training techniques and describe how they link to improving team reliability. MacPhail and Edmondson (Chapter 7) suggest that the work context is a key determinant in identifying how best to learn from errors. Given that errors are inevitable, it is critically important that individuals, units, and organizations effectively learn from these errors. This not only feeds into future error prevention efforts but also can lead to the development of improved error management systems.

Chapter 8 shifts our focus to the top floor of the organization. In this chapter, Shimizu and Hitt discuss errors made at the strategic level. These errors are particularly difficult to identify because of the inherent uncertainty in the broader context of the decision. This context often provides unclear signals, long time lags, and no clear benchmarks against which to judge the outcome of the decision. Hollnagel (Chapter 9) describes three ages of safety—the ages of technology, human factors, and safety management—and how each relates to the attribution of failure and, relatedly, the way in which each age dealt with these failures. The last age, the age of safety management, is characterized by the concept of the resilient organization, which becomes increasingly necessary as systems become more complex and less tractable. Ramanujam and Goodman (Chapter 10) take up the issue of the relationship between organizational errors and adverse consequences to the organization. One primary focus is the notion of latent errors or organizational deviations from rules and operating procedures that can potentially generate adverse outcomes when they interact with triggering events. They further describe, from a systems viewpoint, the antecedents of organizational feedback processes for error correction and error amplification. In Chapter 11, Gelfand, Frese, and Salmon examine the cross-cultural implications for both error prevention and error management. A number of cross-cultural dimensions (e.g., uncertainty avoidance, humane orientation, individualism, tightness-looseness) are discussed and how they might have an impact on error prevention, error detection, and error management. Discussed are a number of cultural paradoxes that result from such a conceptualization.

We conclude the volume with Chapter 12 in which we recap the various contributions, provide some observations about different points of connection, and describe a way to move forward with research on errors in organizations by providing an integrative view of the actor's behavior and the organization's response.

Acknowledgments

We would first like to thank our chapter authors, who have contributed creatively and constructively to this volume. We believe that the chapters herein move the field forward and provide a number of interesting ideas for future research. We would also like to thank the Fulbright Scholar

Program administered by the Council for International Exchange of Scholars, which enabled David to spend 4 months in Germany working with Michael at the University of Giessen. We would also like to thank Anne Duffy, senior editor at Psychology Press/Routledge of the Taylor & Francis Group. Her patience and gentle prodding to keep the volume moving forward were both necessary and extremely critical. Andrea Zekus, also of Psychology Press, provided invaluable editorial assistance as the volume traveled down the home stretch toward publication. Finally, we would like to thank the Society for Industrial and Organizational Psychology for sponsoring the Frontiers Series. Over the years, these volumes have had a tremendous influence on the field. Our hope is that this volume can continue the trend.

Reference

Reason, J. (1990). *Human error*. New York: Cambridge University Press.

About the Editors

David Hofmann is Area Chair and Professor of Organizational Behavior at the University of North Carolina at Chapel Hill's Kenan-Flagler Business School. His research—focused on leadership, organizational climate, multi-level theory/methods, safety, and human error—has appeared in *Academy of Management Journal*, *Academy of Management Review*, *Journal of Applied Psychology*, *Journal of Management*, *Organizational Behavior and Human Decision Processes*, and *Personnel Psychology* and other outlets. He teaches courses in organizational behavior and leadership and was formerly Associate Dean for the full-time MBA program.

David's research on leadership, safety, and human error in organizations led to a corporate partnership with Behavioral Science Technologies where he helped develop a cultural assessment tool that has now been completed by over 200,000 employees in 1,000 companies (including all NASA employees after the Space Shuttle Columbia accident in 2003). In 2006, he received the American Psychological Association's Decade of Behavior Research Award. The APA honored the practical application of his research investigating leadership issues in high-risk industries, and he presented his findings at a Congressional briefing. Dr. Hofmann also received the Yoder-Heneman Research Award from the Society of Human Resource Management and was a Fulbright Senior Scholar. Currently, he is a member of the National Research Council/National Academy of Science's committee investigating the BP Deepwater Horizon accident.

Dr. Hofmann earned his PhD in industrial and organizational psychology from Pennsylvania State University, a master's degree in industrial and organizational psychology from the University of Central Florida, and a bachelor's degree in business administration from Furman University.

Prof. Dr. Michael Frese received his Diploma and Doctorate from the Free University of Berlin and Technical University Berlin, respectively, and holds a joint appointment at the National University of Singapore, Business School and Leuphana University of Lueneburg (Germany). Before that, he held a chair for work and organizational psychology at University of Giessen and also taught at London Business School.

Prof. Frese's research spans a wide range of basic and applied topics within organizational behavior and work psychology. Most important are his longitudinal studies on psychological effects of unemployment, impact of stress at work, predictors of personal initiative, as well as psychological success factors of entrepreneurs. He is also known for his cross-national research on innovation. His field studies on errors, error management, and error management culture also have received wide attention. In addition, he is studying training—most importantly the concept of error management training, leadership training, and psychological training for increasing entrepreneurial success and personal initiative. Most recently, he has conducted studies on cultural factors in organization and across nations, research that looks at psychological success factors in entrepreneurs in developing countries (Africa, Latin America, and Asia) and in Europe.

Frese has authored more than 250 articles and was editor/author of more than 20 books and special issues. He was elected Fellow of the Society for Industrial and Organizational Psychology (Division 14 of the American Psychological Association). He is Germany's most frequently cited work and organizational psychologist and business and management scientist and among the most frequently cited Europeans. He has given more than 25 invited keynote addresses, including presentations at the International Congresses.

Additionally, Prof. Frese was President of the International Association of Applied Psychology and President (speaker) of the division Work and Organizational Psychology in Deutsche Gesellschaft für Psychologie. He also served as Editor of *Applied Psychology: An International Review*, Co-Editor of *Psychologische Rundschau*, as an editorial board member of various book series (Entrepreneurship Series in Germany, SIOP Organizational Frontiers Series, Organization and Management Series [Routledge]), and member of several boards of journals.

Prof. Frese now serves as field editor for *Journal of Business Venturing*. He is also a consultant and lecturer to the management of many companies (among others banks, technology firms, automobile, utilities, telecommunication, and computer industry) with more than 200 talks given and consulting jobs done.

Contributors

Wendy L. Bedwell
Department of Psychology, and
  Institute for Simulation &
  Training
University of Central Florida
3100 Technology Parkway
Orlando, Florida

Bradford Bell
Industrial and Labor Relations
  School
Cornell University
Ithaca, New York

Amy Edmondson
Harvard Business School
Harvard University
Boston, Massachusetts

James Farr
Department of Psychology
The Pennsylvania State
  University
University Park, Pennsylvania

Michael Frese
Department of Management &
  Organization
NUS Business School
National University of Singapore
Singapore

Michele J. Gelfand
Department of Psychology
University of Maryland
College Park, Maryland

Gerd Gigerenzer
Center for Adaptive Behavior and
  Cognition
Max Planck Institute for Human
  Development
Berlin, Germany

Paul Goodman
Graduate School of Industrial
  Administration
Carnegie Mellon University
Pittsburgh, Pennsylvania

Michelle Hammond
Department of Personnel &
  Employment Relations
Kemmy Business School
University of Limerick
Limerick, Ireland

Michael A. Hitt
Department of Management
Mays Business School
Texas A&M University
College Station, Texas

David A. Hofmann
Kenan-Flagler Business School
University of North Carolina at
  Chapel Hill
Chapel Hill, North Carolina

Erik Hollnagel
Institute of Public Health
University of Southern Denmark
Odense, Denmark
and
MINES ParisTech
Sophia Antipolis, France

Nina Keith
Technical University of Darmstadt
Organizational and Business
    Psychology
Darmstadt, Germany

Steve W. J. Kozlowski
Department of Psychology
Michigan State University
East Lansing, Michigan

Lucy MacPhail
Robert F. Wagner Graduate School
    of Public Service
New York University
New York, New York

Shabnam Mousavi
Department of Finance
J. Mack Robinson College of
    Business
Georgia State University
Atlanta, Georgia

Rangaraj Ramanujam
Owen Graduate School of
    Management
Vanderbilt University
Nashville, Tennessee

Eduardo Salas
Department of Psychology, and
Institute for Simulation &
    Training
University of Central Florida
Orlando, Florida

Elizabeth D. Salmon
Department of Psychology
University of Maryland
College Park, Maryland

Katsu Shimizu
Graduate School of Business
    Administration
Keio University
Yokohama, Japan

Sallie J. Weaver
Department of Psychology, and
Institute for Simulation &
    Training
University of Central Florida
Orlando, Florida

1

Errors, Error Taxonomies, Error Prevention, and Error Management: Laying the Groundwork for Discussing Errors in Organizations

David A. Hofmann and Michael Frese

Every organization is confronted with errors; these errors can result in either positive (e.g., learning, innovation) or negative (e.g., loss of time, poor-quality products) consequences. On the positive side, errors can lay the foundation for outcomes such as innovation and learning (e.g., Sitkin, 1992). For example, both Edmondson (1996) and van Dyck, Frese, Baer, and Sonnentag (2005) found that a positive and constructive approach to errors is associated with organizational outcomes such as learning and performance. With regard to the negative aspects of errors, the majority of the attention within the organizational sciences has focused on the investigation of highly salient and visible organizational failures (e.g., *Challenger*, *Columbia*, Chernobyl; Perrow, 1984; Reason, 1987; Starbuck & Farjoun, 2005; Starbuck & Milliken, 1988a; Vaughan, 1996). These investigations have taught us a great deal about how many seemingly independent decisions, actions, and organizational conditions can become interconnected and create extreme failure.

These extreme examples, however, do not really capture the lion's share of errors occurring within organizations. Individuals working in organizations make errors every day and every hour and (sometimes) make multiple errors in the span of a minute. Researchers, for example, have estimated that for some computer tasks, up to 50% of work time is spent on error recovery (Hanson, Kraut, & Farber, 1984; Kraut, Hanson, & Farber, 1983; Shneiderman, 1987), and Brodbeck, Zapf, Prümper, and Frese (1993) found that 10% of computer work time is spent handling and recovering from errors. Other computer-based research suggested that individuals average 18 unnecessary cursor movements per hour (Floyd & Pyun, 1987).

Research investigating the use of spreadsheets within organizations also provides evidence regarding the large number of errors contained in these applications. For example, it has been suggested that between 20% and 40% of all spreadsheets in use within organizations contain errors (Panko, 1988, 2005). As a case in point, Davies and Ikin (1987) found—after inspecting 19 spreadsheets used in 10 different firms (dealing with issues such as project cost, payroll, loan schedules, and short-term money market investment analysis)—that 25% contained serious errors. Two such errors were a mistaken transfer of $7 million in funds between divisions and inconsistent currency conversions. Along similar lines, Lawrence and Lee (2004) audited 30 spreadsheets used to justify the financing of projects. They found, on average, that 7% of the spreadsheets contained errors, and that it took an average of six iterations before they were fully error free (see Panko, 2005).

Although many of these "smaller" errors occurring within organizations are quickly handled and rectified, sometimes they can create significant negative consequences. For example, Smelcer (1989) estimated that an error in the command code of the Structured Query Language resulted in a loss of $58 million per year (based on the estimated time for error recovery). The loss of the $125 million *Orbiter* spacecraft provides another example. In this case, the postincident investigation board determined the underlying reason for the loss was a failure to convert several calculations from English measures of force to newtons.

Despite the importance and prevalence of errors within organizations, there has been no attempt within the field of industrial and organizational psychology or organizational behavior to create a single source that both summarizes what we know regarding errors in organizations and provides a focused effort toward identifying future directions for research. The goal of this volume is to address this gap by providing a forum for researchers who have conducted a considerable amount of research in the error domain to discuss how to extend this research and provide researchers who have not considered the implications of errors for their domain of organizational research an outlet to do so. Our goal of this first chapter is to provide those not familiar with the error literature an overview of this literature. We begin by defining errors and differentiating errors from other related terms. We then describe a goal-directed view of behavior within organizations. Next, we turn to a discussion of an error taxonomy specifying which types of errors might be expected at the different stages of goal-directed behavior. After discussing the different types of errors, we consider the challenges involved in error detection. Following this, we transition into a discussion of collective errors. We conclude by discussing the distinction between error prevention and error management along with some thoughts regarding how to implement each of these activities.

Errors, Inefficiencies, Violations, and Risk

The *Merriam-Webster Online Dictionary* (2006) defines an *error* as an act or condition of ignorant or imprudent deviation from a code of behavior; an act involving an unintentional deviation from truth or accuracy; and an act that through ignorance, deficiency, or accident departs from or fails to achieve what should be done. There are several key ideas nested within this definition worth highlighting. First, an error only occurs when there is a deviation from something else. In other words, classifying something as an error implies "an error compared to what"; the "what" in this case is some external goal, standard of behavior, or truth. Second, an error is an *unintended* deviation. Third, an error can come about through different mechanisms. For example, an individual may not know the standard (ignorance) or may fail to enact his or her intention successfully (e.g., I intended to hit the nail with the hammer but erroneously hit my finger).

As described in more detail in this chapter, we assume that behavior in organizations consists of goal-oriented action. Thus, errors imply a nonachievement of goals; the successful accomplishment of these goals would be the intention. Reason (1990) noted that actions should not be classified as an error if they are brought about by some chance agency. For example, if a person is prevented from achieving a goal due to a lightning strike that results in a temporary power outage, this should not be classified as an error. If errors are not brought about by some chance agency, then the individual was—at least theoretically—in control of his or her action; therefore, errors are potentially avoidable. Taking into consideration these various aspects of errors, we have chosen to define *actions* as erroneous when they unintentionally fail to achieve their goal if this failure was potentially avoidable (i.e., did not arise from some unforeseeable chance agency; Reason, 1990; Zapf, Brodbeck, Frese, Peters, & Prümper, 1992).

In light of this definition of errors, we can now discuss several related concepts, such as inefficiencies, violations, and risk. Errors can be differentiated from inefficiencies because inefficient pursuits do in the end reach the goal. Thus, inefficient actions seem not to meet our definition of an error. However, if we assume that actions occurring in organizations typically include efficiency as part of the broader goal, then inefficiencies would be errors. In other words, if the goal of an action is to achieve some end result efficiently, then inefficient routes do reflect deviations from this standard and therefore would be classified as erroneous.

Viewing errors as unintentional deviations differentiates them from intentional deviations from standards or goals. Specifically, we operationalize *violations* as *intentional* deviations from task goals, rules, or some standard (Reason, 1990). Although when viewed in isolation it is difficult to imagine purposely deviating from a standard or goal, when one considers the pursuit of various goals simultaneously it is easier to imagine such intentional deviations. This is particularly the case with actions in organizational settings in which any given action often involves the pursuit of multiple (and often contradictory) goals simultaneously. A chemical plant worker, for example, may seek to repair a faulty electrical system with, at its most basic level, the goal to get the system working. Yet, this overall goal actually involves the pursuit of multiple goals, such as performance (get it working), quality (so that it works not only in the short term but also in the long term), efficiency, and safety. We believe that most violations within organizations occur when a lower-priority goal is sacrificed to pursue more vigorously a higher-priority goal. For example, the maintenance worker may maximize the efficiency goal by *intentionally* violating safety standards (e.g., by not following accepted protocol regarding lockout and tag-out procedures). Similarly, an individual late for a meeting across town may *intentionally* violate the highway speed limit to arrive at the meeting on time (a higher-priority goal). Of course, it is possible for individuals to engage in violations for other reasons as well, with this deviance having a more malicious intent (Griffin, O'Leary-Kelly, & Pritchard, 2004) or sensation seeking being one of the contributing factors. That said, however, we believe that intentional violations based on differential goal priority (e.g., performance receiving a higher priority than safety) will be the most frequent cause of violations.

It is also worth mentioning that simple observation of actions often does not allow differentiating errors from violations. For example, the observation of cars speeding on the highway in violation of the speed limit might be indicative of either a violation or an error. The drivers may, in fact, be late for an important meeting and be intentionally violating the speed limit, which they view as a lower-priority goal when compared to arriving on time. Or, they may simply be unaware that they are violating the speed limit because they are engaged in conversation with another passenger (i.e., an error due to inattention).

A number of investigations at the individual level have reinforced this distinction between errors and violations (e.g., Kontogiannis, Kossiavelou, & Marmaras, 2002; Lawton, 1998; Reason, Manstead, Stradling, Baxter, & Campbell, 1990). Yet, even though errors and violations are distinct, they can and often do interact with each other. A number of investigations of large-scale organizational accidents—for example, investigations of British Rail accidents, Chernobyl, among others—have revealed that many of these incidents involved a combination of both errors and violations (Reason, 1987, 1990).

Finally, we mention here the concept of *risk*. One could ask whether a well-thought-out, calculated risk that subsequently turns out to result in harm is an error. Suppose a broker does the normal due diligence research and, based on this research, invests in a particular stock with the goal of making a positive return on the investment. Now, suppose the stock subsequently goes down in price such that the observed outcome (negative return on investment) deviates from the original goal. Would this be classified as an error? There is clearly a deviation from the original goal that is unintended, so in these two respects, this outcome fulfills our definition of error. But, the third aspect of our error definition—that it should have been potentially avoidable—seems to be something that differentiates errors from risks. Errors are things that after they occur give us the feeling of "we should have known better," whereas risks that turn out to be harmful are more likely to give us the feeling of "given the same information at that same time and in the same context, I would make the same decision."

In light of this definition of risk, we believe it is possible to draw distinctions between errors and risks as well as discuss the relationship between violations and risks. Errors occur when there is an unintended deviation from a goal or standard; the factors causing this deviation were potentially avoidable (i.e., under the control of the individual). Risks seem to reside in the objective situation. In principle, risks can be analyzed before an action is started or decision is made. Thus, individuals engage in actions that involve risk knowing that the situation has the potential to result in harm, but they believed the probability of this harm actually occurring was lower than the gain. However, it is possible that people can miscalculate the risks inherent in the situation. This may be the case as technology and other improvements reduce the objective risk of the situation. In particular, risk homeostasis theory suggests that often individuals increase their risky behavior as technology and other system improvements reduce the objective risk (e.g., as technological improvements increase the safety of automobiles, individuals drive faster; Pfafferott & Huguenin, 1991; Stetzer & Hofmann, 1996; Wilde, 1982, 1988).

This seems to be the point at which violations and risk begin to interrelate. In other words, engaging in intentional violations often seems to involve assessments of risk. For example, individuals may intentionally violate traffic laws because they believe that the risk of being caught is low. In other words, they recognize that factors outside their control (a police officer) might result in a negative outcome (a fine), but they view the likelihood of this factor occurring to be relatively small in light of the potential benefits brought about by achieving the higher-priority goal (arriving on time). This highlights the fact that often individuals engage in violations when they perceive the risks of negative outcomes to be minimal (Reason, 1990). Of course, nonviolation behavior can also carry certain risks as well. The decision to invest in stocks carries with it the assumption of risk that sometimes bad outcomes (deviations from goals) occur due to the inherent risk in the chosen alternative. Negative outcomes resulting from the assumption of risk are not errors. Of course, if the risks are calculated incorrectly—due to, say, an error in a spreadsheet—and more risk is assumed than the actor believes is the case, then an error has occurred. We, like others, assume that the interaction of incorrect risk calculations (which is an error) coupled with violations can produce catastrophes more easily. The Chernobyl disaster is a case in point. In this disaster, highly skilled operators conducted an experiment in the middle of the night (a high-risk situation) that involved several other decisions that violated accepted safety protocol. This risky experiment coupled with several violations and other errors of judgment resulted in the most significant accident in nuclear power plant history (Dörner, 1996; Reason, 1987, 1990).

Action Processes as the Foundation of Individual Behavior in Organizations

Now that we have defined errors and related terms, we turn our attention to an integrative error taxonomy. As noted, actions can only be defined as erroneous when there is some referent goal against which to compare the outcome (i.e., an error compared to what?). This brings us to what we believe is the defining feature of behavior occurring within organizations: goal-directed action (Dörner & Schaub, 1994; Frese & Zapf, 1994; Hacker, 1985, 2003; Locke & Latham, 2002; Norman, 1981). This goal-directed action involves several distinct steps (Dörner & Schaub, 1994; Frese & Zapf, 1994; Locke & Latham, 2002): (a) goal development and decisions among competing goals; (b) information search to understand the environment in which one acts, and the prognosis of future states (this is particularly important in dynamic systems that change over time even if a person does not intervene, e.g., an organization); (c) generation of a plan to execute the goal; (d) execution of the plan coupled with monitoring the actions involved in this plan execution; and (e) processing feedback emerging from these monitoring activities as well as feedback that provides information at the end of an action cycle (e.g., regarding the success or failure of the action). These action steps do not imply that there is an immutable sequence; people may jump ahead and start an action before planning, notice that they have to redevelop their plans of action or have to modify their goals, and so forth.

The execution of any goal requires the generation and coordination of a large number of subactions and tasks. The coordination of these subactions requires higher-order regulation or some higher-level grammar of action (Miller, Galanter, & Pribram, 1960). In other words, the specification of a goal triggers a set of hierarchically ordered actions such that when all of the subactions are successfully completed, the overall goal is achieved. For example, the goal of driving to the market to purchase bread would entail a number of subactions, such as starting the car, driving to the market, identifying bread to be purchased, purchasing the bread, returning home. Each of these subactions (e.g., starting the car) would be broken down into a series of subactions that, once successfully completed, would achieve the subgoal. Of course, ongoing feedback could result in adjustments and changes to these action sequences along the way. While all of these hierarchically ordered actions are being executed, the overall pattern of actions and progress toward the macrogoal is monitored via a higher-order regulatory process (e.g., periodic attentional checks to make sure all is progressing according the overall plan).

This hierarchy of actions and regulatory processes can be decomposed into four different levels (displayed in Figure 1.1). The three lower levels of regulation address how we initiate actions and subactions in pursuit of the goal. They vary from a conscious level of regulation to a nonconscious level of regulation in which automatized operations are performed (sensorimotor level). These three levels of regulation have been suggested by a number of authors in slightly differing ways (Ackerman & Cianciolo, 2000; Anderson, 1983; Frese & Zapf, 1994; Hacker, 1998; Rasmussen, 1983; Shiffrin & Schneider, 1977). The lowest level—that is, the nonconscious—is the *skill level*. At this level, skill-based scripts, or automatic movement sequences, are executed without ongoing and active conscious attention. The ongoing regulation of these actions takes place predominantly at the nonconscious level through proprioceptive and exterioceptive feedback.

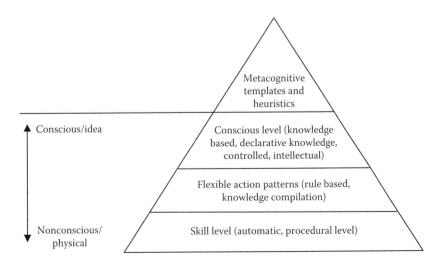

FIGURE 1.1
Four levels of action regulation.

Even though there is no ongoing, conscious regulation, there may be the need for periodic, purposeful (i.e., conscious) attentional checks to either ensure that the actions are consistent with the overall plan or interrupt the dominant response for a less-dominant response at the appropriate time. The latter type of attentional check would be necessary, for example, if an individual intends to make a stop at the store on the way home from work. In other words, for this less-dominant response to occur, the skill-based actions would need to be interrupted so that a left turn is made into the store parking lot instead of the more dominant right turn. It is also the case, however, that conscious attention—particularly if mistimed—may have a negative impact on skill-based actions because it can disrupt the elegance and ease of the ongoing automated performance (Kahneman, 1973).

In contrast to the largely nonconscious regulation occurring at the skill-based level of regulation, the *conscious level* of regulation is defined by more effortful and active cognitive processing. Actions occurring at this level typically involve active processing and consideration of overall goals and how subactions need to be organized to accomplish the goal. Thus, this level of regulation is enacted during the development of new action patterns and when individuals are faced with novel, unfamiliar situations for which behaviors must be planned "online" using conscious, analytical reasoning based on stored knowledge (Reason, 1990). Processing at this level is effortful, slow, and resource limited and proceeds serially, interpreting feedback step by step (Kahneman, 1973).

We refer to the level between the nonconscious skill level and the more effortful conscious level as the level of *flexible action patterns*. These flexible action patterns—or, as Norman (1981) described them, schemata—are ready-made action sequences triggered by situational cues and flexibly adjusted to the situation. Reason (1990) discussed these flexible action patterns as "if-then" statements for familiar problems; these action patterns are not completely automated routines but rather constitute routine use of rules. Rasmussen (1987) used the term *rule based* for this level of regulation. Unfortunately, rules as if-then statements are needed on each level of regulation (cf. Anderson, 1983), so we prefer the term *flexible action patterns* for this level. An example of a flexible action pattern would be when an individual is initiating a maintenance checklist on a machine. All of the required steps are routine, but they have not and are not executed frequently enough to become fully automated, skill-based routines. So, the individual would be guided by a series of rules or steps (e.g., a checklist of actions) for which conscious information is needed as input to perform these routine actions. Writing routine letters to new customers who require different treatment would be a similar process. Although none of the individual steps requires full conscious processing, new information has to be added; therefore, the action is not fully automated. In sum, these action patterns allow individuals to adapt flexibly to the situation (e.g., a new set of instructions, a checklist that is not frequently performed, cooking a meal) for which many of the subactions involve automated scripts, but the scripts are put together in unique ways to respond to an infrequent or new situation.

In addition to these three levels of regulation, Frese and Zapf (1994) suggested an overarching metacognitive or heuristic level of regulation. This level of regulation controls how individuals go about formulating their goals, information searches, plans, and feedback processes (Brown, 1987; Flavell, 1979). On this level of regulation, processes can be either conscious or nonconscious (automatized). An example might be that when specifying an overall goal, some individuals develop detailed plans while others develop superficial plans. Frese and Zapf (1994) and Frese, Stewart, and Hannover (1987) proposed that many of these individual differences are driven by metacognitive strategies for, in this case, plan development, and that individuals may either be consciously aware of their metacognitive approaches to situations (i.e., "Yes, I know, I rarely develop detailed plans") or be largely not consciously aware of these overarching processes (i.e., tacit knowledge; it is just the way they do things with little conscious thought).

To this point, we have described the action regulation as a top-down process; that is, an individual formulates a goal and then enacts a hierarchically ordered set of actions and subactions to accomplish the goal. It is also the case, however, that there is a bottom-up aspect to action regulation that appears when either things go wrong (e.g., in the case of an error) or feedback suggests that actions are not proceeding according to the plan (i.e., feedback suggests that the plan is not going to be achieved). In such a case, people tend to go upward and become more conscious in their action regulation (Lord & Levy, 1994). In this view of the action process, skill-based actions are proceeding unconsciously with only periodic checks to ensure that the action is still in keeping with the plan and overall goals. If feedback suggests that things are not going according to plan, the typical first step is to try to initiate flexible action patterns (Reason, 1990). Initially, individuals attempt to match the feedback to existing action plans for recovery. For example, if an individual engages an attentional check on the drive home from work and realizes that he or she just missed the turn-off to the grocery store (where the goal was to pick up some bread on the way home), then, there is likely to be an existing flexible action pattern that facilitates recovery from this error (e.g., proceed to the next street, turn left, take back road to store, etc.).

If existing action patterns do not return the individual to a point at which skill-based scripts can be reengaged or when there is no existing action specifically matched to the feedback, then the typical next step is for individuals to use analogies (Reason, 1990). Here—instead of matching existing action patterns to the specific feedback—individuals consider analogous situations and attempt to apply action patterns from analogous situations to see if this resolves the problem and returns them to a path of successful goal pursuit. This application of action patterns from analogous situations may cycle through several different iterations (as the individual considers different, but still analogous in some way, situations).

If attempts to resolve the problem through flexible action patterns are not successful, individuals engage in full-scale conscious regulation (as they do in general when a unique and new response needs to be developed for a situation). This involves a new compilation of actions by which the pursuit of this action remains regulated at the conscious level. If this successfully resolves the problem, then more routinized goal pursuit can be resumed. If it does not, then the individual will continue to develop a unique response (often by seeking help from others). Although we believe that our description captures the typical process discussed in the error literature (Rasmussen, 1987; Reason, 1990), there will be exceptions to this process. For example, if the feedback emerging from a skill-based action is so unique and significant, then an individual may switch directly to full conscious regulation.

Error Taxonomy

To both understand and make practical recommendations on how to prevent or manage errors, it is necessary to identify the various types of errors that individuals are likely to commit at each stage of the goal process and at different levels of regulation. As it turns out, there are many taxonomies of errors in the field, for example, those provided by Rasmussen (1987), Reason (1990), Norman (1984), and Heckhausen and Beckmann (1990). Unfortunately, these taxonomies have not typically been empirically validated.

A good taxonomy should be built on a theoretical foundation and have taxons that are clearly distinguishable from each other. It also should be moderately fine-tuned so that it is theoretically and practically meaningful and that empirical differentiation is possible. From a theoretical perspective, the taxonomy presented here is founded on action theory and the cognitive regulation of these actions (Frese & Zapf, 1994); it also builds on and extends error taxonomies presented by Rasmussen (1983), Reason (1990), and Norman (1984).

Frese and Zapf (1994) and Zapf et al. (1992) have developed and validated a taxonomy based on the action theory described here (cf. Figure 1.2). This taxonomy is structured in two-dimensional space. One dimension contains the different levels of cognitive regulation discussed previously in this chapter (i.e., ranging from nonconscious to conscious with the

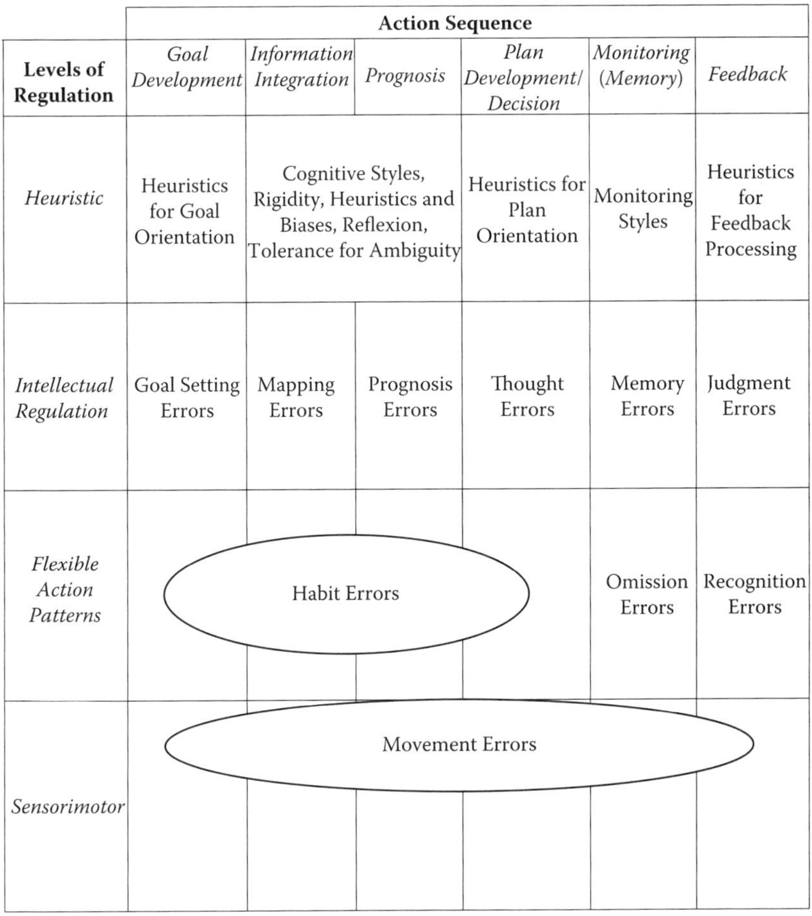

	Action Sequence					
Levels of Regulation	Goal Development	Information Integration	Prognosis	Plan Development/ Decision	Monitoring (Memory)	Feedback
Heuristic	Heuristics for Goal Orientation	Cognitive Styles, Rigidity, Heuristics and Biases, Reflexion, Tolerance for Ambiguity		Heuristics for Plan Orientation	Monitoring Styles	Heuristics for Feedback Processing
Intellectual Regulation	Goal Setting Errors	Mapping Errors	Prognosis Errors	Thought Errors	Memory Errors	Judgment Errors
Flexible Action Patterns		Habit Errors			Omission Errors	Recognition Errors
Sensorimotor		Movement Errors				

FIGURE 1.2
A General Taxonomy of Errors. (Adapted from: Frese, M. & Zapf, D. Action as the core of work psychology: A German approach. In H. C. Triandis, M. D. Dunnette, & L. M. Hough (Eds.), *Handbook of industrial and organizational psychology* (2nd ed., Vol. 4, pp. 271–340). Palo Alto, CA: Consulting Psychologists Press. Used by permission.)

addition of the metacognitive or heuristic level). These different levels of regulation can be crossed with the different steps involved in goal specification and execution (i.e., goal development, information search, prognosis, plan generation, monitoring of the execution, and processing of feedback). In addition to these different levels of regulation, a knowledge base is needed for successful regulation.

Perhaps the most straightforward type of error is a knowledge error. Errors at the knowledge level of regulation occur because an individual does not know certain facts, does not have adequate knowledge of appropriate procedures, or lacks an adequate cognitive representation of the situation (i.e., an inadequate or inaccurate mental model). Given that this knowledge is used in every aspect of the action cycle (i.e., goal development to feedback), knowledge errors can occur at any point of the action process.

We think of the metacognitive and heuristic level of the cognitive level of regulation as responsible for what kind of goals and plans are developed, which information search strategies are used, and how feedback is processed. The metacognitive and heuristic level of processing dictates the superordinate rules or processes that individuals use in the development of goals, searching for information, and processing feedback. So, for example, an individual might approach all problems using a similar macroprocess of information search and problem diagnosis. In particular, one individual might enact a detailed information search process, whereas another individual might simply rely on their "gut" feeling or intuition. Another distinction might be the extent to which individuals engage in self-reflection during problem-solving activities (Dörner, Kreuzig, Reither, & Stäudel, 1983). We believe also that some of the decision-making heuristics and biases would operate at this level (e.g., framing and anchoring heuristics [Kahneman, Slovic, & Tversky, 1982]; promotion versus prevention focus [Higgins, 1997]; and action styles [Frese et al., 1987]). It is important to emphasize that these metacognitive and heuristics of style and approach are different from the *content* of the knowledge used in the action cycle, which resides in the knowledge base available for regulation.

With regard to errors at the other three levels of regulation, errors at the conscious level are largely due to bounded rationality (March & Simon, 1958) and resource limitations of conscious information processing (Norman, 1981). These errors typically take the form of goals and plans that are inadequately developed, wrong decisions made regarding the required subactions, or misinterpretations of environmental data that lead to the development of inadequate plans. Errors in goal setting occur when goals are inadequately developed, for example, when vague criteria are used to decide whether a goal is achieved. Mapping and prognosis errors occur when the gathering, integration, and elaboration of information is not done correctly (Dörner, 1996). This might involve plans that are inadequately developed or incorrect decisions are made regarding the assignment of plans and subplans. Thought errors (i.e., incorrect thoughts regarding goal execution) occur when the long-term and side effects are not considered (e.g., not considering long-term opportunities or problems). Memory errors—or errors of remembering or forgetting—occur when a certain part of the plan is forgotten and not executed, although the goals and plans were initially correctly specified. Judgment errors appear when one cannot understand or is not able to interpret a certain aspect of the feedback.

Errors at the level of flexible action patterns occur when well-known actions are performed. Errors at this level of regulation typically involve misclassifying the situation, which results in the initiation and application of the wrong action sequence or when the action sequence that is specified is incorrectly recalled (i.e., the wrong action sequence is correctly performed or the right action sequence is incorrectly performed; Reason, 1990). Habit errors imply that a correct action is performed in a wrong situation. A person who switches to a new computer system, for example, often shows habit errors (correct procedures for the first system are incorrect for the second one). Another example of a habit error is the tendency to use routines even if they are not adequate or appropriate (e.g., in cross-cultural situations when a person uses responses that are adequate in their home culture but inadequate in a different culture). Frequency gambling, as discussed by Reason (1990), is an important contributor to habit errors. In other words, individuals enact responses that have been correct most of the time and look like they "may work" in a new situation, but in fact they turn out not to address the new situation adequately. Omission errors appear when a person does not execute a well-known subplan. An example may be that a person forgets a step of a routinized plan after he or she was disturbed by a telephone call. Recognition errors occur when a well-known message or some feedback is not noticed or is confused with another one.

At the nonconscious level of regulation, errors largely consist of movement errors. Reason (1990) noted that most errors at this level occur due to inattention—failure to monitor actions at certain key junctures—or overattention, which entails attending to action at the wrong time and therefore inappropriately interrupting the action sequence (Reason, 1990). There is only one category here because it is difficult to differentiate the steps of the action sequence on this level (goals, information search, etc.).

To provide a few more ideas regarding the form that these different error types may take within organizations, Table 1.1 provides examples of each type of error for the design and development of a new widget.

Reliability and Validity of the Error Taxonomy

Using an observational methodology, Zapf et al. (1992) investigated white-collar employees using computer software in their normal day-to-day work. One of the key findings was that both knowledge errors and errors at the intellectual level of regulation required more error-handling time. This is consistent with theory as this level of regulation implies conscious processes that appear in a slow sequential mode. Processes at this level are also often complex; therefore, the errors are based on complex

TABLE 1.1

Examples of Errors at Each Level of Regulation for the Design and Development of a New Widget

Level of Regulation	Error Type	Error Examples
Heuristic		The heuristic level of regulation describes general "approaches" to goal accomplishment. For example, in initiating a research paper, some individuals' "default" approach might consist of writing a detailed outline, whereas the default approach of other individuals might be to jump in and start writing. These heuristics maybe at the conscious level at which there is awareness of this default approach or at the nonconscious level. As can be seen in the widget example, errors occurring at the heuristic level often feed into errors at the intellectual level of processing. For example, a default cognitive style of a product designer suggesting that the designer "knows best" results in little, if any, market research. This lack of market research in turn results in the development of an erroneous goal, for example, the development of a product that consumers do not want to buy or that cannot be manufactured at a reasonable cost.
	Goal orientation: Errors resulting from typical approach to goal development	Project designer always starts with existing widget model, which means the goal is always incremental improvement. This can result in errors when market demands transformational change.
		Product designer always evaluates cost as the most important goal, resulting in missing opportunities to add luxury features.
	Cognitive styles: Errors resulting from typical cognitive styles (e.g., degree of self-reflection, impulsivity)	Product designer does not engage in any reflection regarding the last several widgets developed, the learning from this experience, or the market reaction. Instead, the product designer (without reflection) calls up the existing model (goal orientation) and immediately initiates incremental improvements through the same product development cycle.
		Product designer adopts a cognitive style of "I know best" what the consumer wants. This leads to a lack of market of analysis and can result in downstream "goal errors" (see following entries).
	Plan orientation: Errors resulting from typical approach to plan development	Product designer uses default product development procedures and structure, which can result in errors if the new widget is fundamentally different (i.e., transformational change) and requires a different process and structure.

	Monitoring styles: Errors resulting from typical ways in which progress of action cycle is monitored	After the initiation of the product development cycle, the only metric that is monitored is the production schedule, resulting in errors in quality because these metrics are either not collected or not the primary focus.
	Feedback processing: Errors resulting from typical ways of processing feedback	Rough estimates of scheduled launch date serve as fixed anchor and adjusting away from it becomes extremely difficult. Errors can result when these anchors take on too much weight, leading to the explaining away or ignoring of deviations suggesting that revisions to these estimates are necessary.
Intellectual	Errors generally occur at the intellectual level of regulation because the information-processing capacity is limited or the information is uncertain or equivocal, such that the meaning of the situation is difficult to ascertain.	
	Goal error: Errors resulting from the choice of a wrong goal or the goal is inadequately developed (e.g., unclear goals, vague criteria for success)	Perhaps due to errors at the heuristic level (e.g., nonconscious decisions for incremental improvement), the product designer chooses to focus on incremental improvements to reduce costs when the market demands more innovation. Product designer adopts the goal of designing the best widget that the world has ever seen coupled with an "I know best" cognitive style. This can result in (a) an unclear end to the design process such that the designer continually tweaks the design in perpetuity, (2) a final design that cannot be produced at a cost level the market will bear, or (3) a final design that the consumer does not want.
	Mapping error: Errors in the information needed to achieve goals (e.g., relevant information is ignored or information is processed or integrated incorrectly)	In light of the correct goal—for example, incremental design with a focus on cost reduction—the product designer does not perceive that subtle changes in the design are increasing costs significantly. These signals could be missed because each change is a one-off change occurring in a different aspect of the design but results in a significant cost increase when aggregated.

(continued)

TABLE 1.1 (Continued)

Examples of Errors at Each Level of Regulation for the Design and Development of a New Widget

Level of Regulation	Error Type	Error Examples
	Prognosis error: Errors occurring in complex environments in which individuals incorrectly predict behavior or events based on insufficient "model" of the environment	In light of the correct goal—for example, radical widget innovation—the product designer uses an incorrect cognitive model of consumer preferences (perhaps based on a heuristic of "we know best") to predict reactions to different features. This erroneous model of consumer preferences leads to features that are overly complicated to use or are not desired. In the context of radical innovation, the product designer significantly underestimates the degree of communication and coordination necessary to develop all aspects of the widget (based on product development processes used for incremental improvement). This underestimation results in the lack of sufficient processes to manage integration, which in turn results in errors of integration or communication.
	Thought error: Errors resulting from inadequate plans or wrong decisions made in the assignment of plans and subplans	To expedite a market research study, the product designer selects the sample using the same protocol as for the last three widget processes. This results in a sample that does not match the target market for the new design. Errors regarding the attractiveness of the design ensue.
	Memory error: Correct goal chosen and correct plan developed, but errors occur because a certain part of the plan is forgotten and not executed	Assuming a brand new widget design, there is a marketing research check scheduled halfway through the product development process. This research study is never conducted. This could come about because the product designer is distracted due to a competitor's launch of a similar product when this study is to occur or because the designer is busy focusing on other metrics (cost, schedule) and simply "forgot."
	Judgment error: Errors resulting from a misinterpretation or inability to understand feedback	During the marketing research of their new widget design, the product designer receives feedback from some consumers that they love the new design and from others that they dislike it. The product designer mistakenly interprets this feedback as a green light to continue the development based on the positive feedback because the designer was unable to clearly understand the reasons why individuals disliked the design.

Errors, Error Taxonomies, Error Prevention, and Error Management

Flexible action patterns	Errors on this level occur when well-known actions are not performed correctly.	
	Habit error: Errors resulting from a correct action performed in the wrong situation	In computing the overall cost to produce the new widget, the product designer inadvertently uses the part costs from current widgets instead of updated cost figures.
	Omission error: Errors resulting from not executing a well-known subplan	The product development process has nested in it a routine check with the legal department to ensure that the design does not violate any legal specifications. This is a well-known and routine process. But, as the product designer is sitting down to send the design to legal, the designer is interrupted by a phone call detailing a family emergency. After several other calls and conversations regarding the emergency, the product designer continues with the next step in the process following the step specifying sending the design to legal for review. As a result, the design is never sent to legal, resulting in an omission error (subplan omitted due to interruption and rejoining in the wrong part of the process).
	Recognition error: Errors occurring when a well-known feedback message is not noticed or is confused with another	Product designer fails to notice that one metric nested in the quality control testing on the prototype design is outside the allowable range.
Sensorimotor	Errors at this level of regulation involve psychomotor errors (e.g., typographical errors, movement errors while typing).	
	Movement errors: Errors resulting from wrong psychomotor actions	While recording the manufacturing specifications, the product designer intends to convert a 1-inch measurement to 2.5 centimeters. But instead, the designer writes 2.5 inches.

decisions that take longer to recognize, understand, and respond to than errors occurring at lower levels of regulation. In addition to errors at the conscious level taking more time, knowledge errors may require more handling time as individuals need additional time to look up information or engage in other exploration efforts to find the correct procedures (i.e., gain the correct knowledge).

Thought and memory errors also appeared more frequently in more complex tasks (Zapf et al., 1992) because they required more regulation at the intellectual level. Other research suggested that errors occurring at the conscious level and within the knowledge base used for regulation can often only be corrected through the use of some external support, such as help from coworkers (Brodbeck et al., 1993). In addition, knowledge errors appear more frequently in novices because novices do not have sufficient knowledge of the system. This lack of knowledge can in turn lead to more thought and judgment errors because it is more difficult for them to adequately understand the context and apply the correct action cycle to this context.

In contrast to novices, whose errors resided at the conscious level with their knowledge base, experts were more likely to make habit errors that occurred within their flexible action patterns. This is expected because experts have had more practice to develop routines and therefore are able to perform actions at a lower level of regulation. In other words, given the knowledge and experience of the experts, much of the task was well known and posed little subjective complexity. As a result, much of the task execution involved flexible action patterns, and when errors occurred, the majority of these errors were quickly corrected without any external help.

We have opted to focus on the Zapf et al. (1992) error taxonomy because it is one of the few error taxonomies that have been empirically validated. That notwithstanding, there are significant overlaps with the other well-known taxonomies of Reason (1990), Rasmussen (1987), and Norman (1981). Rasmussen differentiated between knowledge-based, rule-based, and skill-based levels of regulation. We agree with Zapf et al. that the differentiation between intellectual level and some kind of knowledge base is important because there are different processes on these two levels, and the remedies for frequent errors in these two taxons are different. Theoretically and practically, there is a difference between someone who is having difficulty in a particular context because of a lack of knowledge (knowledge error) and someone who has adequate knowledge but is unable to build an adequate plan (conscious level of regulation planning or thought error). Reason (1990) and Norman (1981) classified errors into mistakes, lapses, and slips. There is overlap in these error classes with the differentiation between the different levels of regulation. Lapses and slips seem to be more on the level of flexible action patterns and the sensorimotor level, while mistakes seem to be more on the intellectual level and knowledge base for regulation.

Error Detection, Error Signals, and Error Consequences

Modes of Error Detection

Now that we have defined errors, described their prevalence within organizations, and identified the different types of errors that can occur, we can turn our attention to the factors influencing error detection. A detailed consideration of error detection is important because it constitutes the necessary first step in managing and correcting errors. Error detection has been defined as the realization that an error has occurred independently from understanding the nature or cause of the error (Zapf et al., 1994; Zapf & Reason, 1994). We refer to information about the occurrence of an error as an "error signal," and we differentiate these error signals from error consequences next.

There have been several studies investigating the different modes of error detection (Allwood, 1984; Sellen, 1994; Zapf et al., 1994). The most basic mode of error detection occurs during the action itself (Sellen, 1994). This type of error detection occurs in the absence of any external error signals. In other words, these errors provide the individual a "caught-in-the-act" feeling such that the individual senses that "something is not right" based on the actions in process. Examples might be the intention of calling one person and dialing another person's number when, before the other person answers, you realize that you dialed the wrong number, hang up, and dial the correct number. Sellen (1994) cited several examples of this type of error detection (e.g., an individual providing directions to someone telling them that they need to turn right while holding up their left hand). Other examples from sports and music would be a golfer realizing during his or her swing that something is not right and then hitting the ball into the woods. Playing music (and similarly typing on a keyboard) can be prone to these types of errors—where the mere actions of an individual's fingers signal an error is occurring prior to receiving any external feedback.

Moving beyond errors detected during the action process, it is important to note that two pieces of information are required for error detection to be successful: (a) feedback regarding the action outcome and (b) information about the goals of the action. In the following section, we distinguish errors detected by the actor (internal goal comparison), errors detected with the assistance of external sources, and errors that are detected because next steps in the action sequence are limited in some way.

Many errors are detected by the initiator of an action sequence through an internal comparison between feedback signals and the goal of the action (i.e., *internal goal comparison*). This type of error detection can occur at all levels of regulation. For example, it may range from realizing that the wrong letter was just typed (skill level) to the more macro planning level, where feedback signals that a specific action was in error and that the broader plan was also in error. This type of internal goal comparison has often been discussed in the context of control theories of motivation (Kanfer, 1990).

In addition to individuals detecting their own errors through internal goal comparison, errors can be detected by external sources, for example, colleagues detect others' errors or errors in other external systems (e.g., information system error messages). Errors can be externally detected through two primary modes. The first is through external communication from colleagues or other organizational systems. The special case of error detection by others has been discussed by Hutchins (1994) and Kontogiannis (1999), who both noted that four conditions are necessary for an observer to identify errors committed by another person: access, attention, knowledge/expectation, and perspective. First, the observer must have access to the performance of the actor such that the observer has exposure to the actions of the actor. In addition to access, the observer needs to devote sufficient attention to the monitoring of these actions. Third, the observer must have sufficient knowledge of the actor task so that the observer can adequately judge whether the task has been performed correctly (i.e., the observer needs some expectation regarding correct performance). Fourth, and finally, the observer must have an appropriate perspective or, in other words, understand the goals of the actor (Kontogiannis, 1999).

The difference between these last two conditions can be illustrated by the observation of an actor traveling from City A to City B. Suppose that the actor has the goal of taking the most scenic, not the most direct, route between these two cities. Assume also that an observer is following the actor in his or her own car. The observer will see the actor take a "wrong" turn and try to correct the error. But, the observer has reached the conclusion that this is an error only because he or she has assumed that the goal of the actor was to take the fastest, most efficient route. Thus, in the absence of knowledge of the broader goals of the actor, observers can develop inaccurate conclusions regarding what constitutes correct versus erroneous performance by assuming their goals and expectations for the task are the same as the goals and expectations of the actor. The exception to this is when the goal can be reasonably inferred. Zapf et al. (1994) referred to this type of error detection as an evident error. Spelling errors and typographical errors would fall into this category; the external source of error detection does not need to have explicit knowledge regarding the goals of the actor, but rather those goals can be reasonably inferred. For example, a colleague reviewing PowerPoint slides prior to a client meeting does not need to know the overall goals for the presentation to realize that an incorrectly formatted chart or typographical problem is an error.

The final way that errors can be detected is through what both Sellen (1994) and Zapf et al. (1992) referred to as limiting function or planning barriers. Limiting function error detection occurs when some external system prevents one from moving forward. This may be an organizational system, such as a security card-reading machine that fails to unlock the door when the employee inadvertently slides a credit card through the machine or when a person putting together a spreadsheet for discounting cash flow receives a message from the computer that a computational formula is incorrect. Error detection through planning barriers occurs when a plan of action cannot be continued because some external event or action makes it impossible to accomplish the additional elements of the action plan. An example is when an individual does not know the correct command to change the connection to the printer. In this case, error detection is easy, but error correction is often difficult. Another example of error detection through a planning barrier would be when a supervisor tells an individual that he or she needs to stop pursuing the plan of action because it will not achieve the intended goal. For example, a major professor may, on learning of a student's plan for analyzing data, tell the student to adopt a different plan (as the original plan will not achieve the intended goal and is, therefore, in error). The last example of a person blocking the plans of someone else assumes that the four criteria of Hutchin (1994) and Kontogianni (1999) as discussed in this section are in place.

Complexity of Error Detection

Thus far, our discussion of error detection has been pretty straightforward. Error detection requires certain pieces of information, and if this information is available, then errors are detected through several different predictable modes. But, if we probe a little deeper into the error detection process, we realize that error detection is not always this straightforward—particularly in organizational settings that involve multiple action sequences that produce a complex flow of feedback consisting of substantial information for which some of this information may be unclear or uncertain. In these settings, error detection frequently takes the form of a probabilistic statement; that is, one thinks that an error might have occurred but is not sure. Consider, for example, an individual's plan to pursue one career choice over another. After several years, he or she may experience some difficulties in this career as well as some success. Then, he or she meets with a friend who pursued the other career option and is experiencing a great deal of success. Was it an error (or a mistake) for this person to pursue this career? It will be extremely difficult to tell. The individual will have to reflect on all of the feedback suggesting success and the information suggesting this career choice may have been a mistake. At any given moment, he or she will have to weigh this evidence and conclude that either they did or did not make the correct choice. This example is analogous to many organizational tasks and actions. The feedback signaling whether an error has occurred is nested in a rich and complex flow of feedback containing a degree of uncertainty and equivocality, making it difficult to state clearly at any given moment that an error has occurred.

Thus, three potential categories of information exist: (a) feedback that clearly signals an action cycle is progressing correctly, (b) feedback that clearly signals that there is an error in the action cycle, and (c) feedback that does not clearly signal either correct or incorrect actions (i.e., it is open to interpretation). The ambiguity of this third category of information could result from uncertainty or equivocality nested in the feedback itself (e.g., time lags), uncertainty, or equivocality in the goal criteria or other more traditional sources of noise (e.g., measurement unreliability). Figure 1.3a depicts a hypothetical distribution of the information flow within an organizational context. Assuming a relatively complex action cycle with multiple subgoals as well as competing criteria for determining success (e.g., efficient, cost-effective, high quality), then there will likely be some feedback that suggests that part of the action cycle is correct (Area A), other feedback suggests that an error has occurred and needs to be corrected in other parts of the action cycle (Area B), and some information that is open to interpretation (Area C).

Overall, the complexity as well as the success of error detection are influenced by three factors implicit in Figure 1.3a. First, the feedback signaling erroneous actions is necessary, but not sufficient, for the detection of the error. Certainly, the error signal must be present, but that signal must also be perceived and processed. Thus, these error signals will only lead to error detection if sufficient attentional and cognitive resources are applied toward monitoring for errors. Second, Reason (1990) and others (e.g., Kahneman et al., 1982) have discussed the natural tendency of humans to accentuate confirming feedback and to discount or explain away feedback suggesting that our actions are incorrect. Thus, ambiguous feedback (i.e., Area C, Figure 1.3a) presents a challenge psychologically. Specifically, the actors' natural tendency will be to explain away this ambiguity and continue pursuing the chosen course of action. Exceptions to this, of course, may occur. For example, high-reliability organizations often create cultures in which even the slightest hint of an error triggers a rigorous investigation. In other words, these organizations exercise a highly sensitive error response criterion (cf. Davies & Parasuraman, 1982)

Errors, Error Taxonomies, Error Prevention, and Error Management

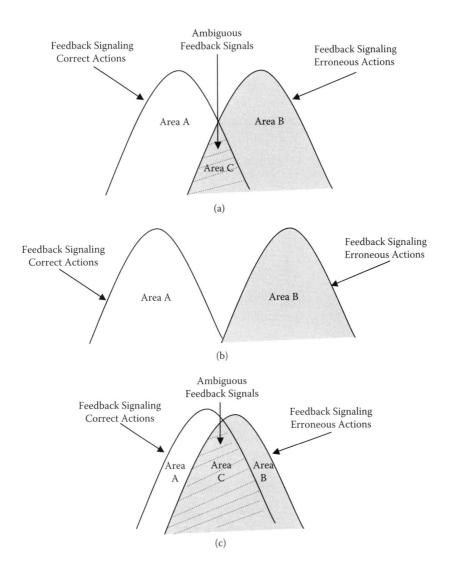

FIGURE 1.3
(a) Hypothetical feedback distributions signaling correct versus erroneous actions. (b) Hypothetical feedback distributions signaling correct versus erroneous actions during skill-based cognitive regulation. (c) Hypothetical feedback distributions signaling correct versus erroneous actions during intellectual cognitive regulation.

such that any feedback remotely signaling an error receives substantial attention. On the other hand, organizations that are not as focused on errors will invoke a much more lenient response criterion such that it needs to be clear that an error has occurred before any significant error correction actions will be invoked.

The third factor influencing the success of error detection is related to the level of cognitive regulation because it has an impact on the degree of overlap of the two data distributions in Figure 1.3a and therefore the ease with which these errors can be detected. At the lowest level of cognitive regulation (sensorimotor), behavior is automatic. These actions occur in familiar environments with clearly specified performance criteria. Given this highly familiar context within which this behavior is performed and the development of clear criteria specifying correct performance, feedback signaling correct versus erroneous actions should be easily recognized and categorized. In other words, the feedback distributions will look like those depicted in Figure 1.3b. Thus, even though many errors occur at this level of regulation, they are typically easy to identify and correct. As a case in point, Zapf et al. (1992) found, in their investigation of computer use, that sensorimotor errors needed the least amount of time to correct, and that 100% of these errors could be corrected without the aid of external support (e.g., from colleagues, a help menu). This provides some empirical support to our notion that errors are salient at the sensorimotor level of regulation, and that errors will be easily distinguishable from correct performance (i.e., feedback falls into either Area A or B, with little, if any, feedback in Area C; Figure 1.3b).

At the level of flexible action patterns, correct and error signals will be slightly less differentiated. At this level of regulation, individuals typically commit habit errors, omission errors, and errors of recognition. Essentially, these errors suggest that a correct action is undertaken in the wrong situation, a subaction is not executed, or a signal specifying a particular action is not noticed, confused with another signal (which triggers a habit error), or explained away (an error signal is not noticed as a real error signal and is explained away). Zapf et al. (1992) found that with flexible action patterns—while still highly correctable—there was a slight increase in the need to obtain external support for correction.

In contrast to the sensorimotor and flexible action pattern levels of regulation, intellectual and knowledge errors exhibit a much less differentiated error and correct performance distributions. In other words, much of the feedback will be difficult to interpret in terms of error detection. Figure 1.3c depicts what these feedback distributions might look like. Consistent with this notion, Zapf et al. (1992) found that intellectual regulation errors occurred more frequently when performing complex tasks, it took longer to correct the errors, and the correction of these errors required significant use of more external resources.

Action Theory, Action Cycles, and Errors at the Collective Level

Thus far, we have grounded our entire discussion at the individual level, albeit individuals working within organizations. But, a number of authors have suggested that groups, teams, and organizations also function as information-processing entities. Much of this research has used concepts similar to those presented here. For example, Gersick and Hackman (1990) discussed habitual routines at the group level in similar terms to our discussion of actions occurring at the lowest level of cognitive regulation (nonconscious, sensorimotor). At the organizational level, concepts such as rules, programs, standard operating procedures, and routines (Cyert & March, 1963; Galbraith, 1977; Knight & McDaniel, 1979; Levitt & March, 1988; Nelson & Winter, 1982) have been used in a similar manner. Perhaps the most direct linkage is Nelson and Winter's (1982, p. 124) discussion of the automatic patterns of behavior as the skill of the organization.

Pentland and Rueter (1994) provided a collective-level example of flexible action patterns when they investigated the problem-solving behavior of computer software customer service representatives. The actions of these representatives were guided by broader "grammars of action." In other words, their performance was guided by a set of preexisting flexible action patterns that determined the processing of particular problem types or classes (i.e., a series of if-then rules). Pentland and Rueter (1994) described them as

> not a single pattern but, rather a set of possible patterns—enabled and constrained by a variety of organizational, social, physical, and cognitive structures—from which organizational members enact performances. ... [This definition] points to similar, but not fixed patterns that emerge through interaction, what is fixed, to some extent, is the space of possibilities for action. (p. 491)

At the organizational level, flexible action patterns are also initiated—similar to the individual level—when problems are encountered in skill-based actions. Both Mintzberg (1973) and Nelson and Winter (1982) noted that problems encountered in ongoing organizational routines occasion the involvement of managers and supervisors. Specifically, Nelson and Winter (1982) stated that the intervention of management "in the detailed functioning of lower levels is ordinarily symptomatic ... of difficulties with the functioning of existing routines—just as conscious awareness ... [is] symptomatic ... of trouble in the case of individual skills" (p. 125). This parallels the discussion by Louis and Sutton (1991) regarding when organizations must "shift cognitive gears" by moving from skill-based actions to more controlled actions that actively draw on collective memory and knowledge stores (Walsh & Ungson, 1991).

We can see from this brief review that similar language has been used to describe both individual and collective information processing. But, before we go too far down this path, we need to discuss what will be similar and what will be different about individual and collective action cycles. To do this, we must make a distinction between the function and structure of constructs (Morgeson & Hofmann, 1999). The function of a construct is simply the effects or outputs of the construct. In other words, the function deals with the nomological network of relationships between the construct and other constructs of interest. In contrast to the functional aspects of constructs, the structure of a construct focuses on the processes through which these outputs and effects occur. This is not "structure" in the sense of structural models (i.e., measurement, validity, reliability), but rather it is more about the conceptual notions of structure; namely, it deals with the "stuff" that the construct is made of and the processes that produce the functional outcomes. For example, information processing across levels enables entities to perceive, interpret, and respond to stimuli. Moving across levels, however, the underlying structural processes, or underpinnings, of information processing change substantially. At the individual level, the structure of information processing is scripts, schemas, and so on (i.e., a cognitive foundation), whereas the structure of collective information processing consists of social interaction (i.e., a social foundation; see Morgeson & Hofmann, 1999). Other excellent examples of this distinction are theoretical treatments of the threat-rigidity responses across levels (Staw, Sandelands, & Dutton, 1981) and organizational memory (Walsh & Ungson, 1991). So, we are arguing here that the functional aspects of these action cycles are similar across levels. The processes, however, are substantially different when moving from individuals to collectives. At the individual level, these action cycles are intrapersonal, and at the collective level they are interpersonal.

We have argued thus far that individual action cycles involve several different levels of cognitive regulation, and that certain types of errors occur within each of these different levels of regulation. Given the functional parallels in discussions regarding the different levels of regulation occurring at the individual and collective levels, it seems possible to think about a collective-level analog of our individual error taxonomy (of course, there are many degrees of freedom and more variance in organizational design than in action design within individuals).

Collective Action Cycles

To do this, we must first consider how organizations might engage in functionally similar action cycles to those described at the individual level. One way to view this would be to think of a traditionally functionally designed organization in which the organization is clustered into functions (e.g., sales and marketing, manufacturing, research and development). In this type of organization, one could imagine a process occurring that parallels the individual action process. For example, one could envision top management establishing a strategic plan with associated goals. These goals create subgoals and subplans for individuals at the top management level leading each of the functional divisions. These goals in turn provide the input to a planning cycle that begins at the next level down. The plans and goals at this level not only develop subactions within this level in the organization but also provide the input for planning at the next level and so forth. So, in this example one could envision organizations as a cascading series of action cycles, with each cycle taking on the following characteristics: (a) goal development and decisions among competing goals; (b) information search to understand the environment in which one acts; (c) generation of a plan to execute the goal; (e) execution and monitoring of the plan execution; and (f) processing of feedback that is both ongoing during action and provides information on the accomplishment of the plan. It is important to realize that as these action cycles cascade down across the different levels of the hierarchy within the organization, separate action cycles are also initiated to manage the coordination and integration function between the action cycles occurring in different, but yet interdependent, parts of the organization.

Just as these organizational action cycles may be initiated through a top-down process (strategy initiates a cascading series of goals and plan development), one could also imagine a bottom-up process in which errors and problems are encountered. For example, when faced with problems in skill-based routines, the organization may first attempt to match the current problem with existing rules or routines. If such rules cannot be found or they are unsuccessful, then the collective might look for analogous situations and apply the rules or routines from that situation. For example, they may attempt to review their past experience in an effort to identify a similar situation, they may identify similar units or organizations that faced the same problem and attempt to learn from that experience (i.e., benchmarking), or they may contract outside consultants in an effort to expand both their collective knowledge base and the number of possible situations that could be considered analogous (Levitt & March, 1988; Starbuck, 1989; Starbuck, Greve, & Hedberg, 1978).

If the organization cannot find a suitable analog for the current situation, it will have to develop new action sequences by reformulating existing skills. The development of these new action sequences will, however, require considerable effort aimed at both assessing and understanding the situation and developing a response. Assuming that an environmental interpretation is agreed on, the resulting actions require, as noted by Nelson and Winter (1982), the development of "new combinations of existing routines" (p. 130). Specifically, Nelson and Winter (1982) described the process in this way:

> Consider the foreman of a work team responsible for a particular operation (set of routines) who observes that a machine is not working properly. He routinely calls in to the maintenance department, which in turn routinely sends out a machine repairman. The machine repairman has been trained to diagnose in a particular way the troubles that such a machine might have. He goes down a list of possible problems systematically, and finds one that fits the symptoms. He fixes the part so that the machine again may play its role in the overall work routine. He may also, however, report to the foreman that this particular kind of trouble has become very common since the supplier started using aluminum in making the part in question and that perhaps the machine should be operated in a different manner to avoid the difficulty. (p. 129; see also Chapple & Sayles, 1961)

Thus, the interruption of one action cycle by an error initiates a second error-handling action cycle similar to the bottom-up process discussed at the individual level. Either the problem will be solved through this process or a new action cycle eventually will generate a new routine and so forth.

Collective Failure Modes

We have established that discussions of both individuals and collectives (teams, groups, organizations) seem to use similar information-processing-based language and seem to engage in functionally equivalent action cycles. If this is the case, then we should be able to observe similar types of collective errors at the respective levels of regulation. So, for example, we should expect to see both individuals and collectives experience goal-setting errors when involved in the intellectual level of regulation, habit errors at the level of flexible action patterns, and movement errors at the skill-based level of processing.

To illustrate how some of the same types of errors that we discussed at the individual level might "look" at the collective level, we provide a few brief examples. Gersick and Hackman (1990) described an airline accident in which the crew did not engage the anti-ice mechanism even though the weather would have deemed it necessary. From our vantage point, this describes a *habit error*. In describing the accident, Gersick and Hackman (1990) noted that the captain had only flown eight takeoffs in wet, freezing weather since being hired by Air Florida. The first officer had only two such takeoffs in his career with the organization. Their example is a clear case of an omitted attention check at a critical choice point in the action sequence, with this omission resulting in the initiation of the strongest, most frequently enacted action. Given the weather conditions, either the first officer or the captain should have invoked conscious attention when anti-ice was called out in the preflight checklist to ensure that the anti-ice mechanism was engaged. This attentional check was even more critical given the past experience of the captain and first officer (i.e., virtually all of their takeoffs were in conditions for which the anti-ice mechanism was not required) and the resulting "strength" of the "anti-ice—off" response pattern. This seems to be a clear case in which an omitted attentional check resulted in pursuit of the strongest path. In other words, the lack of attention resulted in engagement of the most typical response (i.e., a habit error). It is important to point out that *both* interacting parties contributed to the failure. Had either invoked an appropriately timed attentional check, the failure would likely have been averted. Thus, it is a collective failure, not merely a single individual's error.

Weick's (1990) description of the crash of a KLM 747 and a Pan Am 747 in Tenerife, Spain, provides an example of a collective *omission error*. Weick described a situation in which questions about the return trip interrupted an ongoing routine (i.e., taxi and takeoff). While devoting their attentional resources to this higher-order plan (i.e., a legal return trip), the pilots inadvertently initiated their routine takeoff procedures prior to gaining clearance from the tower (Weick, 1990). Viewing it in the current context, this failure could have resulted from an unintended interruption of an ongoing organizational routine. Specifically, the pilots had engaged a well-practiced routine (takeoff procedures), which activated associated subroutines (e.g., taxiing and takeoff). But, this routine was interrupted by an unexpected problem (i.e., whether they would be able to make a legal return trip), which captured their collective attentional resources. After devoting a great deal of attentional resources to the legal return trip, the cockpit crew may have simply resumed the interrupted automatic routine several steps further along than they actually were (i.e., an omission failure due to a unintended interruption or "a place-losing error"). Specifically, the pilots might have simply resumed their automatic routine at the "taxi-and-takeoff" stage instead of contacting the control tower for clearance stage (e.g., "Now where were we. ... Oh, yes, we're ready to go").

Other examples of collective errors that parallel errors in our error taxonomy (Figure 1.2) include Kiesler and Sproull's (1982) as well as Starbuck and Milliken's (1988b) discussions of a number of cognitive biases that can infiltrate the interpretative and cognitive processes of senior management teams, such as bounded rationality (prognosis and thought errors), inaccurate mental representations of the environment (mapping and prognosis errors), and the tendency to discount, or explain away, disconfirming evidence (heuristics for feedback processing). Shrivastava et al. (1987) noted how organizational frames of reference (i.e., the conceptual models of organizational members, mapping, and prognosis errors related to overarching heuristics) can produce failures by constraining and making more rigid organizational responses to external conditions. Similarly, Hall (1984) discussed how inaccurate, or diminished, causal maps (mapping and prognosis errors) used in the development of strategic policies led to the downfall of the *Saturday Evening Post*.

An example of errors occurring at the heuristic level is Danny Miller's (1990, 1993) discussion of how strategic-level dysfunction can occur when organizations become too simple over time. At the individual level, we described the heuristic level of regulation as a level that specifies how action cycles are created (e.g., how to go about setting goals, planning, and so forth). Miller (1990, 1993) argued that, over time, organizations become too simple; that is, they depend too much on overarching, superordinate routines for how to gain competitive advantage. These routines are eventually applied without reflection and become the default and accepted way of doing things. These collective assumptions and routines at the collective level conceptually parallel the discussion of heuristics at the individual level.

Conclusions and Generalizations Regarding Collective Error

Our goal here is not to be exhaustive or to answer all the possible questions regarding the application of action theory to collectives. Instead, our goal is to introduce the possible functional similarities between action theory at the individual level and organizational actions—both top-down and bottom-up processes and the coordinating mechanism between functional groups. Considering the functional similarity of action cycles across levels opens up possibilities for one to see and understand problems occurring within organizations through a different lens. But, this application of action theory to collectives must provide more than a new language to describe the same types of observations. To justify a new look at organizational and collective problems through the lens of action theory, it must do more than provide a new language for the same phenomenon. It must allow us to go further. It must allow us to think of implications that emerge from this viewpoint that might not emerge otherwise.

We believe that there might be at least two areas in which the application of action theory to collective action might better help us either understand or develop implications for organizations. In terms of understanding organizations better, we think that not only is it instructive to view only the actions within a particular collective from an action perspective, but also it is equally instructive to view the coordination of interdependent collectives as regulated via the same action control processes with the same types of errors occurring in the *coordination of actions* of groups and teams within the organization. This raises the possibility that two interdependent collectives each might not commit an error within their respective action cycles, yet the process as a whole may fail due to errors in the coordination action cycle. For example, it is interesting think about not only what habit errors or thought errors might look like within individuals and within collectives but also what these errors might look like when they occur during the action cycle designed to coordinate the actions *between* interdependent collectives.

The second area in which we believe this application of action theory to organizational behavior—both the behavior of individuals within organizations and the actions of collectives as a whole—is likely to lead to new insight is in discussions of error prevention and error management actions. It is to this discussion that we now turn.

Error Management and Error Prevention in Individual and Collective Action Cycles

Figure 1.4 describes two strategies used to deal with errors: (a) error prevention for which the focus is on the reduction of the occurrence of errors and (b) error management for which the focus is disconnecting the error from error consequences (Frese, 1991; van Dyck et al., 2005). Examples of employing an error management approach can be found in software systems (e.g., the Undo function), physical setups (e.g., the containment egg around nuclear power plants), individual and team behaviors (e.g., cross-checking in the cockpit that leads to "error trapping," catching the error before its negative consequences can unfold; see Helmreich & Merritt, 2000), and organizational practices (in the sense of error management culture; van Dyck et al., 2005). Although error prevention efforts (e.g., human factors, system engineering) can clearly reduce the number of errors that occur within organizational settings, they will not be able to eliminate all errors (Reason, 1997). This is because whenever there are humans involved in an system, one gains both the advantages of human cognition—characterized by efficient processing in uncertain environments (Reason, 1990)—and the associated limitations, such as bounded

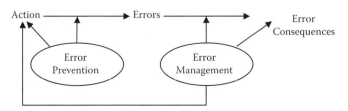

FIGURE 1.4
Error prevention and error management.

rationality (March & Simon, 1958), errors due to limited working memory (Norman & Bobrow, 1975), and errors that appear as a result of wanting maintain one's self-confidence (Dorner, 1996). Thus, a pure error prevention approach cannot adequately deal with the fact that errors are ubiquitous at both the individual and the collective/organizational levels. In addition to not being able to fully eliminate errors, an error prevention focus can reduce the learning potential of errors and the possibility that some errors may result in long-term positive consequences (Sitkin, 1992).

Despite this fact, the first and most frequent response of professionals working in complex organizations (and most engineers and economists) is to adopt an error prevention focus. Similarly, most laypeople seem to think that once an error has occurred the negative consequences are unavoidable. But, both error prevention and error management share a common goal: to reduce the negative consequences of errors. Error prevention tries to achieve this goal by preventing the error from occurring in the first place, whereas error management accepts that errors will occur and is designed to provide a second layer of defense, specifically one that attempts to intercept and rectify the error prior to the accrual of significant negative consequences. In our view, organizations should use both error prevention and error management approaches because this dual-prong approach is likely to have a better chance of preventing error consequences than either approach alone.

Discussions of error management suggest that there is a clear differentiation between the error itself and the consequences of the error. Both Reason (1990) and Frese and Zapf (1994) suggested that errors occur when a plan does not achieve its intended outcomes (goal), which implies that errors may not be noticed until an outcome (i.e., the wrong outcome) is observed. The question then emerges regarding what defines the *outcome* or signal that an error has occurred versus the *consequence* of the error. When viewed within the broader perspective of the action cycle, we believe that a useful way to distinguish between the outcomes and consequences of errors is to differentiate between the indicators of an error *during* the action cycle versus errors that remain *after* the conclusion of the action cycle. Clearly, errors are frequently identified when the outcomes of actions are inconsistent with intentions. For example, a spreadsheet programming error can be clearly identified when a test of the programming statement produces a software error message. We consider these opportunities for feedback during the action cycle to be error signals (i.e., signs pointing to the possibilities of errors).

These opportunities for feedback regarding the success of an action cycle, however, are qualitatively different from what we consider the potential system-level consequences of errors. These system-level consequences are more likely to occur after the action cycle is completed, when these errors have a negative impact on the performance of other entities within the organization. In other words, significant error consequences can accrue when errors remain resident within the system after an action cycle is completed (Ramanujam & Goodman, 2003). For example, NASA (National Aeronautics and Space Administration) *Orbiter* engineers' failure to discover the wrong basis of force calculations remained resident in the system after the action cycle had been completed. These specifications were then used by other parts of the organization as a basis for design and engineering specifications. In such situations, serious error consequences can emerge. When errors remain after the action sequence is completed, such errors begin to have an impact on the performance and functioning of *other, interdependent* entities. It is these types of ripple effects that we consider to be the consequence of an error (as opposed to its signal).

Goals of Error Management

With the distinction between error signals and error consequences established, we want now to mention briefly several primary goals of error management. Specifically, error management is designed to (a) control the potential damage of errors, (b) reduce the potential of error cascades, and (c) facilitate secondary error prevention.

Quick Damage Control

Reason (1990, 1997) convincingly argued that there are latent errors in organizations (see also Ramanujam & Goodman, 2003). The longer they are unchecked in an organization, the higher is the probability that these latent errors interact with local trigger conditions to produce accidents (Maurino, Reason, Johnston, & Lee, 1995). Moreover, in dynamic systems (e.g., organizations dealing with a competitive market), it is necessary to detect errors quickly so that any potential damage can be controlled (Dörner, 1996). There are several factors that can work against quick damage control: punishing errors, blaming as the dominant response to errors, and not discussing errors openly within the organization (van Dyck et al., 2005). Drawing on a number of business cases, Peters (1987, p. 230) eloquently argued why problems accumulate when there is a fear of revealing errors. If errors are not openly discussed, then small errors are not revealed, such that minor adjustments and corrections to the action cycle are not enacted. When this happens, managers do not receive the necessary information (or are even misled), no learning occurs, testing and evaluation of error signals are delayed, and the potential ramifications of the error continue to grow. The end result is a much more serious long-term error consequence.

Reducing Error Cascades

The second goal of error management is to help minimize the potential of one error to result in other errors (i.e., an error cascade). When errors occur, people have to start to regulate routines on the conscious level of regulation (see Figure 1.1). When actions have been regulated on the lower level of regulation for a long time, it is not only effortful to start regulating them on the higher level but also there is little regulatory power on the higher level in such cases (remember the fable of the centipede that, when asked how it coordinated its many feet, fell over and could not move any more when it tried to think consciously about this task). Moreover, the negative emotions that often accompany errors (e.g., based on negative internal dialogues) also use conscious capacity; therefore, capacity limits are often reached when errors occur (Meijman & Mulder, 1998). Because error management leads to a more relaxed attitude toward errors and a more task-oriented approach to deal with them quickly and efficiently, the tendency of error cascades may be reduced.

Secondary Error Prevention

Figure 1.4 displays that error management may lead to secondary error prevention. As individuals learn more about the reasons for and the potential problems of a particular error, they will tend to prevent this particular error from appearing again in the future. Of course, other learning processes may appear as well; they may, for example, learn that they tend to make more errors in a particular task arena and then develop strategies for catching errors more quickly (e.g., learning leads to quicker and more efficient error handling). Thus, secondary gains of error management lead not only to better error prevention but also to better error handling.

Factors Influencing Error Management

Effective error management processes are influenced by a number of factors. Several of these factors are more general in that they influence the error management processes across all of the different stages of the action cycle. Others are more specific, such that they are more critical in error management activities occurring at certain phases of the action cycle.

With respect to general influences on the error management process, errors will be better managed if individuals have a general expectation that errors will occur. This general expectation facilitates error management through several processes. One such process is that individuals have a tendency to look for confirming instead of disconfirming evidence (Kahneman et al., 1982; Reason, 1990). Thus if there is no hypothesis regarding the possibility of an error, evidence will not be sought in search of errors. In fact, one indicator that groups are falling into "groupthink" (Janis, 1982) is the perception that erroneous decisions are not possible. As a result, decisions are made with little, if any, effort put forth to see if an error has occurred. Put simply, if there is no expectation that errors are possible, then both individuals and collectives will not "see" errors when they occur.

Another process through which a general expectancy regarding errors can come about is through either individual values or the broader organizational culture. Rybowiak, Garst, Frese, and Batinic's (1999) developed an assessment of individual error orientation, that is, the degree to which individuals have a positive and constructive view of errors (e.g., "Mistakes help me improve my work; when I make a mistake at work, I tell others about it so that they do not make the same mistake"; "Employees who admit to their errors make a big mistake"). Similar to the work of O'Reilly, Chatman, and Caldwell (1991), these individual values also have been used to assess organizational culture. With regard to error cultures, van Dyck et al.'s (2005) assessment of error culture focused on shared perceptions regarding error aversion (e.g., "People in this organization feel stressed or embarrassed after making a mistake," "People in this organization are often afraid of making errors") and error management (e.g., "Errors are useful for improving work processes"). Van Dyck et al. (2005) found that error cultures were positively related to firm profitability. Although not specifically error focused, Edmondson's (1999) assessment of psychological safety similarly assessed, in part, the extent to which problems and mistakes can be openly discussed. Hofmann and Mark (2006) also found that error culture—as part of a broader safety orientation—was significantly related to a number of health-related outcomes (e.g., medication errors, nurses' back injuries). We believe that shared organizational practices and values focused on communicating about errors, sharing error knowledge, helping in error situations, and quickly detecting and handling errors will help have an impact on error detection time by increasing both individual and collective error vigilance.

Another factor influencing error management is the degree to which organizational or technical systems are structured in a transparent way that allows individuals to develop a better conceptualization or mental model of the system. This facilitates a deeper understanding of the system and, as a result, an improved capacity for error management (Norman, 1988, talks about "visibility," which is a similar principle). Although understanding is necessary, it is not sufficient. One must also receive feedback as this feedback helps in understanding an organization or a technical system better. An individual can receive better feedback by asking more questions and by working harder on a better understanding of an organization (e.g., developing an explicit hypothesis, including a hypothesis on where one can fail within the system).

In addition to these general influences on the error management process, there are some influences on the error management process that are more likely to be effective during certain phases of the action cycle. Recall that the action cycle consists of (a) goal development and decisions among competing goals; (b) information search to understand the environment in which one acts and the prognosis of future states, which is particularly important in dynamic systems that change over time even if a person does not intervene (e.g., an organization); (c) generation of a plan to execute the goal; (d) execution of the plan coupled with monitoring the actions involved in execution of this plan; and (e) processing feedback emerging from these monitoring activities as well as feedback that provides information at the end of an action cycle (e.g., regarding the success or failure of the action).

Error management during goal development is facilitated by having both clear goals and goals that are not in conflict or potential conflicts are thoughtfully addressed. Goal conflicts are more likely to occur at the higher levels of cognitive regulation (e.g., conflict between superordinate goals, such as being an outstanding teacher and establishing one's reputation as a scholar). As a result of these potential conflicts and the uncertainty around goals, errors are more likely to happen, and effective error-handling strategies are necessary (e.g., expectations that errors will occur, clear guidelines for help in error handling, a culture that rewards open discussion of errors, and so forth).

With respect to information search and prognosis activities, error management activities are supported by good analogies and metaphors. As noted, complexity (i.e., the amount and the interaction of the system parameters; Dörner, 1981) can influence the clarity of the signals indicating that an error has occurred. Although this clarity regarding error signals will make error detection a significant challenge during information search and prognosis activities, error management is improved using good analogies and metaphors. The more a person knows about the situation in which he or she works, the better and richer analogies and metaphors they will have available to them to bring to bear on the situation and therefore the more effectively they will be able to deal with errors. Thus, even though errors with respect to information search and prognosis will be more difficult to detect in complex systems, error management can be facilitated by transparent organizational and technical systems and well-developed cognitive maps.

Plan development and decision making are also supported by a high degree of transparency and consistency. Interestingly, however, some of the recommendations that are offered to prevent errors in plan development can actually work against error management. For example, one of the best ways to reduce the complexity of plans is to reduce the time scope (i.e., short-range plans are less complex and therefore more transparent).

However, short-range plans frequently do not mirror the requisite complexity of the situation. Only long-range plans (and proactive plans) effectively deal with complex and dynamic environments because they offer more degrees of freedom when handling errors (Frese et al., 2007). Therefore, an error prevention focus within organizations often works against the development of complex long-range plans, while an error management focus would encourage the development of these plans.

Monitoring, and error management during monitoring, is dependent on memory. Memory is supported by technical systems that provide the kinds of information that one needs. To a certain extent, memory function is also helped by checklists, rules of thumb, procedures, and the like. Often, memory errors occur when one is interrupted (e.g., by a telephone call). Having relevant checklists and procedures available can help prevent and manage these errors as they can facilitate the ease with which individuals can leave a task, successfully return to it later, and not skip or repeat steps.

As noted, error detection related to the feedback process is complicated by unclear error signals. The more unclear the error signals are, the more difficult it is to evaluate correctly the feedback and to detect errors. The more complex and dynamic the environment is, the more unclear the error signals will be. Within organizational contexts, there is an added complexity due to organizational hierarchies. Individuals at the base of the organization (e.g., those who have direct customer and production contact) receive feedback. Yet, if they think that negative feedback has to be suppressed because of an error prevention culture that punishes errors, feedback will not be given to higher levels of the organization, and the error management process will be hindered significantly. The necessary feedback for error management will only be communicated to the upper levels within the organization if an error management culture is created. Although this more constructive culture around errors will help, there is still a time lag between the reception of the feedback and its communication up the hierarchy. This argues for two approaches to error management that should be used simultaneously. First, there must be processes in place for the quick communication of the most serious errors up the various levels of the hierarchy. Second, more minor errors—in keeping with recommendations from sociotechnical system design (Clegg, 2000)—should be dealt with as close to the source of the error as possible. Although this will require leaders to establish clear goals and a strong, compelling vision so that the right decision frames are used in responding to errors, it will significantly improve the error management process.

Conclusions

In this chapter, we have discussed what we believe to be some of the foundational concepts related to errors within organizations. We started by defining and drawing distinctions among several related concepts: errors, violations, inefficiencies, and risk. We then presented action processes as the foundation for behavior within organizations. Emerging from these action processes was a taxonomy of errors occurring at the different levels of cognitive regulation. Once we established that different types of errors can occur during the various phases of the action cycle, we turned our attention to error detection; we discussed both the various modes of error detection and the implications of signal detection theory for error detection activities. Finally, we transitioned into a discussion of the notion of collective errors as well as some thoughts on error management versus error prevention. We now turn over the rest of the volume to our chapter authors. Our goal, hope, and desire are that these chapters further flesh out these foundational concepts as they consider the implications that errors have for their domain of organizational research.

Addendum

After all of the chapters were written and the book was well into final preparations, two significant events reinforced the importance of human errors in organizations and, in particular, the importance of effective error management systems. On April 20, 2010, the Deepwater Horizon drilling rig exploded resulting in the largest accidental marine oil spill in the history of the petroleum industry. Then, on March 11, 2011, Japan experienced an earthquake and tsunami that resulted in an extremely serious nuclear accident at the Fukushima Nuclear Power Plant. As the investigatory reports emerge on the various causes of these accidents, our hope is that the chapters in the current volume will help contribute to building systems that are designed to prevent these types of accidents in the future and better manage and control the unfolding events following similar accidents.

References

Ackerman, P. L., & Cianciolo, A. T. (2000). Cognitive, perceptual-speed, and psychomotor determinants of individual differences during skill acquisition. *Journal of Experimental Psychology: Applied, 6*, 259–290.

Allwood, C. M. (1984). Error detection processes in statistical problem solving. *Cognitive Science, 8*, 413–437.

Anderson, J. R. (1983). *The architecture of cognition*. Cambridge, MA: Harvard University Press.

Brodbeck, F. C., Zapf, D., Prümper, J., & Frese, M. (1993). Error handling in office work with computers: A field study. *Journal of Occupational and Organizational Psychology, 66*, 303–317.

Brown, A. L. (1987). Metacognition, executive control, self-regulation, and other more mysterious mechanisms. In F. W. Weinert & R. H. Kluwe (Eds.), *Metacognition, motivation, and understanding* (pp. 65–116). Hillsdale, NJ: Erlbaum.

Chapple, E. D. and Sayles, L. R. (1961). *The measure of management: Designing organizations for human effectiveness*. New York: Macmillan.

Clegg, C. W. (2000). Sociotechnical principles for system design. *Applied Ergonomics, 31*, 463–477.

Cyert, R. M., & March, J. G. (1963). *A behavioral theory of the firm*. Englewood Cliffs, NJ: Prentice.

Davies, D. R., & Parasuraman, R. (1982). *The psychology of vigilance*. London: Academic Press.

Davies, N., & Ikin, C. (1987). Auditing spreadsheets. *Australian Accountant, December*, 54–56.

Dörner, D. (1981). Über die Schwierigkeiten menschlichen Umgangs mit Komplexität. *Psychologische Rundschau, 31*, 163–179.

Dörner, D. (1996). *The logic of failure*. Reading, MA: Addison-Wesley.

Dörner, D., Kreuzig, H. W., Reither, F., & Stäudel, T. (1983). *Lohhausen. Vom Umgang mit Unbestimmtheit und Komplexität*. Bern, Switzerland: Huber.

Dörner, D., & Schaub, H. (1994). Errors in planning and decision-making and the nature of human information processing. *Applied Psychology: An International Review, 43*, 433–454.

Edmondson, A. (1999). Psychological safety and learning behavior in work teams. *Administrative Science Quarterly, 44*, 350–383.

Edmondson, A. C. (1996). Learning from mistakes is easier said than done: Group and organizational influences on the detection and correction of human error. *Journal of Applied Behavioral Science, 32*, 5–28.

Flavell, J. H. (1979). Metacognition and cognitive monitoring: A new area of cognitive-developmental inquiry. *American Psychologist, 34*, 906–911.

Floyd, B. D., & Pyun, J. (1987). *Errors in spreadsheet use* (October Working Paper 167). New York: Center for Research on Information Systems, Information Systems Department, New York University.

Frese, M. (1991). Error management or error prevention: Two strategies to deal with errors in software design. In H.-J. Bullinger (Ed.), *Human aspects in computing: Design and use of interactive systems and work with terminals* (pp. 776–782). Amsterdam: Elsevier Science.

Frese, M., Krauss, S., Keith, N., Escher, S., Grabarkiewicz, R., Luneng, S. T., et al. (2007). Business owners' action planning and its relationship to business success in three African countries. *Journal of Applied Psychology, 92,* 1481–1498.

Frese, M., Stewart, J., & Hannover, B. (1987). Goal-orientation and planfulness: Action styles as personality concepts. *Journal of Personality and Social Psychology, 52,* 1182–1194.

Frese, M., & Zapf, D. (1994). Action as the core of work psychology: A German approach. In H. C. Triandis, M. D. Dunnette, & L. Hough (Eds.), *Handbook of industrial and organizational psychology* (Vol. 4, pp. 271–340). Palo Alto, CA: Consulting Psychologists Press.

Galbraith, J. R. (1977). *Organization design.* Menlo Park, CA: Addison-Wesley.

Gersick, C. J. G., & Hackman, J. R. (1990). Habitual routines in task-performing groups. *Organizational Behavior and Human Decision Process, 47,* 65–97.

Griffin, R. W., O'Leary-Kelly, A., & Pritchard, R. D. (2004). *The dark side of organizational behavior.* San Francisco: Jossey-Bass.

Hacker, W. (1985). Activity: A fruitful concept in industrial psychology. In M. Frese & J. Sabini (Eds.), *Goal directed behavior: The concept of action in psychology* (pp. 262–283). Hillsdale, NJ: Erlbaum.

Hacker, W. (1998). *Allgemeine Arbeitspsychologie.* Bern, Switzerland: Huber.

Hacker, W. (2003). Action regulation theory: A practical tool for the design of modern work. *European Journal of Work and Organizational Psychology, 12,* 105–130.

Hall, R. I. (1984). The natural logic of management policy making: Its implications for the survival of an organization. *Management Science, 30,* 905–927.

Hanson, S. J., Kraut, R. E., & Farber, J. M. (1984). Interface design and multivariate analysis of UNIX command use. *ACM Transactions on Office Information Systems, 2*(1), 42–57.

Heckhausen, H., & Beckmann, J. (1990). Intentional action and action slips. *Psychological Review, 97,* 36–48.

Helmreich, R. L., & Merritt, A. C. (2000). Safety and error management: The role of crew resource management. In B. J. Hayward & A. R. Lowe (Eds.), *Aviation resource management* (pp. 107–119). Aldershot, UK: Ashgate.

Higgins, E. T. (1997). Beyond pleasure and pain. *American Psychologist, 52,* 1280–1300.

Hofmann, D. A., & Mark, B. A. (2006). An investigation of the relationship between safety climate and medication errors as well as other nurse and patient outcomes. *Personnel Psychology, 59,* 847–869.

Hutchins, E. (1994). *Cognition in the wild.* Cambridge, MA: MIT Press.

Janis, I. L. (1982). *Groupthink: A psychological study of policy decisions and fiascoes.* Boston: Houghton Mifflin.

Kahneman, D. (1973). *Attention and effort.* Englewood Cliffs, NJ: Prentice Hall.

Kahneman, D., Slovic, P., & Tversky, A. (1982). *Judgment under uncertainty: Heuristics and biases.* Cambridge, UK: Cambridge University Press.

Kanfer, R. (1990). Motivation theory and industrial/organizational psychology. In M. D. Dunnette and L. Hough (Eds.), *Handbook of industrial and organizational psychology. Volume 1. Theory in industrial and organizational psychology* (pp. 75–170). Palo Alto, CA: Consulting Psychologists Press.

Kiesler, S., & Sproull, L. (1982). Managerial response to changing environments: Perspectives on problem sensing from social cognition. *Administrative Science Quarterly, 27*, 548–570.

Knight, K. E., & McDaniel, R. (1979). *Organizations: An information systems perspective*. Belmont, CA: Wadsworth.

Kontogiannis, T. (1999). User strategies in recovering from errors in man-machine systems. *Safety Science, 32*, 49–68.

Kontogiannis T., Kossiavelou Z., & Marmaras N. (2002). Self-reports of aberrant behaviour on the roads: Errors and violations in a sample of Greek drivers. *Accident Analysis and Prevention, 34*, 381–399.

Kraut, R. E., Hanson, S. J., & Farber, J. M. (1983). Command use and interface design. *Proceedings of CHI '83 Conference on Human Factors in Computing Systems*, 120–123.

Lawrence, R. J., & Lee, J. (2004, August). Financial modeling of project financing transactions. *Institute of Actuaries of Australia Financial Services Forum*, pp. 26–27.

Lawton, R. (1998). Not working to rule: Understanding procedural violations at work. *Safety Science, 28*, 77–95.

Levitt, B., & March, J. G. (1988). Organizational learning. In W. R. Scott & J. Blake (Eds.), *Annual review of sociology* (Vol. 14, pp. 319–340). Palo Alto, CA: Annual Reviews.

Locke, E. A., & Latham, G. P. (2002). Building a practically useful theory of goal setting and task motivation. *American Psychologist, 57*, 705–717.

Lord, R. G., & Levy, P. E. (1994). Moving from cognition to action: A control theory perspective. *Applied Psychology: An International Review, 43*, 335–366.

Louis, M. R., & Sutton, R. I. (1991). Switching cognitive gears: From habits of mind to active thinking. *Human Relations, 44*, 55–75.

March, J., & Simon, H. A. (1958). *Organizations*. New York: Wiley.

Maurino, D. E., Reason, J., Johnston, N., & Lee, R. B. (1995). *Beyond aviation human factors*. Hants, UK: Ashgate.

Meijman, T. F., & Mulder, G. (1998). Psychological aspects of workload. In P. J. D. Drenth, H. Thierry, & C. J. De Wolff (Eds.), *Handbook of work and organizational psychology* (2nd ed., Vol. 1, pp. 5–33). London: Psychology Press.

Merriam-Webster Online Dictionary. (2006). Retrieved June 2006 from http://www.m-w.com

Miller, D. (1990). *The Incarus paradox: How exceptional companies bring about their own downfall*. New York: HarperBusiness.

Miller, D. (1993). The architecture of simplicity. *Academy of Management Review, 18*, 116–138.

Miller, G. A., Galanter, E., & Pribram, K. H. (1960). *Plans and the structure of behavior*. London: Holt.

Mintzberg, H. (1973). *The nature of managerial work*. New York: Harper & Row.

Morgeson, F. P., & Hofmann, D. A. (1999). The structure and function of collective constructs: Implications for research and theory development. *Academy of Management Review, 24*, 249–265.

Nelson, R. R., & Winter, S. G. (1982). *An evolutionary theory of economic change.* Cambridge, MA: Harvard University Press.

Norman, D. A. (1981). Categorization of action slips. *Psychological Review, 88,* 1–15.

Norman, D. A. (1984). Stages and levels in human-machine interaction. *International Journal of Man-Machine Studies, 21,* 365–375.

Norman, D. A. (1988). *The psychology of everyday things.* New York: Basic Books.

Norman, D. A., & Bobrow, D. G. (1975). On data-limited and resource-limited processes. *Cognitive Psychology, 7,* 44–64.

O'Reilly, C. A., Chatman, J. A., & Caldwell, D. F. (1991). People and organizational culture: A profile comparison approach to person-organization fit. *Academy of Management Journal, 34,* 487–516.

Panko, R. R. 1988. What we know about spreadsheet error. *Journal of End User Computing, 10,* 15–21.

Panko, R. R. 2005. *What we know about spreadsheet errors.* Unpublished manuscript. Retrieved June 2006 from http://panko.cba.hawaii.edu/ssr/Mypapers/whatknow.htm

Pentland, B. T., & Rueter, H. H. (1994). Organizational routines as grammars of action. *Administrative Science Quarterly, 39,* 484–510.

Perrow, C. (1984). *Normal accidents: Living with high risk technologies.* New York: Cambridge University Press.

Peters, T. (1987). *Thriving on chaos.* New York: Harper & Row.

Pfafferott, I., & Huguenin, R. D. (1991). Adaptation nach Einführung von Sicherheitsmaßnahmen. *Zeitschrift für Verkehrssicherheit, 37,* 71–83.

Ramanujam, R., & Goodman, P. S. (2003). Latent errors and adverse organizational consequences: A conceptualization. *Journal of Organizational Behavior, 24,* 815–836.

Rasmussen, J. (1983). Skills, rules, and knowledge; signals, signs, and symbols, and other distinctions in human performance models. *IEEE Transactions on Systems, Man, and Cybernetics, SMC-13,* 257–266.

Rasmussen, J. (1987). Cognitive control and human error mechanisms. In K. Rasmussen, J. Duncan, & J. Leplat (Eds.), *New technology and human error* (pp. 53–61). London: Wiley.

Reason, J. (1987). The Chernobyl errors. *Bulletin of the British Psychological Society, 40,* 201–206.

Reason, J. (1990). *Human error.* New York: Cambridge University Press.

Reason, J. (1997). *Managing the risks of organizational accidents.* Aldershot, UK: Ashgate.

Reason, J. T., Manstead, A. S., R., Stradling, S. G., Baxter, J. S., & Campbell, K. A. (1990). Errors and violations on the road: A real distinction? *Ergonomics, 33,* 1315-1332.

Rybowiak, V., Garst, H., Frese, M., & Batinic, B. 1999. Error Orientation Questionnaire (EOQ): Reliability, validity, and different language equivalence. *Journal of Organizational Behavior, 20,* 527–547.

Sellen, A. 1994. Detection of everyday errors. *Applied Psychology: An International Review, 43,* 475–498.

Shiffrin, R. M., & Schneider, W. (1977). Controlled and automatic human information processing: II. Perceptual learning, automatic attending, and a general theory. *Psychological Review, 84,* 127–190.

Shneiderman, B. (1987). *Designing the user interface: Strategies for effective human-computer interaction.* Reading, MA: Addison-Wesley.

Shrivastava, P., Mitroff, I. I., & Alvesson, M. (1987). Nonrationality in organizational actions. *International Studies of Management and Organization, 17,* 90–109.

Sitkin, S. B. (1992). Learning through failure: The strategy of small losses. *Research in Organizational Behavior, 14,* 231–266.

Smelcer, J. B. (1989). *Understanding user errors in database query.* Unpublished dissertation thesis, University of Michigan, Ann Arbor.

Starbuck, W. H. (1989). Why organizations run into crises ... and sometimes survive them. In K. C. Laudon & J. A. Turner (Eds.), *Information technology and management strategy* (pp. 11–33). Englewood Cliffs, NJ: Prentice Hall.

Starbuck, W. H., & Farjoun, M. (2005). *Organization at the limit: Lessons from the Columbia disaster.* New York: Wiley.

Starbuck, W. H., Greve, A., & Hedberg, B. L. T. (1978). Responding to crisis. *Journal of Business Administration, 9,* 111–137.

Starbuck, W. H., & Milliken, F. J. (1988a). *Challenger*: Fine-tuning the odds until something breaks. *Journal of Management Studies, 25,* 319–340.

Starbuck, W. H., & Milliken, F. J. (1988b). Executives' perceptual filters: What they notice and how they make sense. In D. C. Hambrick (Ed.), *The executive effect: Concepts and methods for studying top managers* (pp. 35–65). Greenwich, CT: JAI.

Staw, B. M., Sandelands, L. E., & Dutton, J. E. (1981). Threat rigidity effects in organizational behavior: A multi-level analysis. *Administrative Science Quarterly, 26,* 501–524.

Stetzer, A., & Hofmann, D. A. (1996). Risk compensation: Implications for safety interventions. *Organizational Behavior and Human Decision Processes, 66,* 73–88.

van Dyck, C., Frese, M., Baer, M., & Sonnentag, S. (2005). Organizational error management culture and its impact on performance: A two-study replication. *Journal of Applied Psychology, 90,* 1228–1240.

Vaughan, D. (1996). *The Challenger launch decision: Risky technology, culture, and deviance at NASA.* Chicago: University of Chicago Press.

Walsh, J. P., & Ungson, G. R. (1991). Organizational memory. *Academy of Management Review, 16,* 57–91.

Weick, K. E. (1990). The vulnerable system: An analysis of the Tenerife air disaster. *Journal of Management, 16,* 571–593.

Wilde, G. J. S. (1982). The theory of risk homeostasis: Implications for safety and health. *Risk Analysis, 2,* 209–225.

Wilde, G. J. S. (1988). Risk homeostasis theory and traffic accidents: Propositions, deductions and discussion of dissension in recent reactions. *Ergonomics, 31,* 441–468.

Zapf, D., Brodbeck, F. C., Frese, M., Peters, H., & Prümper, J. (1992). Errors in working with computers: A first validation of a taxonomy for observed errors in a field setting. *International Journal of Human-Computer Interaction, 4,* 311–339.

Zapf, D., Maier, G. W., Rappensperger, G., & Irmer, C. (1994). Error detection, task characteristics and some consequences for software design. *Applied Psychology: An International Review, 43,* 499–520.

Zapf, D., & Reason, J. T. (1994). Human errors and error handling. *Applied Psychology: An International Review, 43,* 427–432.

2

Learning Through Errors in Training

Nina Keith

Traditionally, errors are ascribed a negative role for learning during training. A well-known example is the Skinnerean approach, which equates errors with punishment that can inhibit behavior but does not contribute to learning. In a similar vein, social-cognitive theory supports a guided and error-free learning environment: Learners should be "spared the costs and pain of faulty effort" (Bandura, 1986, p. 47). Yet, in the past decades, training research has increasingly acknowledged that errors can actually be beneficial for learning; therefore, errors should explicitly be incorporated in training procedures (e.g., Frese et al., 1991; Heimbeck, Frese, Sonnentag, & Keith, 2003; Ivancic & Hesketh, 1995–1996). The underlying idea is that errors provide negative but informative feedback as they pinpoint needs for further improvement in knowledge and skills. Also, errors are a by-product of active learning and of adaptation to environmental changes. Whenever new skills are required to master a novel task or whenever work requirements change and established routines become obsolete, people will inevitably make errors (Bauer & Mulder, 2007). It is therefore worthwhile to employ training methods that prepare employees for coping with these changes. Training methods that incorporate errors may be better suited than error-avoidant training methods to make trainees flexible and ready for change (cf. Frese, 1997; Smith, Ford, & Kozlowski, 1997).

The present chapter reviews existing research on approaches that explicitly incorporate errors into training. The emphasis is on one particular training method, called error management training, because for this training method there is the relatively largest body of research available. Special consideration is given to the issue of transfer, that is, to the question if and under what circumstances error management training can lead to better transfer performance than error-avoidant training approaches. Specifically, I review empirical evidence on the notion that error management training, compared to error-avoidant training, may impede immediate performance during training but will ultimately lead to better performance, particularly on tasks that require the development of a novel solution. A further section deals with the question of which psychological mechanisms (i.e., processes) underlie the effectiveness of the error management training method. I start by describing the concept of error management as well the design of training methods that incorporate ideas of error management.

Strategies of Dealing With Errors During Training: Error Avoidance Versus Error Management

As mentioned, errors are traditionally regarded as negative events that should be avoided during training (e.g., Bandura, 1986; Skinner, 1953). A measure to protect trainees from committing errors during training is to provide tight structure and step-by-step guidance through the training material and the procedures to be learned, as is done in conventional tutorials. In fact, for rehabilitation of memory impairment in clinical populations, error-avoidant training has been shown to be an effective technique (Kessels & de Haan, 2003). In the domain of computer skill acquisition, some studies demonstrated shorter training times for training methods that prevented trainees from making errors during training (e.g., by means of blocking some functions of the computer program; Carroll & Carrithers, 1984). It seems that error-avoidant training methods may be useful for the acquisition of some well-defined skills. For organizational training, however, I argue and review evidence suggesting that training methods that do not prevent errors but explicitly incorporate them in training are most promising to boost transfer of skills (cf. Frese, 1995; Frese et al., 1991; Heimbeck et al., 2003; Hesketh, 1997; Ivancic & Hesketh, 1995–1996).

In contrast to error-avoidant approaches, an error management approach acknowledges that errors are ubiquitous. Despite all efforts, errors cannot be completely avoided, even among experts (Prümper, Zapf, Brodbeck, & Frese, 1992)—and this applies all the more to performance during training, when knowledge and skills of trainees are still to be developed. As a consequence, strategies of error avoidance should be supplemented by strategies of error management, which are directed at quickly reacting to and effectively dealing with errors after they have occurred (Frese, 1995). In addition, an error management approach proposes that errors are not only negative events but also may lead to positive consequences, such as innovation and learning—as long as they can be dealt with effectively (van Dyck, Frese, Baer, & Sonnentag, 2005). In other words, while an error-avoidant approach seeks to erect a barrier between the action and the potential error, an error management approach seeks to erect a barrier between the error and its potentially negative consequences (Frese, 1995). A more thorough discussion of the two organizational strategies of error prevention and error management is provided by Hofmann and Frese (Chapter 1, this volume).

Design of Training Methods That Incorporate Errors

Incorporating errors in training implies that trainees should be given ample opportunities to make errors during training. One such training method, called error management training, implements this idea by providing only minimal task information and by reducing direct instruction to a minimum. Trainees are encouraged to independently explore the training tasks without a tight structure imposed by a trainer. In computer software training, for example, trainees of error management training would receive only basic information on the functions of the software and how these can be activated (e.g., a list of commands, Frese et al., 1991; an introduction to the most important toolbars as well as the Undo and Delete functions, Keith & Frese, 2005). Based on this minimal information, trainees would be asked to work on relatively complex training tasks without additional guidance by the trainer—a procedure that inevitably leads to many errors and impasses.

With regard to the minimal guidance, error management training is similar to exploratory or discovery learning, which stresses the importance of allowing the learner to actively explore ideas and to test them (Bruner, 1966). Unlike pure discovery methods, however, error management training gives explicit training tasks as learning objectives, thereby providing at least a low degree of structure during training. In addition, there is greater emphasis on errors in error management training than in discovery learning. In most applications, so-called error management instructions are given that inform trainees about the positive function of errors during training to reduce potential frustration in the face of errors and impasses. Also, the trainer explicitly encourages trainees to make errors as they work independently on the training tasks and to use these errors to learn more about the task. The main ideas of this approach are summarized in brief statements that frame errors positively, such as "You have made an error? That is great because now you can learn something new!" (Figure 2.1; Frese et al., 1991; Heimbeck et al., 2003; Keith & Frese, 2005, 2008). These statements are repeatedly presented by the trainer throughout the training. When trainees make errors or face problems while working on the training tasks, the trainer would encourage them to solve the problem by themselves. Only in extreme cases in which trainees

> **Whenever you make an error, remember ...**
> "Errors are a natural part of the learning process!"
> "There is always a way to leave the error situation!"
> "Errors inform you about what you still can learn!"
> "The more errors you make, the more you learn!"

FIGURE 2.1
Positive error framing in error management instructions (From Frese et al., 1991, *Human-Computer Interaction, 6*, 77–93; Heimbeck et al., 2003, *Personnel Psychology, 56*, 333–361; Keith & Frese, 2005, *Journal of Applied Psychology, 90*, 677–691; Keith & Frese, 2008, *Journal of Applied Psychology, 93*, 59–69.)

cannot recover from the error by themselves would the trainer intervene (e.g., when a novice computer user accidentally closed the working file and does not know how to reopen it).

In sum, there are two characteristics that distinguish error management training from training methods that seek to avoid errors during training: (a) active exploration with minimal guidance and (b) explicit encouragement to make errors during training along with positive error framing in error management instructions (note that some authors refer to this type of training as "error training," e.g., Dormann & Frese, 1994; Heimbeck et al., 2003; other authors have used terms such as "error-filled training," Ivancic & Hesketh, 1995–1996, or "error-based training," Smith et al., 1997). Other training methods that incorporate the idea of error management provided the opportunity to make errors but without the explicit error encouragement and positive framing of errors (e.g., Ivancic & Hesketh, 2000, Study 1). Still others presented trainees with errors that were systematically preselected (by the trainer or the researchers) rather than exposing trainees to errors that they make unsystematically during active exploration (Ivancic & Hesketh, 2000, Study 2; Joung, Hesketh, & Neal, 2006; Lorenzet, Salas, & Tannenbaum, 2005; Rogers, Regehr, & MacDonald, 2002).

Effectiveness of Error Management Training and the Issue of Transfer

To evaluate training effectiveness, several studies compared outcomes of error management training and alternative training methods, such as error-avoidant training or exploratory training without the explicit encouragement and positive framing of errors that characterizes error management training. Early studies were mostly conducted in the domain of computer software skills and used rather small participant samples (e.g., Frese et al., 1991). These studies found error management training to lead to better posttraining performance (an overview of some early studies is provided by Frese, 1995). Later studies replicated the effect in larger samples (e.g., Chillarege, Nordstrom, & Williams, 2003; Heimbeck et al., 2003) or in other task domains, such as driving simulator training (Ivancic & Hesketh, 2000); decision making (Gully, Payne, Koles, & Whiteman, 2002); surgical skills training (Rogers et al., 2002); and firefighting (Joung et al., 2006). A meta-analysis identified 24 studies that compared error management training with an alternative training method and found a mean Cohen's d effect size of 0.44 in favor of error management training (Keith & Frese, 2008).

Despite the empirical evidence that generally supports the effectiveness of error management training, in some studies alternative training methods were equally effective or even better than error management training (Debowski, Wood, & Bandura, 2001; Gully et al., 2002). In addition, there are theoretical reasons why error management training cannot be expected to be effective for all types of training outcomes alike. In particular, two aspects of training outcomes need to be considered. First, error management training aims at improving performance after training as opposed to immediate training performance. This distinction between immediate within-training performance and posttraining performance is crucial as manipulations that positively affect training performance may negatively affect performance in the long run and vice versa (Bell & Kozlowski, 2008; Goodman & Wood, 2004; Hesketh, 1997; Keith & Frese, 2005, 2008; Schmidt & Bjork, 1992). This phenomenon is illustrated in a study from the domain of motor learning (Shea & Morgan, 1979). Participants learned three movement tasks in either blocked-practice or random-order condition. During training, the blocked-practice group performed better than the random-order group. On a posttraining transfer task, however, the pattern reversed, and the random-order group performed better, particularly if the transfer task was presented in random order.

During error management training, trainees are explicitly encouraged to explore the task and to make errors. As a result, immediate training performance in terms of error rate, efficiency, or training time may be depressed as trainees explore and experiment and sometimes arrive at suboptimal solutions. The benefits of error management training are expected to unfold on posttraining tasks, when errors are no longer encouraged and trainees are aware that their performance is now being evaluated. It is therefore important to distinguish training performance conceptually and operationally from posttraining performance when conducting studies to evaluate training effectiveness (cf. the distinction of acquisition vs. retention performance as indicators of learning by Schmidt & Bjork, 1992). In line with this reasoning, the meta-analysis by Keith and Frese (2008) found that the evaluation phase moderated the magnitude of the effect sizes. Of the 24 studies, most (but not all) used a study design that included a separate posttraining test phase to evaluate performance; some studies, however, did not include a separate test phase but used training performance to evaluate performance. The former studies yielded significantly larger effect sizes than the latter. Also, the studies that used training performance for evaluation yielded a mean effect size that did not differ from zero. In other words, across studies, error management training led to better posttraining performance than alternative training methods, whereas for immediate training performance, error management training and alternative training methods fared equally well.

The second aspect pertains to the type of transfer affected by error management training. Transfer implies that knowledge and skills are "transferred from one task or job to another" (Hesketh, 1997, p. 318). Depending on the similarity between training tasks and transfer tasks, two types of transfer can be distinguished (Ivancic & Hesketh, 2000). *Analogical transfer* refers to situations for which problem solutions are familiar or analogous; *adaptive transfer* comprises "using one's existing knowledge base to change a learned procedure, or to generate a solution to a completely new problem" (Ivancic & Hesketh, 2000, p. 1968; see Barnett & Ceci, 2002, for a similar distinction between near and far transfer). Adaptive transfer tasks are, beyond being more difficult, structurally distinct from training tasks. For solving adaptive transfer tasks, rote application of procedures learned in training is insufficient as they require the development of new procedures. In computer training that teaches a spreadsheet program, for example, if the training task were to produce and format a table, then producing and formatting a different table would constitute analogical transfer; creating and formatting a diagram, however, would constitute adaptive transfer.

In some situations, promoting analogical transfer may be the training goal (e.g., when the job requires a limited behavioral repertoire that can be taught within the allotted training time). In many organizational applications, however, adaptive transfer is probably more important because not all potential work-related problems and their solutions can be explained and practiced during the limited training time (Hesketh, 1997). For example, not all functions and potential pitfalls of a new computer program can be taught within a 1-day training session. Back on the job, trainees may encounter problems with the software that they have to solve independently without the additional assistance that was available in the protected training situation.

Training researchers have argued that training methods that encourage independent exploration and making errors during training, such as error management training, may be particularly suited to promote adaptive transfer (Ivancic & Hesketh, 1995–1996; Smith et al., 1997) because error management training resembles the transfer situation more than error-avoidant training methods. When errors and impasses occur in the transfer situations, trainees of error management training may be better prepared as they have learned to deal with errors during training. Trainees of error-avoidant training, in contrast, face these difficulties for the first time. This issue is captured in the principle of transfer-appropriate processing, which postulates that those processes required on transfer tasks should be practiced in training (Morris, Bransford, & Franks, 1977). In a way, error management training reduces the psychological distance between the training and the transfer situation (Heimbeck et al., 2003).

For promoting analogical transfer, error management training may also be useful because errors made during training potentially facilitate the retrieval of similar problems and their solutions (Ivancic & Hesketh, 2000). In fact, research suggests that cues for which an individual makes incorrect predictions are learned better than cues for which the initial predictions were correct and that prediction errors trigger attentional resources (Wills, Lavric, Croft, & Hodgson, 2007). However, error-avoidant training methods, which directly teach the correct strategy whose rote application is sufficient to solve the analogous problem, may be effective as well. As a result, a comparison of both training methods would not reveal any performance differences. In line with this argument, meta-analytical results found type of transfer to moderate the magnitude of effect sizes. Studies that used adaptive transfer tasks to evaluate training effectiveness found a large mean Cohen's d effect size of 0.80, whereas studies that used analogical transfer tasks yielded a small (but significant) mean effect size of 0.20 (Keith & Frese, 2008).

The previous sections have argued on a rather general level why incorporating errors during training may be useful for learning and performance. The next sections are more detailed about the psychological processes that may be instigated in error management training and that, in turn, may be associated with enhanced transfer performance. Also, empirical evidence concerning these processes is reviewed.

Processes in Error Management Training That Promote Transfer

Several processes to explain the effectiveness of error management training have been proposed in the literature, although only a few studies directly tested these potentially effective mechanisms. The approaches can be grouped based on whether they focus on cognitive or on emotional/motivational processes. In addition, self-regulatory approaches focus on the regulation of both cognitions and emotions.

Cognitive Processes in Error Management Training

Error management training may facilitate the development of a good mental model of the subject to be learned (Frese, 1995; Heimbeck et al., 2003). Errors, particularly if they were unexpected, may attract learners' attention (cf. Wills et al., 2007), which is then devoted to understanding the cause of the error and to finding a solution to the problem (Ivancic & Hesketh, 1995–1996). Also, errors during exploration provide negative but informative feedback as they point out areas of misunderstanding and help to detect incorrect assumptions (Heimbeck et al., 2003; Ivancic & Hesketh, 1995–1996). Action theory stresses the importance of action-oriented mental models for successful work-related actions; these mental models may develop best if learners actively tackle a problem rather than passively receive correct instructions (Frese & Zapf, 1994; Hacker, 1998; Keith & Frese, 2005, 2008). In addition, as learners explore, experiment, and make errors, premature automatization of actions is disrupted (Frese, 1995; Frese & Zapf, 1994). Learners' attention is constantly triggered, and they are encouraged to switch from automatic processing to conscious and deeper-level processing (cf. Heimbeck et al., 2003; Ivancic & Hesketh, 1995–1996). Error-avoidant training, on the other hand, in which participants receive detailed instructions on task solution, requires less-deep processing because one can, in principle, follow these instructions without really reflecting on what one is doing. As a result, the development of a rich mental model may be impeded.

In line with the idea that exploration during error management training is important for its effectiveness, Dormann and Frese (1994) found the extent of exploration during training to be related to posttraining performance (for methodological reasons, mediation could not be tested in their study as the measure of amount of exploration differed between the error-avoidant and error management training conditions). Also, Bell and Kozlowski (2008) investigated effects of several elements of active learning approaches and found exploration (vs. step-by-step guidance) to benefit performance.

Emotional and Motivational Processes in Error Management Training

Experiencing errors is usually associated with negative emotions, such as anger or anxiety, as well as with reduced motivation and frustration. Learning through errors can only occur if these potential negative consequences of errors can be reduced (Frese, 1995; Heimbeck et al., 2003;

Ivancic & Hesketh, 1995–1996; Keith & Frese, 2005). From a resource allocation perspective, motivational processes determine how much attentional effort is devoted to task-related cognitive activities (Kanfer & Ackerman, 1989). Also, if off-task processes such as negative emotional reactions absorb one's attentional capacities, task-related processing and learning are unlikely to occur (Kanfer & Ackerman, 1989; Kluger & DeNisi, 1996). It therefore seems worthwhile to consider emotional and motivational processes in error management training. Studies examining such processes, however, have provided inconsistent results. Wood, Kakebeeke, Debowski, and Frese (2000) hypothesized that intrinsic motivation mediates the effectiveness of error management training for learning an electronic search task. Yet, they found intrinsic motivation to be negatively related to the quality of participants' search strategy and unrelated to performance. Building on earlier studies on error management training (Frese et al., 1991), Nordstrom, Wendland, and Williams (1998) investigated the effect of error management training on frustration. They found that, over the course of training and posttraining transfer phases, frustration decreased in participants of error management training, whereas frustration increased in participants of error-avoidant training (this study did not test mediation effects of frustration). In a replication study, however, no such group differences were found (Chillarege et al., 2003).

In line with the notion that a positive error-tolerant attitude, as is conveyed in error management instructions, is important for the effectiveness of error management training, the meta-analysis of Keith and Frese (2008) identified both active exploration and error management instructions to be effective elements. Similarly, a study that varied both the provision of error management instructions and the type of training (exploratory training vs. step-by-step instructions) found both elements to be effective (Bell & Kozlowski, 2008).

Self-Regulation in Error Management Training and the Issue of Transfer-Appropriate Processing

The cognitive and emotional/motivational approaches can be integrated within a self-regulatory perspective that acknowledges the significance of both (Bell & Kozlowski, 2008; Keith & Frese, 2005). *Self-regulation* refers to processes "that enable an individual to guide his or her goal-directed activities over time" and can be directed at the "modulation of thought, affect, behavior, or attention" (Karoly, 1993, p. 25). Self-regulation may be expected to be particularly important for error management training because this training method offers only a low degree of structure and little external guidance (Keith & Frese, 2005). Also, it may be argued that error management training fosters self-regulatory skills, and that these skills in turn prove valuable when trainees are confronted with novel problems that were not practiced in training. In an attempt to test this notion, Keith and Frese (2005) found two self-regulatory skills, emotion control (i.e., self-regulation of emotions) and metacognition (i.e., self-regulation of cognitions), to mediate effects of error management training on adaptive transfer performance (see also Bell & Kozlowski, 2008). Their conceptual model is depicted in Figure 2.2.

Apparently, participants of error management training learned to exert emotion control aimed at reducing negative emotional reactions to errors and setbacks (Kanfer, Ackerman, & Heggestad, 1996). Participants also engaged in metacognitive activities of planning, monitoring, evaluation, and revision of task strategies during training (Brown, Bransford, Ferrara, & Campione, 1983). Error management training stimulates such metacognitive activities because "errors prompt learners to stop and think about the causes of the error" (Ivancic & Hesketh, 2000, p. 1968). They then have to come up with potential solutions to the problem, implement them, and monitor their effectiveness (Ivancic & Hesketh, 2000).

It is important to note that in the study by Keith and Frese (2005), metacognition *during training* (assessed based on think-aloud protocols) explained performance differences on *posttraining* adaptive transfer tasks. This finding is in line with the principle of transfer-appropriate processing (Morris et al., 1977) mentioned in this chapter. The idea is that during error management training, participants not only acquire specific knowledge and skills that are directly related to the task but also acquire general self-regulatory skills. These skills in turn are useful in the transfer situation when no external guidance is available and self-regulation is required (cf. Ford, Smith, Weissbein, Gully, & Salas, 1998).

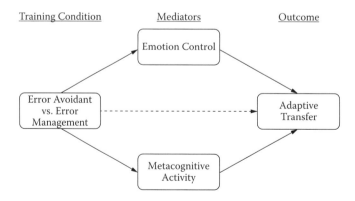

FIGURE 2.2
Self-regulatory processes as mediators of performance effects (conceptual model adapted from Keith & Frese, 2005, *Journal of Applied Psychology, 90,* 677–691). The dashed line denotes the direct effect that is accounted for by the mediator variables.

Individual Differences and Error Management Training

The general advantages of error management training notwithstanding, trainees may benefit differentially from this training method depending on certain trainee characteristics (cf. Snow, 1986). It is conceivable that not all trainees benefit from independent exploration and error encouragement alike. For some trainees, independent exploration with the prospect to make errors may be an anxiety-provoking situation or a task that is cognitively too demanding (cf. Kirschner, Sweller, & Clark, 2006). Also, trainees can widely vary with regard to the types of exploration strategies in which they spontaneously engage (e.g., systematic exploration vs. trial and error), and the choice of strategies may determine their learning and performance (Van der Linden, Sonnentag, Frese, & Van Dyck, 2001). Finally, the same training instructions can have differential impacts on trainee behavior depending on trainee characteristics (e.g., Schmidt & Ford, 2003).

Existing studies have investigated moderating effects of motivational and cognitive variables on training outcomes and on process variables. Heimbeck et al. (2003) investigated moderating effects of dispositional goal orientations (Dweck & Leggett, 1988; VandeWalle, 1997), that is, an individual's tendency to adopt the goal of competence improvement and mastery (i.e., learning goal orientation), the goal of demonstrating competence (i.e., prove goal orientation), or the goal of avoiding to appear incompetent (i.e., avoidance goal orientation). Heimbeck et al. expected trainees with a high-prove or a high-avoid goal orientation to benefit more from error-avoidant than from error management training and found partial support for this expectation. For trainees high in learning goal orientation, however, the study did not find the expected advantage of error management training. Overall, performance was best in error management training—irrespective of individuals' goal orientations. In other words, whether trainees benefited from error-avoidant training depended on their goal orientations, whereas all trainees benefited equally well from error management training. Gully et al. (2002) found that trainees with high cognitive ability or more openness benefit more from error encouragement than trainees with low cognitive ability or less openness (however, their study did not include a posttraining test phase but used performance during training as a criterion). Finally, Bell and Kozlowski (2008) found that trainees with higher cognitive ability show more metacognitive activity when they had received an exploratory rather than a guided error-avoidant training; for lower-ability trainees, the level of metacognitive activity did not differ between training conditions. Also, error encouragement instructions led to higher state learning orientation only among trainees with a low dispositional learning orientation.

In sum, studying aptitude–treatment interactions (i.e., interactions between training method and trainee characteristics) has the potential to advance theoretical knowledge about cognitive and motivational processes instigated in training. It also has practical implications as the results can specify which trainees may benefit most from specific training methods. In interpreting the existing research, however, it should be noted that research on such interactions is entirely based on samples of college and university students. To the extent that students differ from the general population (e.g., above-average cognitive abilities, different pattern of motivational goal orientations), generalizations to the organizational context need to be made with caution (cf. Bell & Kozlowski, 2008).

Implications for Training Research and Practice

The research on error management training reviewed in this chapter has several implications for training research and practice. First, the existing research shows that explicitly incorporating errors in training benefits learning. This finding is in contrast to traditional approaches that focus on teaching correct behaviors during training and that deny any positive functions of errors for learning (e.g., Bandura, 1986; Skinner, 1953). Trainers may consider integrating modules of exploration and error encouragement within their training programs and should resist the temptation to offer immediate help when participants face errors and impasses during training. By having participants solve such problems independently, they are given the opportunity to develop self-regulatory skills (i.e., emotion control and metacognition) that are useful when confronted with novel tasks—provided that they view errors positively and do not react with anger and frustration. To prevent such negative emotions, error management instructions and positive error statements (see Figure 2.1) may be integrated in practically any type of training as these instructions are effective and inexpensive, simple, and easy to administer (cf. Beier & Campbell, 2007).

Research on error management training has supported the claim made by some training researchers (e.g., Hesketh, 1997; Schmidt & Bjork, 1992) that interventions that improve immediate within-training performance may be detrimental to learning and, conversely, that interventions that slow or impede immediate training performance may be beneficial for learning and performance in the long run. In error management training, immediate within-training performance may be worse than in training methods that directly teach correct behaviors; on posttraining transfer tasks, however, participants of error management training fare better than their counterparts from error-avoidant training, particularly if these tasks involve the development of new solutions to problems that were not practiced during training (i.e., adaptive transfer). This should be kept in mind both by training researchers conducting studies on training effectiveness and by practitioners delivering training. Trainers should not mistakenly be led to the conclusion that their intervention produced real learning only because immediate within-training performance is enhanced or, conversely, that participants are not learning just because they are not performing well on a training task. As Schmidt and Bjork (1992) point out, trainers should also not be misled by the subjective reaction of their own trainees, who may favor a training method that in fact produces less learning. In the case of error management training, it is imaginable that some trainees may prefer receiving correct task instructions to taking the effortful "detour" of independent exploration and errors—irrespective of the empirical evidence showing that taking this detour may be worthwhile. In some cases, it may be a challenge for trainers to motivate and enable trainees to benefit from error management training.

Trainers may opt to use error management training whenever the training goal is to promote adaptive transfer, such as when the skills needed on the job are too diverse and extensive to be covered completely during training time (i.e., adaptive transfer). If, however, the training goal is to learn a specific procedure that is required in the same manner on the job (i.e., analogical transfer), it may be suitable to teach and practice this procedure directly to establish a routine. Ivancic and Hesketh (1995–1996) argued that the training goal should determine whether an error-avoidant or error management training is used. The meta-analytical finding that the effectiveness of error management training was larger for adaptive than for analogical transfer tasks provides empirical support for this suggestion. An example is the study by Debowski et al. (2001), who found that a guided training group performs better than an error management training group. In this study, both the training and the posttraining transfer tasks were to conduct electronic literature searches concerning a particular research question. This task was solved best using a specific search strategy (identify relevant keywords, connect keywords to form a search statement, etc.). In the guided group, this strategy was taught directly, and participants were interrupted and corrected whenever they deviated from the prescribed steps. Participants in the error management group (called enactive exploration in this study) were informed about the eight steps but otherwise worked independently and without any corrective feedback. On the transfer task, guided participants probably performed better because they had developed a routine of using the prescribed steps, which they had practiced extensively during training and whose rote application was the best strategy to solve the new but analogical transfer task (cf. Debowski et al., 2001).

The present chapter focused on error management training that involves active exploration and encouragement to make errors and to learn from them because for this training method a considerable amount of research exists. As mentioned, however, the theoretical ideas pertaining to learning from errors apply to other ways to incorporate errors in training as well. An example is the study by Joung et al. (2006), who trained firefighters with a "vicarious" error management training method. Participants did not explore and make errors by themselves but were presented with actual or potential errors made by others in firefighting scenarios (i.e., firefighting stories based on real incidents). Participants of the group who received vicarious error management training outperformed their counterparts who received errorless scenarios on new scenarios that were presented posttraining. Vicarious error management training of this kind may be better suited than active error management training when the prospect of making errors oneself is too threatening and stressful or in which there is no safe environment available (e.g., high-resolution simulators) that allows for active exploration and experimentation by trainees (cf. Joung et al., 2006). Trainers may come up with even more ways of integrating errors in training—ways that are best suited for their particular training content and goals.

Open Research Questions and Further Considerations

Since its development in the 1980s, there has been a growing body of research concerning the effectiveness of error management training, its underlying mechanisms, and the role of personal characteristics in training. Most of the existing research, however, was concerned with the acquisition of computer skills using error management training. This popularity of computer tasks in error management training research is probably due to several reasons. First, computer skills are relevant for many tasks, both on the job and in everyday life (Debowski et al., 2001; Quinones & Ehrenstein, 1997). Second, many computer tasks are complex enough to elicit errors during training while providing a safe environment in which most errors can be corrected. Also, most computer tasks provide task-generated feedback. Trainees are, in principle, able to observe changes of the system (e.g., visual changes on the display) in response to their actions and to use this information as feedback to evaluate and improve one's task strategies (Frese & Zapf, 1994; Neubert, 1998). While feedback is generally considered important for learning and performance (e.g., Ilgen, Fisher, & Taylor, 1979), task-generated feedback may be of particular importance for error management training as trainees work independently without having a trainer who constantly monitors the correctness of their actions. In addition, task-generated feedback implies that errors can be detected in the first place—a condition that is prerequisite for learning from errors.

The positive function of errors for learning and the mechanisms involved in learning from errors may apply to a wide range of skills. Empirical evidence for this proposition, however, remains scarce. The meta-analysis by Keith and Frese (2008) included only studies that either taught computer skills or delivered computer-based training (e.g., decision-making tasks delivered on the computer). Other studies that explicitly incorporated errors involved driving simulation training (Ivancic & Hesketh, 2000) and training of firefighting skills (Joung et al., 2006). Future research could continue to use error management training in other domains. Also, all existing studies used some kind of a posttraining performance test; none of the existing studies actually measured on-the-job performance to evaluate training effectiveness (cf. Ghodsian, Bjork, & Benjamin, 1997), a research gap that should be filled. From a theoretical perspective, there is reason to expect error management will benefit transfer of skills on the job. First, despite all efforts, errors will occur on the job as well, and there is usually no trainer available who immediately assists with error handling. Error management training probably prepares for this situation better than guided trainings because trainees of error management training learn to deal with errors independently from the beginning (Frese, 1995; Ivancic & Hesketh, 1995–1996). Second, research indicated that participants of error management training learn to apply self-regulatory skills of emotion control and metacognition when confronted with new tasks (Keith & Frese, 2005). These skills, which help to remain focused on the task and to select and adjust one's task strategies according to task requirements, may serve as generalizable skills that benefit a wide range of tasks on the job (cf. Ford et al., 1998). For the same reason, it may be speculated that error management training can contribute to general flexibility and adaptability of employees. Changing work requirements always imply the possibility to make errors. Employees who adopt a positive attitude toward errors and who are able to use self-regulatory skills to cope with new demands may be better prepared for change situations than employees who rely on external guidance or routinized skills (cf. Smith et al., 1997).

Studies dealing with the effectiveness of error management training used a comparative training design in which error management training was contrasted with an alternative training method. In many cases, this alternative training was an error-avoidant training that mimicked conventional tutorials. This training provided detailed written instructions that guided participants to correct task solutions. Compared with this particular type of error-avoidant training, error management consistently fared better (e.g., Chillarege et al., 2003; Frese et al., 1991; Heimbeck et al., 2003; Keith & Frese, 2005; Nordstrom et al., 1998). It may be argued, however, that this type of error-avoidant training is not representative of all guided training approaches, and that there may be other error-avoidant training methods that are more effective—even more effective than error management training. In the organizational training literature, behavior-modeling training, which is derived from social-cognitive theory, is a popular and effective training method that stresses the importance of guidance and that promotes error avoidance (Bandura, 1986; a meta-analysis on behavior-modeling training is provided by Taylor, Russ-Eft, & Chan, 2005). In behavior modeling, trainees watch a model displaying the correct behaviors, they practice these behaviors, and they receive feedback and social reinforcement following practice (e.g., Latham & Saari, 1979). Future research could contrast error management training with behavior-modeling training or, even better, go beyond mere comparisons and develop training methods that combine advantages of both approaches (Joung et al., 2006). As described, one study that explicitly used a training method based on behavior modeling found this training method to be more effective than error management training on an analogical transfer task (Debowski et al., 2001). Given this research finding, it is conceivable that a combination of behavior-modeling and error management training could be useful in some applications. For example, behavior-modeling training could be used early in training to establish some basic routine skills; in later training phases, error management training could be used to boost self-regulatory skills that ultimately lead to adaptive performance.

When comparing error management training with behavior-modeling training, two more aspects should be kept in mind. First, while most of the error management training studies taught technical skills (i.e., computer skills), many behavior-modeling training studies taught social skills (e.g., supervisory skills; Taylor et al., 2005). Given the differences between these two skill domains (e.g., in terms of trainees' pretraining skill levels, clarity of task-generated feedback), the question remains regarding the extent that the existing findings apply to the other skill domain. Second, although Bandura's (1986) social-cognitive theory originally promotes error avoidance, a number of applications did in effect incorporate errors in training—"based on trainer intuition" (Baldwin, 1992, p. 152) rather than on social-cognitive theory. These applications included so-called negative models that displayed undesired behavior in addition to positive models that displayed desired behavior. In fact, in a study by Baldwin (1992), training with positive-only models led to better performance on tasks analogical to training tasks, whereas training with both positive and negative models led to better performance on a generalization task. In a meta-analysis, studies including both positive and negative models yielded larger overall effect sizes for job behavior criteria than studies using positive models only, while the pattern was reversed for declarative knowledge tests (Taylor et al., 2005). These findings are consistent with the view promoted here, that guided error-avoidant training approaches (such as behavior-modeling training) may be appropriate to instigate reproduction on analogical tasks, but that training methods that include errors in training are more powerful in promoting adaptive transfer to novel tasks and situations (cf. Tannenbaum & Yukl, 1992).

Finally, the research summarized in this chapter focused on how novices learn through errors in training. Yet, the potential positive functions of errors may apply for skilled performers and even experts. Research from expert performance indicates that individuals who eventually become experts in a particular task domain (e.g., chess, music, sports) have devoted several years of time and effort to specific practice activities—called deliberate practice—that are explicitly directed at improving one's performance (Ericsson, Krampe, & Tesch-Römer, 1993). These practice activities are designed, for example, by trainers or teachers, according to weaknesses and errors in one's current performance. Research further indicates that skilled individuals independently engage in similar activities as they pursue everyday activities such as typing (Keith & Ericsson, 2007) and work-related activities (Sonnentag & Kleine, 2000; Unger, Keith, Hilling, Gielnik, & Frese, 2009), and that the amount and quality of these activities are related to performance. In a way, deliberate practice at work may be conceived of as a self-regulated training on the job that incorporates errors. Future research may continue to investigate how errors at work, that is, performance errors that occur outside formal training activities, can be used by individuals and organizations to promote learning (cf. Bauer & Mulder, 2007).

References

Baldwin, T. T. (1992). Effects of alternative modeling strategies on outcomes of interpersonal-skills training. *Journal of Applied Psychology, 77*, 147–154.

Bandura, A. (1986). *Social foundations of thought and action: A social cognitive theory.* Englewood Cliffs, NJ: Prentice Hall.

Barnett, S. M., & Ceci, S. J. (2002). When and where do we apply what we learn? A taxonomy for far transfer. *Psychological Bulletin, 128*, 612–637.

Bauer, J., & Mulder, R. H. (2007). Modelling learning from errors in daily work. *Learning in Health and Social Care, 6*, 121–133.

Beier, M. F., & Campbell, M. (2007). *Age and learning in technology training: The importance of motivation and self-regulation.* Paper presented at the 22nd annual conference of the Society for Industrial and Organizational Psychology, April, New York.

Bell, B. S., & Kozlowski, S. W. J. (2008). Active learning: Effects of core training design elements on self-regulatory processes, learning, and adaptability. *Journal of Applied Psychology, 93*, 296–316.

Brown, A. L., Bransford, J. D., Ferrara, R. A., & Campione, J. C. (1983). Learning, remembering, and understanding. In J. H. Flavell & E. M. Markman (Eds.), *Handbook of child psychology* (Vol. 3, pp. 77–166). New York: Wiley.

Bruner, J. S. (1966). *Toward a theory of instruction*. Cambridge, MA: Harvard University Press.

Carroll, J. M., & Carrithers, C. (1984). Blocking learner error states in a training-wheels system. *Human Factors, 26*, 377–389.

Chillarege, K. A., Nordstrom, C. R., & Williams, K. B. (2003). Learning from our mistakes: Error management training for mature learners. *Journal of Business and Psychology, 17*, 369–385.

Debowski, S., Wood, R. E., & Bandura, A. (2001). Impact of guided exploration on self-regulatory mechanisms and information acquisition through electronic search. *Journal of Applied Psychology, 86*, 1129–1141.

Dormann, T., & Frese, M. (1994). Error management training: Replication and the function of exploratory behavior. *International Journal of Human-Computer Interaction, 6*, 365–372.

Dweck, C. S., & Leggett, E. L. (1988). A social-cognitive approach to motivation and personality. *Psychological Review, 95*, 256–273.

Ericsson, K. A., Krampe, R. Th., Tesch-Römer, C. (1993). The role of deliberate practice in the acquisition of expert performance. *Psychological Review, 100*, 363–406.

Ford, J. K., Smith, E. M., Weissbein, D. A., Gully, S. M., & Salas, E. (1998). Relationships of goal orientation, metacognitive activity, and practice strategies with learning outcomes and transfer. *Journal of Applied Psychology, 83*, 218–233.

Frese, M. (1995). Error management in training: Conceptual and empirical results. In C. Zucchermaglio, S. Bagnara, & S. U. Stucky (Eds.), *Organizational learning and technological change, Series F: Computer and systems sciences* (Vol. 141, pp. 112–124). Berlin: Springer.

Frese, M. (1997). Dynamic self-reliance: An important concept for work in the twenty-first century. In C. L. Cooper, & S. E. Jackson (Eds.), *Creating tomorrow's organizations* (pp. 399–416). New York: Wiley.

Frese, M., Brodbeck, F. C., Heinbokel, T., Mooser, C., Schleiffenbaum, E., & Thiemann, P. (1991). Errors in training computer skills: On the positive function of errors. *Human-Computer Interaction, 6*, 77–93.

Frese, M., & Zapf, D. (1994). Action as the core of work psychology: A German approach. In H. C. Triandis, M. D. Dunette, & L. M. Hough (Eds.), *Handbook of industrial and organizational psychology* (Vol. 4, pp. 271–340). Palo Alto, CA: Consulting Psychologists Press.

Ghodsian, D., Bjork, R. A., & Benjamin, A. S. (1997). Evaluating training *during* training: Obstacles and opportunities. In M. A. Quinones & A. Ehrenstein (Eds.), *Training for a rapidly changing workplace* (pp. 63–88). Washington, DC: American Psychological Association.

Goodman, J., & Wood, R. E. (2004). Feedback specificity, learning opportunities, and learning. *Journal of Applied Psychology, 89*, 809–821.

Gully, S. M., Payne, S. C., Koles, K. L. K., & Whiteman, J. A. K. (2002). The impact of error management training and individual differences on training outcomes: An attribute–treatment interaction perspective. *Journal of Applied Psychology, 87*, 143–155.

Hacker, W. (1998). *Allgemeine Arbeitspsychologie: Psychische Regulation von Arbeitstätigkeiten* [General industrial psychology: Mental regulation of working activities]. Bern, Switzerland: Huber.

Heimbeck, D., Frese, M., Sonnentag, S., & Keith, N. (2003). Integrating errors into the training process: The function of error management instructions and the role of goal orientation. *Personnel Psychology, 56*, 333–361.

Hesketh, B. (1997). Dilemmas in training for transfer and retention. *Applied Psychology: An International Review, 46*, 317–339.

Ilgen, D. R., Fisher, C. D., & Taylor, M. S. (1979). Consequences of individual feedback on behavior in organizations. *Journal of Applied Psychology, 64*, 359–371.

Ivancic, B., & Hesketh, K. (2000). Learning from error in a driving simulation: Effects on driving skill and self-confidence. *Ergonomics, 43*, 1966–1984.

Ivancic, K., & Hesketh, B. (1995–1996). Making the best of errors during training. *Training Research Journal, 1*, 103–125.

Joung, W., Hesketh, B., & Neal, A. (2006). Using "war stories" to train for adaptive performance: Is it better to learn from error or success? *Applied Psychology: An International Review, 55*, 282–302.

Kanfer, R., & Ackerman, P. L. (1989). Motivation and cognitive abilities: An integrative/aptitude–treatment interaction approach to skill acquisition [Monograph]. *Journal of Applied Psychology, 74*, 657–690.

Kanfer, R., Ackerman, P. L., & Heggestad, E. D. (1996). Motivational skills and self-regulation for learning: A trait perspective. *Learning and Individual Differences, 8*, 185–209.

Karoly, P. (1993). Mechanisms of self-regulation: A systems view. *Annual Review of Psychology, 44*, 23–52.

Keith, N., & Ericsson, K. A. (2007). A deliberate practice account of typing proficiency in everyday typists. *Journal of Experimental Psychology: Applied, 13*, 135–145.

Keith, N., & Frese, M. (2005). Self-regulation in error management training: Emotion control and metacognition as mediators of performance effects. *Journal of Applied Psychology, 90*, 677–691.

Keith, N., & Frese, (2008). Effectiveness of error management training: A meta-analysis. *Journal of Applied Psychology, 93*, 59–69.

Kessels, R. P., & de Haan, E. H. (2003). Implicit learning in memory rehabilitation: A meta-analysis on errorless learning and vanishing cues methods. *Journal of Clinical and Experimental Neuropsychology, 25*, 805–814.

Kirschner, P. A., Sweller, J., & Clark, R. E. (2006). Why minimal guidance during instruction does not work: An analysis of the failure of constructivist, discovery, problem-based, experiential, and inquiry-based teaching. *Educational Psychologist, 41*, 75–86.

Kluger, A. N., & DeNisi, A. (1996). Effects of feedback intervention on performance: A historical review, a meta-analysis, and a preliminary feedback intervention theory. *Psychological Bulletin, 119*, 254–284.

Latham, G. P., & Saari, L. M. (1979). Application of social-learning theory to training supervisors through behavioral modeling. *Journal of Applied Psychology, 64,* 239–246.

Lorenzet, S. J., Salas, E., & Tannenbaum, S. I. (2005). Benefiting from mistakes: The impact of guided errors on learning, performance, and self-efficacy. *Human Resource Development Quarterly, 16,* 301–322.

Morris, C. D., Bransford, J. D., & Franks, J. J. (1977). Levels of processing versus transfer appropriate processing. *Journal of Verbal Learning and Verbal Behavior, 16,* 519–533.

Neubert, M. J. (1998). The value of feedback and goal setting over goal setting alone and potential moderators of this effect: A meta-analysis. *Human Performance, 11,* 321–335.

Nordstrom, C. R., Wendland, D., & Williams, K. B. (1998). "To err is human": An examination of the effectiveness of error management training. *Journal of Business and Psychology, 12,* 269–282.

Prümper, J., Zapf, D., Brodbeck, F. C., & Frese, M. (1992). Errors of novices and experts: Some surprising differences in computerized office work. *Behaviour and Information Technology, 11,* 319–328.

Quinones, M. A., & Ehrenstein, A. (Eds.). (1997). *Training for a rapidly changing workplace.* Washington DC: American Psychological Association.

Rogers, D. A., Regehr, G., & MacDonald, J. (2002). A role for error training in surgical technical skill instruction and evaluation. *The American Journal of Surgery, 183,* 242–245.

Schmidt, A. M., & Ford, J. K. (2003). Learning within a learner control environment: The interactive effects of goal orientation and metacognitive instruction on learning outcomes. *Personnel Psychology, 56,* 405–429.

Schmidt, R. A., & Bjork, R. A. (1992). New conceptualizations of practice: Common principles in three paradigms suggest new concepts for training. *Psychological Science, 3,* 207–217.

Shea, J. B., & Morgan, R. L. (1979). Contextual interference effects on the acquisition, retention, and transfer of a motor skill. *Journal of Experimental Psychology: Human Learning and Memory, 5,* 179–187.

Skinner, B. F. (1953). *Science and human behavior.* New York: Free Press.

Smith, E. M., Ford, J. K., & Kozlowski, S. W. J. (1997). Building adaptive expertise: Implications for training design strategies. In M. A. Quinones & A. Ehrenstein (Eds.), *Training for a rapidly changing workplace* (pp. 89–118). Washington, DC: American Psychological Association.

Snow, R. E. (1986). Individual differences and the design of educational programs. *American Psychologist, 41,* 1029–1039.

Sonnentag, S., & Kleine, B. M. (2000). Deliberate practice at work: A study with insurance agents. *Journal of Occupational and Organizational Psychology, 73,* 87–102.

Tannenbaum, S. I., & Yukl, G. (1992). Training and development in work organizations. *Annual Review of Psychology, 43,* 399–441.

Taylor, P. J., Russ-Eft, D. F., & Chan, D. W. L. (2005). A meta-analytic review of behavior modeling training. *Journal of Applied Psychology, 90,* 692–709.

Unger, J. M., Keith, N., Hilling, C., Gielnik, M. M., & Frese, M. (2009). Deliberate practice among South African small business owners: Relationships with education, cognitive ability, knowledge, and success. *Journal of Occupational and Organizational Psychology, 82*, 21–44.

van der Linden, D., Sonnentag, S., Frese, M., & van Dyck, C. (2001). Exploration strategies, performance, and error consequences when learning a complex computer task. *Behavior and Information Technology, 20*, 189–198.

VandeWalle, D. (1997). Development and validation of a work domain goal orientation instrument. *Educational and Psychological Measurement, 57*, 995–1015.

van Dyck, C., Frese, M. Baer, M., & Sonnentag, S. (2005). Organizational error management culture and its impact on performance: A two-study replication. *Journal of Applied Psychology, 90*, 1228–1240.

Wills, A. J., Lavric, A., Croft, G. S., & Hodgson, T. L. (2007). Predictive learning, prediction errors, and attention: Evidence from event-related potentials and eye tracking. *Journal of Cognitive Neuroscience, 19*, 843–854.

Wood, R. E., Kakebeeke, B. M., Debowski, S., & Frese, M. (2000). The impact of enactive exploration on intrinsic motivation, strategy, and performance in electronic search. *Applied Psychology: An International Review, 49*, 263–283.

3

The Role of Errors in the Creative and Innovative Process

Michelle M. Hammond and James L. Farr

You can encourage and teach young people to observe, to ask questions when unexpected things happen. You can teach yourself not to ignore the unanticipated. Just think of all the great inventions that have come through serendipity, such as Alexander Fleming's discovery of penicillin, and just noticing something no one conceived of before.

Patsy O'Connell Sherman

Perhaps that morning in 1953 started out as any other day for Patsy O'Connell Sherman, a young chemist working for 3M in Minneapolis, Minnesota. She began working on her assigned task—to develop a synthetic rubber material for aircraft hoses—when her lab assistant accidentally dropped a glass bottle of synthetic latex on her white canvas tennis shoe. Perhaps she was initially upset at the careless assistant, whom she might later thank for the mistake that changed her life. After being unable to remove the spill and discovering that it resisted further staining, Sherman and her partner, Sam Smith, realized its commercial application. They spent several years perfecting the compound, which later became known as Scotchgard™, a household name in protective coating used on furniture and carpets.

For Sherman, an error in the lab led to a great innovation. She is not alone, for many times great ideas, inventions, and innovations come through identifying problems or making mistakes. Although creativity may emerge through errors, experiencing failure or committing errors is much more likely to happen when pursuing innovation. In this chapter, we identify processes and conditions in which errors are most likely to be used for the good of innovation, discuss the common types of errors that may occur in creative and innovative processes, and discuss rich research opportunities regarding the intersection between errors and workplace creativity and innovation.

Creativity and Innovation in the Workplace

Before discussing errors in relation to creativity and innovation, it is important to first define what we mean by these constructs. Creativity and innovation are highly interconnected; however, they are also often, erroneously, used interchangeably. Workplace innovation can be understood as a broader process that may include not only creativity but also the implementation of ideas within the work setting. Both creativity and innovation in the workplace specify that they must be, at least potentially, of use to or derive some benefit for the person, group, or organization (Zhou & Shalley, 2003). Creativity focuses on the generation of novel ideas, whereas innovation includes not only the generation of ideas but also the selection and implementation of the chosen alternatives (Unsworth, 2001). Further, creativity describes the creation of something new; however, innovation may also include the application of a product, procedure, or process already in use elsewhere, provided it is a new application within a particular role, work group, or organization (Anderson, De Dreu, & Nijstad, 2004).

Further, research on creativity and innovation has dissimilar historical roots. Traditionally, creativity has been considered primarily by psychologists, most commonly at the individual level. Alternatively, innovation has been studied frequently by economists, sociologists, and organizational theorists and is more commonly studied at the organizational level. However, more recently, these bodies of literature have been brought together with a number of multilevel and multifocus theories (e.g., Farr, Sin, & Tesluk, 2003; Staw, 1990), empirical studies distinguishing between creativity and innovation (Axtell et al., 2000; Oldham & Cummings, 1996), and meta-analyses comparing the two (Hammond, Neff, Farr, Schwall, & Zhao, in press).

Several models of the components of the creativity and innovation process have been proposed (Patterson, 2002). Although there are differences in terminology, Patterson noted that generally the models propose an initial "creativity" or ideation stage in which the task or problem is identified and further specified with alternative approaches or ideas developed and then an "implementation" stage in which alternatives are assessed with regard to the situational context and selected alternatives are implemented. We follow this distinction, further delineating each stage, similar to Farr et al.'s (2003) model of team innovation and also used by Hammond et al. (in press). The creativity stage includes a preparatory phase during which issue interpretation and problem identification (problem/opportunity identification) take place and then an action phase in which alternative ideas and solutions are generated (idea generation). Within the innovation implementation stage, the preparatory phase includes evaluation and

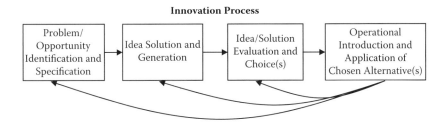

FIGURE 3.1
The innovation process.

selection of ideas (idea evaluation) and then an action stage that involves the actual implementation and application of the chosen solutions in the work situation (implementation). Figure 3.1 illustrates this process model.

We now briefly review the research literature related to each of the four stages shown in Figure 3.1. Although we discuss each component of the model independently, we acknowledge their interdependencies. Figure 3.1 presents a linear sequence of the components simply for the sake of parsimony. Rather, the process of creativity and innovation is more aptly described as nonlinear (Anderson et al., 2004), iterative, and even chaotic (Drazin, Glynn, & Kazanjian, 1999). Following our discussion of relevant research on the stages of creativity and innovation, we discuss how errors may both facilitate and hinder successful innovation in organizations.

Problem/Opportunity Identification

In the typical work setting, many employees may prefer stability, following preestablished norms and habitual routines, and avoiding changes to their work environment (Ford, 1996). As such, unless a task has an explicit component to be creative or innovative, creativity may be the exception rather than the norm. Something is needed to spark a need or desire to engage in innovation. In most cases, an area is identified as a potential venue for innovation. This identification may be proactive in that a potential opportunity for creativity or innovation is identified or more reactive, in which a problem is recognized that needs a creative/innovative solution. With regard to creativity, Unsworth (2001) recommends distinguishing creativity arising from problem or opportunity identification. Specifically, she developed a typology of four creativity types depending on the driver for engagement in the creativity process (internal or external demands) and the type of the problem (open and discovered by the individual or closed and presented to the individual). The four types include responsive (closed, external), expected (open, external), contributory (closed, internal), and proactive (open, internal). Open problems require more creativity than those that are closed or "predefined" for the individual. Whether the motivation for engaging in the creative process is internal or external to the person affects the factors that predict successful problem identification or definition.

The identification of organizationally relevant problems or opportunities requires an initial recognition that some effectiveness standard or goal is not being achieved (a problem) or could be improved or extended in some way (an opportunity). Such identification is based on processes such as contextual scanning, information search and processing, and inference (Cowan, 1986). Amabile (1996) notes that domain knowledge (expertise) and intrinsic motivation are factors that predict overall creativity and innovation in work settings, and the stage of problem identification is no exception. With regard to domain knowledge, Amabile further notes that it is not only the amount of relevant information that an individual possesses that is related to successful problem identification but also the structural organization of that information. A structure is optimal when it facilitates many easily accessed associations among components of that knowledge so that meaning can be imposed and inferences made that integrate the various pieces of information (Mumford, Blair, Dailey, Leritz, & Osburn, 2006). Intrinsic motivation to engage in scanning and complex information processing is especially important for successful problem identification when the problem type is open and discovered by the individual (Unsworth, 2001).

Although problem or opportunity identification is the first step toward creative idea generation and implementation of ideas, finding problems at work can lead to a great deal of stress and frustration. These types of negative attitudes and emotions may actually be important catalysts in the identification of problems or opportunities. Specifically, negative moods were positively related to creative performance when employees felt recognized and rewarded for their creative performance and when clarity of feelings was high (George & Zhou, 2002). Further, job dissatisfaction may foster engagement in the creative process when paired with the right amount of support from the organization or coworkers (Zhou & George, 2001).

Idea Generation

Idea generation is often considered the crux of creativity. Many studies of creativity, in particular laboratory studies, use idea generation as a proxy for the measurement of creativity (Hammond et al., in press). Ideas can be quantified based on several criteria, such as the frequency (i.e., total number of ideas), flexibility (i.e., how different the ideas are from one another), originality or novelty, feasibility, and usefulness (Zhou & Shalley, 2003). Most commonly, studies in the workplace have used Amabile's (1996) consensual assessment technique; several raters serve as expert judges and assess the generated ideas on any of the aforementioned dimensions (e.g., Zhou & Oldham, 2001).

A large body of literature exists investigating brainstorming, which is a formalized session for the generation of ideas, both at individual and group levels of analysis. In the workplace, suggesting new ideas is more strongly related to personal and job characteristics than group or environmental factors (Axtell et al., 2000). Individual factors that contribute to offering idea generation include some (but not extremely high) expertise (Amabile & Conti, 1999), job complexity (Farmer, Tierney, & Kung-McIntyre, 2003), intrinsic motivation (Shin & Zhou, 2003), self-efficacy, and autonomy (Axtell et al., 2000).

In a meta-analysis of the mood-creativity relation, Baas, De Dreu, and Nijstad (2008) reported that the regulatory focus associated with mood states and the specific nature of the creativity facet examined may be critical factors. Specifically, when mood state facilitated a promotion focus (i.e., primary concern for gains), individuals exhibited expanded attentional scope and broader access to less-accessible mental representations, in contrast to a prevention focus (primary concern for threats), which resulted in more constricted scope of attention. Expanded attention and more access to less-accessible cognitive structures are expected to result in generation of more ideas. Baas et al. also argued that promotion-focused mood states enhance creativity via cognitive flexibility, while prevention-focused mood states can enhance creativity by resulting in greater cognitive persistence. They also noted that the nature of the focal task may affect whether cognitive flexibility or persistence is more important for achieving creative results.

Idea Evaluation

After ideas are generated, they must be evaluated against some criteria, selected, and often modified before they are put into action. *Idea evaluation* refers to the stage of the innovation process in which ideas are (a) evaluated for feasibility and resource needs, (b) communicated to others, (c) selected, and (d) modified (Mumford, Lonergan, & Scott, 2002; Zhou & George, 2003). In general, there is less theory and empirical attention given to the idea evaluation stage than other stages, particularly idea generation. However, work has highlighted the importance of this stage in the innovation process (Dailey & Mumford, 2006; Runco, 2003). The success of the idea evaluation stage lies in the ability to select the best idea given time, resources, or support constraints. These types of issues are not usually given much attention in the idea generation stage, when the goal is to create as many solutions as possible. Because the goal of idea evaluation is to assess the viability of each solution, the context of the problem, the need for innovation, and the organizational resources available, organizational-level factors play a much greater role in this stage than for idea generation. In addition, the two stages are linked as the success of the idea evaluation stage hinges on the success of the idea generation stage (Basadur, Runco, & Vega, 2000). It is also important to acknowledge that a certain amount of evaluation occurs within idea generation, both as individuals evaluate their own ideas before sharing them with others and as members collectively evaluate ideas suggested within group interactions. Again, the innovation process is dynamic, cyclical, and messy.

Mumford et al. (2002) proposed a model for understanding the cognitive aspects involved in idea evaluation. The model specifies that individuals predict the potential outcomes or consequences of implementing each idea and the resources required for successful implementation. These are then applied against a chosen standard to make the decision to implement, drop, or modify the idea. Generally, idea evaluation involves three basic stages: (a) forecasting potential consequences of implementation, (b) appraising the idea against some chosen criteria, and (c) revising the idea to increase the likelihood of its success. Basic support for the model has been garnered (Lonergan, Scott, & Mumford, 2004).

Implementation

Following evaluation, the chosen idea, innovation, or solution must be put into action. The final component of the process, implementation, has been defined as "the transition period during which targeted organizational members ideally become increasingly skillful, consistent, and committed in their use of an innovation" (Klein & Sorra, 1996, p. 1057). Until recently, little research has been conducted examining implementation independently from innovation more broadly. Klein, Conn, and Sorra (2001) were among the first to distinguish between implementation effectiveness and innovation effectiveness. Implementation effectiveness focuses on the successful use of the innovation, whereas innovation effectiveness refers to the benefits of the innovation for the organization. Clearly, ideas must be successfully implemented prior to benefiting the organization.

Because the implementation stage can be full of obstacles and challenges, persistence is necessary. At the individual level, factors such as intrinsic motivation, learning goal orientation, and conscientiousness may help maintain the necessary persistence (Farr et al., 2003). Intrinsic motivation is especially important at the implementation stage because difficulties and complications are likely to be experienced during this stage. Similarly, self-efficacy has been identified as a key factor in implementing individual role innovations (Farr & Ford, 1990). Employees must believe that such changes can actually be implemented successfully to provide the motivation to do so. In addition, individuals are more likely to implement ideas if they receive support from their leaders (Axtell et al., 2000), their mistakes are tolerated (Zhou & Woodman, 2003), and they are more involved in decision making (Axtell et al., 2000).

Whereas the previous stages of the innovation process may occur at the individual or collective levels, implementation is inherently more collective. Individuals need to garner support and buy-in for successful implementation of their ideas. Often, individuals need the expertise of others to execute the development of their ideas successfully (e.g., the manufacturing, marketing, and sales of a product idea). Plus, by the nature of the some innovations, the scope reaches far beyond the individual in the diffusion and transfer of innovation across or even outside the organization (e.g., implementing a new knowledge management system in an organization affects all who might use it). As such, the collective becomes ever more important in the implementation stage.

In general, collective-level processes and factors provide the structure and mechanisms by which innovation is fostered or hindered. Often, top management holds the deciding factor on whether an idea will be implemented. Successful implementation requires garnering the support of others and necessary resources. Kanter (1988) used the term *selling* to refer to the process by which an idea champion attempts to promote his or her idea to team members or management. As such, a strong social network is useful in successful selling. Occupying central positions within networks allows access to diverse expertise, resources, and experiences (Tsai, 2001) and is likely to aid in the selling of an idea for successful implementation. At an organizational level, numerous factors influence the success of the implementation stage. Several policies and practices have been identified as contributing to successful implementation, including training, technical support, communication, and rewards. Further, Klein et al. (2001) identify an implementation climate, which refers to shared perceptions about how much an organization values and supports innovation implementation. Other context factors, such as psychological safety and a supportive climate for innovation, have been identified as necessary for successful implementation (Baer & Frese, 2003).

In sum, successful innovation is a complex process involving problem or opportunity identification, idea generation, idea evaluation, and implementation. As mentioned, the process is not necessarily linear, but rather complex, chaotic, and iterative. Several factors at both the individual and collective levels of analysis have been identified to influence each subprocess. We suggest that, in some situations, errors may also facilitate innovation, along with other positive outcomes.

Facilitation Effect of Errors on Innovation

A growing body of literature has proposed positive consequences of errors, such as identifying risk, enhancing motivation to change, building resilience, and increasing learning (Ellis & Davidi, 2005; Jones & O'Brien, 1991; Sitkin, 1996). As illustrated by a number of useful innovations that arose through errors (e.g., penicillin, Scotchgard™, Post-it™ notes), facilitating innovation should be added to the potentially favorable outcomes of errors. But, how and under which conditions might errors lead to innovations? We propose two pathways through which errors may foster innovation: problem or opportunity identification and serendipity. We further specify several conditions in which we believe innovations are most likely to emerge from errors, both at the individual and organizational levels. In doing so, we simultaneously highlight a number of potential research opportunities. To our knowledge, no studies have empirically investigated the facilitation effect of errors on innovation or the conditions in which it may be most likely to occur.

Pathways From Errors to Innovation

Problem or Opportunity Identification

One of the most frequent ways in which errors may facilitate creativity or innovation is through the opportunities that errors may create for improvements of existing processes, procedures, or products. The familiar adage "success breeds complacency" suggests that the "reinforcement" that is derived from successful performance often leads to repetition of existing behavior and little change. An error or failure may signify a problem with the status quo and prompt a need for change (Farr & Ford, 1990). Failures can prompt a need for change through identifying either an existing problem or an opportunity for future innovation.

Further, innovation may arise from the sense making that occurs when individuals cope with the crises that may result from errors (Drazin et al., 1999; Kazanjian, Drazin, & Glynn, 2000). Crises occur when the current organizational system, procedure, or product does not allow a problem to be solved. Such crises may be related to functionality, cost, or schedule. As such, a discovered error or failure may signal a problem to be solved or an opportunity to be taken and thereby foster creative engagement by employees. Further, crises push individuals to "suspend their existing frames of reference and look at the world differently" (Drazin et al., 1999, p. 296), which cultivates creativity.

For example, an error may arise in an initial attempt to implement an innovation. The negative outcomes from that result may lead to finding or creating a better implementation approach than if the organization had correctly followed the initial implementation plan. Alternatively, although an organization may have developed the "perfect" solution to a problem, it may not have been implemented properly. As so much of implementation success is dependent on the environment, failure in implementation may signify larger environmental problems (Drazin et al., 1999). This in turn may lead to further innovation with regard to the environmental issues.

Serendipity

Further, errors may contribute to creativity and innovation serendipitously or through chance. Creative behavior, especially in the sciences, involves a certain amount of chance, randomness, and unpredictability (Simonton, 2003). Through an error, a new discovery may be made. *Serendipity* refers to an unintended or unexpected discovery made by accident (Foster & Ford, 2003). Serendipity may foster a new idea altogether or an idea for the application of an existing product, procedure, or process to be applied in a new domain. A failure in one domain may be a huge success when applied in a different setting. Learning from failures can help decrease the likelihood of repeating certain mistakes but often can be applied in other contexts (Hargadon, 2008).

Serendipity that occurs through information searches may lead to innovations or improvement on innovations in a several ways. In a qualitative study, Foster and Ford (2003) found that serendipitous information in research projects may have positive effects, such as reinforcing the participant's current conceptualization of the problem or taking the participant in a new direction through reconceptualizing or reconfiguring the problem. As such, errors that provide serendipitous information may lead the employee to better understand or confirm an existing problem or provide an opportunity for an innovation.

Further, new ideas may be created by way of serendipitous juxtapositions, that is, instances in which two seemingly unrelated concepts, images, or events happen to come together to reveal a new or unexpected relationship. This basic idea has been at the heart of conceptions of the creative process for a long time. For example, in Guilford's (1984) original model, creativity was discussed as arising through divergent thinking, which refers to the ability to derive multiple solutions to problems through the unexpected combination of such divergent thoughts. Further, the theory behind the frequently used Remote Associates Test (Mednick, 1968) suggests that creativity arises from the formation of useful new associations or combinations of previously separate concepts. Rothenberg (1971) also referred to this idea as Janusian thinking, defining it as the capacity to simultaneously envision and make use of two or more opposite or contradictory ideas.

TABLE 3.1

Potential Facilitation Effects of Errors on Innovation

	Pathways	Predictors and Situational Conditions
Individual	Problem/opportunity identification Serendipity	Openness to experience Self-efficacy Learning goal orientation Personal initiative
Collective	Problem/opportunity identification Serendipity	Job complexity Autonomy Social networks Error management culture Climate for creativity/innovation Psychological safety

Although trying to predict serendipity appears to be somewhat paradoxical, it may be possible to foster an environment in which capitalizing on seemingly unrelated concepts or events may be most likely to occur. Louis Pasteur declared that "chance favors the prepared mind" (cited by Posner, 1973). We suggest several individual differences that may "prepare the mind" for facilitating innovation through errors in terms of both identifying problems and through serendipitous use of errors as innovation. Similarly, the environment may also need to be prepared to facilitate innovation through errors. Encouraging the right frame of mind, whether it is through individual differences or environmental facilitators, may have an impact on the extent to which individuals identify opportunities for innovation and engage in creative action (Drazin et al., 1999). We also discuss several individual differences and additional collective-level factors that may prepare the social environment to capitalize on errors. Because we are unaware of any empirical evidence of these direct relationships, we pose these as potential areas for future research. Table 3.1 summarizes these proposed predictors.

Individual-Level Predictors

Openness to Experience

Of the big five factors, openness to experience is most clearly linked with innovative behavior and has been the most frequently examined personality factor. A positive association has been empirically demonstrated between openness and divergent thinking (McCrae, 1987). Individuals high on openness have high intellectual curiosity, imagination, independence, and sensitivity to the arts (McCrae, 1987) and, as such, are less likely to shy away from new experiences and change. Consequently, individuals who are more open to new experiences may have been exposed to a much larger variety of influences and display less-predictable behaviors and thought patterns. By adulthood, they may have a greater variety of experiences and a richer and more diverse associative network. For example, openness to experience may predict the engagement in behaviors that may themselves enhance creativity, such as choosing to live in nonnative cultural settings. Maddux and Galinsky (2009) have shown that time spent living (not simply traveling) abroad was positively correlated with creativity on several types of tasks. The authors inferred that the greater immersion in another culture and in-depth processing of cultural information that would occur while living abroad led to a broader range of experiences that facilitated creative task responses. This interpretation was further supported by findings that the extent to which individuals had adapted to the new culture mediated the relation between time spent living abroad and task creativity. Also, Maddux and Galinsky (2009) demonstrated that, while openness to experience was correlated with reaching a creative solution in a dyadic negotiation task, time spent living abroad explained additional variance in the creativity of solutions beyond that of the personality variable. These influences are likely to increase the likelihood of the stochastic behavior necessary for creativity (Simonton, 2003).

Personal Initiative

Because the innovation process so often includes trials, tribulations, and failure, individuals who do not give up in the face of these trials may be more successful in their innovative endeavors. *Personal initiative* refers to self-starting, proactive, and persistent behaviors in the face of barriers (Frese & Fay, 2001). Employees who have greater personal initiative are more likely to anticipate future opportunities that may stem from errors (proactivity), to follow through with an idea that stemmed from an error (self-starting), and persevere in the innovation process through barriers and errors (persistence).

Self-Efficacy

Similarly, individuals with a stronger belief in their competence are more likely to overcome fears associated with facing errors and more likely to persist in the effort required for recovery and learning from errors (Bandura, 1997). Both *job self-efficacy*, which refers to beliefs about one's competence with regard to task performance, and *creative self-efficacy*, which refers to beliefs about one's competence with regard to creative performance (Tierney & Farmer, 2002) have exhibited positive relationships with creative and innovative outcomes (Axtell et al., 2000; Carmeli & Schaubroeck, 2007; Frese, Teng, & Wijnen, 1999). However, an optimal level of self-efficacy may be desired as self-efficacy may also lead to overconfidence and increase the likelihood of errors (Vancouver, Thompson, Tischner, & Putka, 2002).

Learning Goal Orientation

A final individual difference that may affect an individual's motivation to advance innovation from errors is that of *goal orientation*. Goal orientation describes the ways in which individuals approach tasks. Individuals with a stronger learning goal orientation approach tasks with the goal of learning and mastery, rather than primarily demonstrating their competence to others, which can be described as a performance goal orientation (Dweck, 1986; Farr, Hofmann, & Ringenbach, 1993). Individuals who learn for the sake of learning may be more likely to have a wider repertoire of knowledge in various areas that they may connect with the error, thereby facilitating serendipitous connections. Further, as learning goal orientation is related to motivation to learn, individuals higher on this construct may be more likely to admit to errors, learn from them, and capitalize on them in the form of innovations. Plus, this orientation is positively related to innovation (Janssen & Van Yperen, 2004; Moss & Ritossa, 2007).

Collective-Level Predictors

Job Complexity

Jobs that are more complex may demand more innovation in their very nature by allowing individuals to focus simultaneously on multiple aspects of their work (Oldham & Cummings, 1996). This may promote remote association or serendipitous discoveries. Complex jobs include diverse activities and challenges, thereby widening the repertoire from which an individual may make serendipitous connections (Amabile, 1988) that may arise through errors. Further, of all the predictor categories, job characteristics held the most consistent and strongest positive relationships with creativity and innovation (Hammond et al., in press).

Autonomy

Autonomy has been found to relate to both creative and innovative behaviors. Autonomy was positively related both to the generation and testing of ideas (Krause, 2004) and to innovation implementation (Axtell et al., 2000). Jobs with little discretion in how, when, or where work is accomplished may stifle an employee's ability to be innovative. Alternatively, providing employees with freedom and independence to determine which procedures should be used to carry out a task may increase the likelihood that they will be willing to implement them within their job. In addition, autonomy may be an important factor in individual innovation as it provides an individual with freedom to decide how, when, and with whom to work (West & Farr, 1990).

Social Networks

As discussed, social networks may influence innovation in organizations through increasing divergent and flexible thinking, which may be encouraged through social discussion with others (Perry-Smith & Shalley, 2003). It is particularly important that discussions ensue that focus on unique information rather than falling into the trap of focusing on shared information (Stasser & Titus, 1985) to optimize the likelihood of serendipitous innovations through errors. Perry-Smith (2006) found that having a more heterogeneous set of direct contacts through weak ties facilitated cognitive processes, whereas having a number of strong ties may lead to conformity pressures that may inhibit creativity. Promoting interdisciplinary contact through teamwork within the organizations (Klein, 2005) may be one way to promoted creativity through heterogeneous ties, as long as "strong ties" are not developed.

Error Management Culture

Theoretical development has suggested that the ways in which errors are analyzed, communicated, prevented, and managed are often incorporated in the organizational culture. Error management culture, as suggested by Van Dyck, Frese, Baer, and Sonnentag (2005), includes aspects of communicating about errors, sharing error knowledge, helping in error situations, detecting and controlling errors, analyzing errors, and handling errors. In theory, positive error management cultures promote firm performance by simultaneously reducing the negative effects of errors and promoting positive consequences, such as innovation, exploration, experimentation, initiative, and learning. Although theorizing about error management suggests this positive effect of errors on innovation, it is yet to be empirically tested.

Climate for Creativity or Innovation

In addition to a climate that supports proper error management, climates and cultures may promote innovation. An organizational climate that is supportive of innovation may be marked by being open to change, supportive of new ideas from members, tolerant of member diversity, and provide adequate resources necessary for innovation (Scott & Bruce, 1994). Employees who perceive supportive climates and relationships are more likely to be innovative as they believe that creativity is expected and valued by the organization (Tierney & Farmer, 2004). Working in an environment supportive of creativity and innovation may promote employees to be more aware of opportunities for innovation, some of which may arise through errors in other domains.

Psychological Safety

In addition to work climates that are explicitly supportive of innovation or focused on the management of errors, environments that are more generally more interpersonally "safe" may also be helpful in promoting innovations from errors. Edmondson (1996) found that individuals are most likely to be able to learn from mistakes, admit errors, and discuss problems in environments in which they do not feel threatened interpersonally. This construct became known as *psychological safety*, which refers to "shared belief that the team is safe for interpersonal risk taking" (Edmondson, 1999, p. 354). Work has provided evidence that psychological safety is positively related to learning from failures (Carmeli & Gittell, 2008), employee involvement in creative work (Kark & Carmeli, 2009), and organizational process innovation (Baer & Frese, 2003). As such, it is likely that more innovations may occur from errors in psychologically safe environments.

In summary, individuals and organizations may be able to capitalize on errors through problem or opportunity identification or serendipitous discovery. Individuals who are high on openness to experience, self-efficacy, learning goal orientation, and personal initiative may be more likely to foster creativity and innovation through errors. In addition, we propose that organizations may increase the likelihood of benefiting from errors by designing complex and autonomous jobs, promoting individuals to develop diverse social networks, as well as building an organizational climate marked by psychological safety, support for innovation, and positive error management.

Errors in Creative and Innovative Endeavors

Although, given the right conditions, errors may foster innovation, errors may also hinder successful innovation. Indeed, the success of innovation often depends to a considerable extent on the reduction of errors that can occur during the innovation process. In their definition of errors, Hofmann and Frese (Chapter 1, this volume) suggest that errors represent actions in which individuals unintentionally fail to achieve their goals. Similarly, innovation involves intention. For example, West and Farr (1990) specify that "innovation is restricted to intentional attempts to derive anticipated benefits from some change" (p. 9). Therefore, errors in innovation might include any failing to carry forward with a change or to benefit from the change. Further, with regard to creativity, there is a body of work suggesting that individuals may also have goals to be creative. In these instances, the goal may not be to implement and derive benefits from change but more simply to generate ideas that are creative (Shalley, 1991). As such, an error in this instance may include failing to be creative or to produce creative ideas.

In the next section, we discuss errors common to each described component of our process model. Table 3.2 provides a summary. As suggested by Farr et al. (2003), various individual, job, and environmental factors may play more or less-important roles throughout the innovation process, particularly with regard to generating ideas and implementing them. Although we discuss each component of the model independently, we again acknowledge their interdependencies. We do not address in detail all of the common errors shown in Table 3.2 but focus on a few that seem to us to be especially useful directions to follow in future research. We do address both individual-level and collective-level errors for each component of the process model.

Problem Identification

Individual Errors in Problem Identification

Most certainly, not all cases of problem identification will lead to creative solutions. A number of individual-level factors may contribute to errors in effectively identifying or recognizing a problem or opportunity in a given situation. These often restrict the breadth of information gathering and data analysis that lead to the development of problem definitions. Mumford et al. (2006) have described in detail many of the factors noted in Table 3.2 (e.g., information search errors, framing), and those interested are directed to their article. We discuss two individual-level factors here that we believe to be the most relevant.

First, schema-based information processing (George & Jones, 2001) may inhibit the recognition of problems. Schemas are relatively complex sets of knowledge about a situation, concept, or stimulus, including details about its features or attributes and the interconnections among the attributes. Individuals generally develop schemas after repeated experiences with the situation or concept. Once schemas have developed, they are activated in a relatively automatic (unconscious) manner when a similar situation or concept is encountered and are used by the individual to interpret specific information related to the new encounter. Thus, new information is processed and interpreted in a manner consistent with the preexisting

TABLE 3.2
Common Errors Across Stages of Innovation

	Problem Identification	Idea Generation	Idea Evaluation	Implementation
Individual	• Biased information search • Framing • Schema-based information processing • Affect	• Expertise • Functional fixedness and satisficing • Attentional capacity limitations • Focusing on contextual factors	• Underestimation of resources required • Severity • Leniency and overoptimism • Errors of inattention	• Overoptimism • Overanalysis • Failure to commit • Estimation errors
Collective	• Shared mental models • Problematic norms and work-group climates	• Production blocking • Derailment • Evaluation apprehension • Social loafing	• Exacerbation of individual information-processing errors • Group polarization • Conformity • Groupthink • Overreliance on shared information	• Monitoring errors • Resistance to change • Decreased working relationships • Escalation of commitment or noncommitment

schema and related expectations. Schemas thus serve to inhibit change in general and the creative identification of problems more specifically. Discrepancies between new information and the existing schema often result in emotional and cognitive reactions that promote resistance and denial that change is needed. Additional research that examines the impact of new information on both the maintenance and the revision of the existing schema would be especially useful.

Individual differences may also play a role in problem recognition and can bias this stage of the creativity process. In a review of empirical research on the relation of personal characteristics to creativity, Shalley, Zhou, and Oldham (2004) noted that individuals who experience positive moods tend to make more connections among divergent concepts and use more inclusive categories when integrating information than those experiencing negative moods and may thus be more likely to recognize problems. However, Shalley et al. (2004) also concluded that, when negative job affect was related to the perception that the status quo was no longer acceptable, such affect could trigger more extensive problem search processes. Thus, the role of affect may be context dependent, and future research is needed to delineate the contextual variables that moderate the relation of individual differences and problem recognition outcomes.

Collective Errors in Problem Identification

Not all errors in problem identification result from individual-level factors. Social or group-level factors also can contribute to such errors. Included among these are shared mental models, group norms, and work-group climates. Generally, these group-level variables can shape the way that group members recognize and frame problems or opportunities for innovation.

For example, shared mental models (George & Jones, 2001) may inhibit the recognition of problems. Such mental models may be thought of as schemas that are shared by group members. These may be even more resistant to change than individual-level schemas because of the agreement among coworkers about the nature of "reality" represented by the shared mental model. It is also likely that pressure to conform to the mental model will exist for those individuals who note that new information is discrepant with the existing knowledge structure and expectations.

Further, group norms and work-group climates for creativity and task achievement also may have an impact on problem recognition or definition. Norms and climates that promote adhering to past routines and traditions ("do it our way") and encourage satisfactory levels of group performance are less likely to perceive the status quo as a "problem." Work groups with such norms and climates also are less likely to construe a need for radical change or innovation, in contrast to groups whose norms and creativity and achievement climates promote change and high levels of achievement. Groups with norms and climates promoting change and improvement should identify more "problems" and needs for major change and performance improvement.

Idea Generation

Individual Errors in Idea Generation

As mentioned, the idea generation stage has received the greatest amount of research attention as it is sometimes perceived synonymously with creativity. Similarly, a large number of individual-level factors have been suggested as correlates of errors in idea generation (see Mumford et al., 2006, for a review) and are listed in Table 3.2. Similar to errors in problem identification, individual errors in idea generation can also arise from limited views of the problem.

Expertise is often considered a necessary factor for creative idea generation (e.g., Amabile, 1996), but Mumford et al. (2006) noted that expertise can also be related to errors in idea generation. In particular, they argue that experts may ignore or discount atypical aspects of the problem situation and apply well-known solutions to all similar cases in a manner reminiscent of the use of schemas in problem identification. Thus, expertise can serve to limit idea generation in some situations.

The complexity of problems and individuals' limited information-processing capacity jointly encourage the simplification of the problem and needed solution or idea. Simplification processes may result in considering only some of the elements of the problem when generating solutions, frequently the more typical or representative attributes or examples (Mumford et al., 2006), and also imputing the existence of unobserved, but common, attributes for the category or type of problem encountered (the illusory correlation bias). Limited information-processing capabilities may also encourage individuals to set a standard of satisficing rather than maximizing. Satisficing is the tendency to accept an initial solution or idea that appears to deal in a satisfactory manner with the problem rather than continuing to generate additional ideas that may lead to superior outcomes.

Collective Errors in Idea Generation

Although the layperson may believe that brainstorming is best achieved in groups, empirical support of this hypothesis is lacking. Whereas individuals may feel that there is a "synergy" in brainstorming groups, often social errors prevent groups from outperforming individuals in idea generation. Such errors in group brainstorming include production blocking, derailment, and evaluation apprehension (Kerr & Tindale, 2004).

During a "vocal" group brainstorming session, only one person can offer a suggestion at a time. This limits the ability for others to make suggestions, or even think about new ideas, at the same time. This concept is referred to as *production blocking* (Diehl & Stroebe, 1987). More recent work has suggested that is not simply an issue with regard the inability to speak or think at the same time, but rather that individuals may not be able to begin on productive "trains of thought" or may derail thought away from a productive line of thought. This derailment effect refers to a shifting away from productive cognition. Further, groups may underperform individuals because of hesitancy to offer suggestions in front of other group members. Individuals may experience evaluation apprehension. They may be reluctant to put forth ideas for fear of a negative reaction from other group members. Several research studies, however, have suggested that idea generation can be increased via "virtual interaction" among members of a brainstorming group in contrast to face-to-face interactions (e.g., Valacich, Dennis, & Connolly, 1994). Such virtual interactions typically use technology-based communication media that can allow multiple individuals to make suggestions anonymously and simultaneously, with the technology providing lists of alternatives to group members for further evaluation after all ideas are first generated. Evaluation apprehension is lessened since other group members are not aware of the source of each idea.

Social loafing can also occur in group-based idea generation. *Social loafing* refers to the phenomenon that individuals in group settings often exert less effort toward task performance than they would if performing the task on an individual basis (Shepperd, 1993). Social loafing is more likely to occur when accountability for individual performance is limited or low. Thus, social loafing may counteract to some extent the benefit of anonymous idea generation in virtual brainstorming. The use of virtual groups in idea generation is likely to be more effective if procedures for increasing individual accountability for performance are utilized. In the next stage, idea evaluation, sharing information and critically processing ideas become of greatest importance.

Idea Evaluation

Individual Errors in Idea Evaluation

Individuals may not be able to provide the careful attention and deep processing required for successful evaluation of ideas because of attention given to other tasks. The innovation process is often wrought with stress and frustration, and employees' information-processing capacity may be taxed by task difficulty, time pressures, other demands or commitments, fatigue, or stress (Mumford et al., 2006). Other individual errors in the idea evaluation stage tend to focus on incorrectly estimating the quality of the idea or the resources needed for successful implementation. Regarding the former, individuals tend to have preferences for ideas that are clearly understood, have an immediate short-term benefit to many people, and fit within prevailing social norms (Blair & Mumford, 2007). In addition, individuals have a tendency to underestimate the resource requirements of implementation (Josephs & Hahn, 1995), especially with regard to time and money. Further, there is a tendency to overestimate the outcomes of a particular idea, especially when individuals are familiar with the content and outcomes are valued (Dailey & Mumford, 2006).

Conceptual progress has been made by Michael Mumford and colleagues in the understanding of errors in the idea evaluation stages (Blair & Mumford, 2007; Dailey & Mumford, 2006; Licuanan, Dailey, & Mumford, 2007; Mumford et al. 2006). These models focus on the individual information-processing level (e.g., cognitive or thought errors). As suggested in Mumford et al.'s (2006) taxonomy, errors in idea evaluation typically occur when the individual is overly critical of ideas (e.g., errors of overanalysis, rejection of risk, aversion to regret) or overly lenient (e.g., errors of overoptimism, confirmatory bias, underestimation errors). Overly critical evaluation of errors typically happens when individuals engage in overprocessing ideas, overestimate the resources required, or prematurely reject ideas. For example, individuals may be overly critical of uncertain ideas because people tend to be averse to risks (Tversky & Kahneman, 1974) and have a preference against radical change. Ideas that are outside individuals' comfort zones, risky, and original are more likely to be accepted when the evaluation criteria are not stringent and when they are under time pressure (Blair & Mumford, 2007). Alternatively, errors of leniency happen when individuals (a) underestimate the resources required for successful innovation, (b) overestimate the positive outcomes of an idea, or (c) fail to engage in the critical processing necessary for successful evaluation.

There has been some progress in the identification of specific errors, conditions in which these errors are more likely to occur, and individuals who may be more prone to these errors. For example, research on the influence of expertise on creativity suggests that experts may be prone to some errors because of their highly developed mental models of problems within their domain of expertise. They may be unable to consider ideas that fall outside their existing mental models of the problem. As such, these ideas may be prematurely rejected (discussed in Mumford et al., 2006). In addition, there is evidence that experts, or even individuals with some familiarity with an issue, tend to overestimate positive outcomes of ideas and underestimate the resources required to implement the ideas successfully (Dailey & Mumford, 2006). Also, there is some evidence that positive mood may foster errors of illusive optimism, overconfidence, and persistence in nonoptimal behaviors (Seo, Barrett, & Bartunek, 2004).

Collective Errors in Idea Evaluation

In addition to individual cognitive errors, idea evaluation errors occur at a group level. Similarly, groups may make overly critical or overly lenient errors in the appraisals of ideas. There is some evidence that groups may either attenuate or exacerbate individual errors in decision making, depending on the type of bias, the type of group decision process, the strength of the bias, and the individual preference distribution in the group (Kerr & Tindale, 2004). However, future work is needed to better understand these conditions.

Collective errors in idea evaluation can be well informed by considering the vast literature on group decision making. Processes such as group polarization, groupthink, and conformity may provide insight into social errors in idea evaluation. For example, individuals' preferences, such as risk aversions, may be polarized or shifted when in the presence of a group (Moscovici & Zavalloni, 1969). Therefore, groups may be more prone to either extreme conservativeness or riskiness than might individuals when evaluating ideas. Further, research on conformity (Asch, 1955), suggests that individuals may be less likely to voice concern in idea evaluation if others have expressed positive evaluation of an idea. Although empirical evidence of this groupthink has been met with mixed results, it may be important to consider the potentially negative role that cohesiveness and collective efficacy as well as directive leadership and perceptions of urgency may play in idea evaluation (Janis, 1982).

The classic proposition that groups have an inclination to focus on shared information rather than unique information is particularly relevant when considering group idea evaluation (Stasser & Titus, 1985). When evaluating ideas, it is critical for group members to share all relevant and unique information each individual has regarding the idea and the context. Focusing on shared information to the detriment of unique information is likely to result in biased evaluation of ideas. Shared information may be more likely to be discussed early in the evaluation processes of the group. Others have suggested that a focus on shared information may stem from a desire for closure, a preference to be perceived as competent and credible, and a dislike for changing initial preferences. Ideas such as increasing decision-making time, assigning members to be in charge of particular areas of information as experts, assigning group leaders, and training members to integrate more information in discussion may be of some use to avoid these biases (as reviewed in Kerr & Tindale, 2004). Further, the presence of a minority dissenter may increase decision-making accuracy as well as the quality of idea generation (De Dreu & West, 2001). In sum, individual-level errors in idea evaluation focus on critical processing of ideas, whereas collective level errors seem to occur more from a lack of sharing pertinent information. In the next stage, implementation, the role of selling becomes most important.

Implementation

Individual Errors in Implementation

Because of the primacy of the collective in implementation, errors common to this stage are more frequently tied to social errors than individual information-processing errors; however, there may be elements of both. Individual information-processing errors relating to successful implementation are more likely related to the planning of the implementation, which may be, in part, linked more closely with the idea evaluation stage than the implementation stage itself. For example, Mumford et al. (2006) identified several errors common to idea evaluation stages as also common errors in implementation planning and monitoring stages, such as overoptimism, overanalysis, failure to commit, and estimation errors, among others. In general, these errors refer most often to people's reactions to ideas and the implications of these ideas as well as ineffective monitoring of the implementation.

Collective Errors in Implementation

Although individual-level monitoring is important for implementation, monitoring at the group level may be especially critical for successful implementation. Farr et al. (2003) argued that four types of monitoring are critical for implementation: monitoring progress toward goals; monitoring the system, including resources and environmental conditions; monitoring the team itself in relation to task accomplishment; and coordinating action of team members. Social errors therefore may stem from a failure to monitor goal process and group functioning properly. Because innovation is an emotional process in which a great deal of time, energy, and even a sense of identity may be invested, groups may not be accurate in their estimates of implementation effectiveness. In cases like this, problems such as escalation of commitment may occur, in which individuals or groups justify increased investment into an idea when evidence would suggest that it is time to abandon the idea or significantly change directions (Staw, 1981).

The implementation of new ideas may be met with resistance from coworkers, supervisors, subordinates, and other actors in a work environment. Even with regard to small-scale changes within an individual's work role or group, others may resist such changes for a variety of reasons, including their own insecurities, a desire to avoid perceived stress of the change, or their commitment to the "status quo" (Janssen, 2003). Further, implementing changes in one's work role may highlight differences between coworkers and damage their sense of cohesion. This type of conflict has been found to be particularly high for employees who strongly link their work with their identity (i.e., high job involvement; Janssen, 2003). Even successful implementation may be met with decreased workplace relationship quality. Thus, an unintended error might be conflict within groups.

Summary of Errors in the Innovation Process

As described, it is clear that the innovation process has significant opportunity for errors, mistakes, and failures. Even with the most careful planning, bountiful resources, and market research, products often flop, and processes may not become popular. Failures have the potential to be destructive to an individual's sense of self, to relationships, and to the bottom line of the organization. As such, it is imperative that employees and organizations learn from their failures and manage the potential negative consequences of errors. Although error prevention strategies aimed at promoting awareness of common errors in the innovation process may be somewhat successful in reducing the frequency of these types of errors, no strategy can eliminate them completely. For example, rater error training has been moderately successful in reducing the effect of common errors in performance appraisal (Woehr & Huffcutt, 1994). Perhaps a complementary strategy would be to promote the potential positive consequences of errors.

Future Research Directions

Little empirical work has investigated the processes in which individuals and organizations may benefit from errors in organizations. Many have posed the potential for positive outcomes of errors, such as learning, exploration, experimentation, and initiative as well as improved quality of products, services, and procedures (Van Dyck et al., 2005). However, few empirical studies have been conducted specifically on the benefit of errors for innovation. We propose two pathways (problem/opportunity identification and serendipity); however, others may also exist.

As important as identifying the link between errors and innovation, it is important to identify the conditions in which the relationship is likely to hold. We propose that individual differences such as openness to experience, self-efficacy, learning goal orientation, and personal initiative may be more likely to promote innovative benefits from errors. In addition, we suggest that organizations may increase the likelihood of benefiting from errors by designing complex and autonomous jobs, promoting individuals to develop diverse social networks, as well as building organizational climates and cultures marked by psychological safety, support for innovation, and positive error management. Future research is necessary to test these propositions as well as consider potential interactive effects.

In addition to better understanding the potential benefits of errors for innovation, future research is needed to better understand common errors in innovation. Work by Mumford and colleagues has paved the way in identifying many errors across the stages of creativity with regard to individual information processing; however, there is still much opportunity for future research. Specifically, we know little about the relative impact of different types of errors on creativity (Mumford et al., 2006) or the impact of social errors.

Conclusions

The process of creativity and innovation is one wrought with errors and failed experiments. These errors may occur at the individual or collective level and may differ based on the stage of the innovation process. Although errors may be an inevitable part of the innovation process, individuals and organizations may be able to capitalize on errors to identify new problems or opportunities to be developed or resolved creatively. Errors may also stimulate serendipitous discovery, leading to new ways of thinking or new product developments. Individuals who are high on openness to experience, self-efficacy, learning goal orientation, and personal initiative may be better prepared to take advantage of these benefits. In addition, organizations may increase the likelihood of benefiting from errors by designing complex and autonomous jobs, promoting individuals to develop diverse social networks, as well as building an organizational climate marked by psychological safety, support for innovation, and positive error management. This domain deserves greater research attention to better understand these potential relationships.

Acknowledgment

Thanks for the partial support for the writing of this chapter from the project, creation and implementation of Radical and Incremental Innovation: An International Comparison at Multiple Levels of Analysis, funded by VW-Foundation II/82 408.

References

Amabile, T. M. (1988). From individual creativity to organizational innovation. In K. Gronhaug & G. Kaufmann (Eds.), *Innovation: A cross-disciplinary perspective* (pp. 139–166). Oslo, Norway: Norwegian University Press.

Amabile, T. M. (1996). *Creativity in context.* Boulder, CO: Westview Press.

Amabile, T. M., & Conti, R. (1999). Changes in the work environment for creativity during downsizing. *The Academy of Management Journal, 42,* 630–640.

Anderson, N., De Dreu, C. K. W., & Nijstad, B. A. (2004). The routinization of innovation research: A constructively critical review of the state-of-the-science. *Journal of Organizational Behavior, 25,* 147–173.

Asch, S. E. (1955). Opinions and social pressure. *Scientific American, 193,* 31–35.

Axtell, C. M., Holman, D. J., Unsworth, K. L., Wall, T. D., Waterson, P. E., & Harrington, E. (2000). Shop floor innovation: Facilitating the suggestion and implementation of ideas. *Journal of Occupational and Organizational Psychology, 73,* 265–285.

Baas, M., De Dreu, C. K. W., & Nijstad, B. A. (2008). A meta-analysis of 25 years of mood-creativity research: Hedonic tone, activation, or regulatory focus? *Psychological Bulletin, 134,* 779–806.

Baer, M., & Frese, M. (2003). Innovation is not enough: Climates for initiative and psychological safety, process innovation and firm performance. *Journal of Organizational Behavior, 24,* 45–68.

Bandura, A. (1997). *Self-efficacy: The exercise of control.* New York: Freeman.

Basadur, M. S., Runco, M. A., Vega, L.A. (2000). Understanding how creative thinking skills, attitudes, and behaviors work together: A causal process model. *Journal of Creative Behavior, 34,* 77–100.

Blair, C. S., & Mumford, M. D. (2007). Errors in idea evaluation: Preference for the unoriginal? *Journal of Creative Behavior, 41,* 197–222.

Carmeli, A., & Gittell, J. H. (2008). High-quality relationships, psychological safety, and learning from failures in work organizations. *Journal of Organizational Behavior, 30,* 709–729.

Carmeli, A., & Schaubroeck, J. (2007). The influence of leaders' and other referents' normative expectations on individual involvement in creative work. *The Leadership Quarterly, 18,* 35–48.

Cowan, D. A. (1986). Developing a process model of problem recognition. *Academy of Management Review, 11*, 763–776.
Dailey, L., & Mumford, M. D. (2006). Evaluative aspects of creative thought: Errors in appraising the implications of new ideas. *Creativity Research Journal, 18*, 367–384.
De Dreu, C. K. W., & West, M. A. (2001). Minority dissent and team innovation: The importance of participation in decision making. *Journal of Applied Psychology, 86*, 1191–1201.
Diehl, M., & Stroebe, W. (1987). Productivity loss in brainstorming groups: Toward the solution of a riddle. *Journal of Personality and Social Psychology, 53*, 497–509.
Drazin, R., Glynn, M. A., & Kazanjian, R. K. (1999). Multilevel theorizing about creativity in organizations: A sensemaking perspective. *Academy of Management Review, 24*, 286–307.
Dweck, C. (1986). Motivational processes affecting learning. *American Psychologist, 41*, 1040–1048.
Edmondson, A. (1996). Learning from mistakes is easier said than done: Group and organizational influences on the detection and correction of human error. *Journal of Applied Behavioral Science, 32*, 5–32.
Edmondson, A. (1999). Psychological safety and learning behavior in work teams. *Administrative Science Quarterly, 44*, 350–383.
Ellis, S., & Davidi, I. (2005). After event reviews: Drawing lessons from failed and successful events. *Journal of Applied Psychology, 90*, 857–871.
Farmer, S. M., Tierney, P., & Kung-Mcintyre, K. (2003). Employee creativity in Taiwan: An application of role identity theory. *Academy of Management Journal, 46*, 618–630.
Farr, J. L., & Ford, C. M. (1990). Individual innovation. Innovation and creativity at work: Psychological and organizational strategies. In M. A. West & J. L. Farr (Eds.), *Innovation and creativity at work: Psychological and organizational strategies* (pp. 63–80). Oxford, UK: Wiley.
Farr, J. L., Hofmann, D. A., & Ringenbach, K. L. (1993). Goal orientation and action control theory: Implications for industrial and organizational psychology. In C. Cooper & I. Robertson (Eds.), *International review of industrial and organizational psychology* (Vol. 8, pp. 193–232). Chichester, UK: Wiley.
Farr, J. L., Sin, H. P., & Tesluk, P. E. (2003). Knowledge management processes and work group innovation. In L. V. Shavinina (Ed.), *International handbook on innovation* (pp. 574–586). Oxford, UK: Elsevier Science.
Ford, C. M. (1996). A theory of individual creative action in multiple social domains. *Academy of Management Review, 21*, 1112–1142.
Foster, A., & Ford, N. (2003). Serendipity and information seeking: An empirical study. *Journal of Documentation, 59*, 321–340.
Frese, M., & Fay, D. (2001). Personal initiative (PI): A concept for work in the 21st century. *Research in Organizational Behavior, 23*, 133–188.
Frese, M., Teng, E., & Wijnen, C. J. D. (1999). Helping to improve suggestion systems: Predictors of making suggestions in companies. *Journal of Organizational Behavior, 20*, 1139–1155.
George, J. M., & Jones, G. R. (2001). Toward a process model of individual change in organizations. *Human Relations, 54*, 419–444.

George, J. M., & Zhou, J. (2002). Understanding when bad moods foster creativity and good ones don't: The role of context and clarity of feelings. *Journal of Applied Psychology, 87,* 687–697.

Guilford, J. P. (1984). Varieties of divergent production. *Journal of Creative Behavior, 18,* 1–10.

Hammond, M. M., Neff, N. L., Farr, J. L., Schwall, A. R., & Zhao, X. (in press). Predictors of individual level innovation at work: A meta-analysis. *The Psychology of Aesthetics, Creativity, and the Arts.*

Hargadon, A. (2008). Creativity that works. In L. Zhou and C. E. Shalley (Eds.), *Handbook of organizational creativity* (pp. 323–343). New York: Erlbaum.

Janis, I. L. (1982). Decision-making under stress. In L. Goldberger & S. Breznitz (Eds.), *Handbook of stress: Theoretical and clinical aspects* (pp. 69–80). New York: Free Press.

Janssen, O. (2003). Innovative behavior and job involvement at the price of conflict and less satisfactory relations with co-workers. *Journal of Occupational and Organizational Psychology, 76,* 347–364.

Janssen, O., & Van Yperen, N. W. (2004). Employees' goal orientations, the quality of leader-member exchange, and the outcomes of job performance and job satisfaction. *Academy of Management Journal, 47,* 368–384.

Jones, C. F., & O'Brien, J. (1991). *Mistakes that worked.* New York: Bantam Books/Doubleday.

Josephs, R. A., & Hahn, E. D. (1995). Bias and accuracy in estimates of task duration. *Organizational Behavior and Human Decision Processes, 61,* 202–213.

Kanter, R. M. (1988). When a thousand flowers bloom: Structural, collective, and social conditions for innovation in organization. *Research in Organizational Behavior, 10,* 169–211.

Kark, R., & Carmeli, A. (2009). Alive and creating: The mediating role of vitality and aliveness in the relationship between psychological safety and creative work involvement. *Journal of Organizational Behavior, 30,* 785–805.

Kazanjian, R. K., Drazin, R., & Glynn, M. A. (2000).Creativity and technological learning: The roles of organization architecture and crisis in large-scale projects. *Journal of Engineering and Technology Management, 17,* 273–298.

Kerr, N. L., & Tindale, R. S. (2004). Group performance and decision making. *Annual Review of Psychology, 55,* 623–655.

Klein, J. T. (2005). Interdisciplinary teamwork: The dynamics of collaboration and integration. In S. J. Derry, C. D. Schunn, & M. A. Gernsbacher (Eds.), *Interdisciplinary collaboration: An emerging cognitive science* (pp. 25–50). Mahwah, NJ: Erlbaum.

Klein, K. J., Conn, A. B., & Sorra, J. S. (2001). Implementing computerized technology: An organizational analysis. *Journal of Applied Psychology, 86,* 3–16.

Klein, K. J., & Sorra, J. S. (1996). The challenge of innovation implementation. *Academy of Management Review, 21,* 1055–1080.

Krause, D. E. (2004). Influence-based leadership as a determinant of the inclination to innovate and of innovation related behaviors: An empirical investigation. *Leadership Quarterly, 15,* 79–102.

Licuanan, B. F., Dailey, L. R., & Mumford, M. D. (2007). Idea evaluation: Error in evaluating highly original ideas. *Journal of Creative Behavior, 41,* 1–27.

Lonergan, D. C., Scott, G. M., & Mumford, M. D. (2004). Evaluative aspects of creative thought: Effects of idea appraisal and revision standards. *Creativity Research Journal, 16,* 231–246.

Maddux, W. W., & Galinsky, A. D. (2009). Cultural borders and mental barriers: The relationship between living abroad and creativity. *Journal of Personality and Social Psychology, 96,* 1047–1061.

McCrae, R. R. (1987). Creativity, divergent thinking, and openness to experience. *Journal of Personality and Social Psychology, 52,* 1258–1265.

Mednick, S. A. (1968). The remote association test. *Journal of Creative Behavior, 2,* 213–214.

Moscovici, S., & Zavalloni, M. (1969). The group as a polarizer of attitudes. *Journal of Personality and Social Psychology, 12,* 125–135

Moss, S. A., & Ritossa, D. A. (2007). The impact of goal orientation on the association between leadership style and follower performance, creativity, and work attitudes. *Leadership, 3,* 433–456.

Mumford, M. D., Blair, C., Dailey, L., Leritz, L. E., & Osburn, H. K. (2006). Errors in creative thought? Cognitive biases in a complex processing activity. *Journal of Creative Behavior, 40,* 75–109.

Mumford, M. D., Lonergan, D. C., & Scott, G. M. (2002). Evaluating creative ideas: Processes, standards, and context. *Inquiry: Critical Thinking Across the Disciplines, 22,* 21–30.

Oldham, G. R., & Cummings, A. (1996). Employee creativity: Personal and contextual factors at work. *Academy of Management Journal, 39,* 607–634.

Patterson, F. (2002). Great minds don't think alike? Person level predictors of innovation at work. *International Review of Industrial and Organizational Psychology, 17,* 115–144.

Perry-Smith, J. E. (2006). Social yet creative: The role of social relationships in facilitating individual creativity. *Academy of Management Journal, 49,* 85–101.

Perry-Smith, J. E., & Shalley, C. E. (2003). The social side of creativity: A static and dynamic social network perspective. *Academy of Management Review, 28,* 89–106.

Posner, M. I. (1973). *Cognition: An introduction.* Glenview, IL: Scott, Forseman.

Rothenberg, A. (1971). The process of Janusian thinking in creativity. *Archives of General Psychiatry, 24,* 195–205.

Runco, M. A. (2003). Idea evaluation, divergent thinking, and creativity. In M. A. Runco (Ed.), *Critical creative processes* (pp. 69–94). Cresskill, NJ: Hampton Press.

Scott, S. G., & Bruce, R. A. (1994). Determinants of innovative behavior: A path model of individual innovation in the workplace. *Academy of Management Journal, 37,* 580–607.

Seo, M.-G., Barrett, L. F., & Bartunek, J. M. (2004). The role of affective experience in work motivation. *Academy of Management Review, 29,* 423–439.

Shalley, C. E. (1991). Effects of productivity goals, creativity goals, and personal discretion on individual creativity. *Journal of Applied Psychology, 76,* 179–185.

Shalley, C. E., Zhou, J., & Oldham, G. R. (2004). The effects of personal and contextual characteristics on creativity: Where should we go from here? *Journal of Management, 30,* 933–958.

Shepperd, J. (1993). Productivity loss in performance groups: A motivation analysis. *Psychological Bulletin, 113,* 67–81.

Shin, S. J., & Zhou, J. (2003). Transformational leadership, conservation, and creativity: Evidence from Korea. *Academy of Management Journal, 46*, 703–714.

Simonton, D. K. (2003). Scientific creativity as constrained stochastic behavior: The integration of product, person, and process perspectives. *Psychological Bulletin, 129*, 475–494.

Sitkin, S. B. (1996). Learning through failure: The strategy of small losses. In M. Cohen & U. Sproull (Eds.), *Organizational learning* (pp. 541–577). Thousand Oaks, CA: Sage.

Stasser, G., & Titus, W. (1985). Pooling of unshared information in group decision making: Biased information sampling during discussion. *Journal of Personality and Social Psychology, 48*, 1467–1478.

Staw, B. M. (1981). The escalation of commitment to a course of action. *Academy of Management Review, 6*, 577–587.

Staw, B. M. (1990). An evolutionary approach to creativity and innovation. In M. A. West & J. L. Farr (Eds.), *Innovation and creativity at work* (pp. 287–308). Chichester, UK: Wiley.

Tierney, P., & Farmer, S. M. (2002). Creative self-efficacy: Its potential antecedents and relationship to creative performance. *Academy of Management Journal, 45*, 1137–1148.

Tierney, P., & Farmer, S. M. (2004). The Pygmalion process and employee creativity. *Journal of Management, 30*, 413–432.

Tsai, W. (2001). Knowledge transfer in intraorganizational networks: Effects of network position and absorptive capacity on business unit innovation and performance. *Academy of Management Journal, 44*, 996–1004.

Tversky, A., & Kahneman, D. (1974). Judgment under uncertainty: Heuristics and biases. *Science, 185*, 1124–1131.

Unsworth, K. (2001). Unpacking creativity. *The Academy of Management Review, 26*, 289–297.

Valacich, J. S., Dennis, A. R., & Connolly, T. (1994). Idea generation in computer-based groups: A new ending to an old story. *Organizational Behavior and Human Decision Processes, 57*, 448–467.

Van Dyck, C., Frese, M., Baer, M., & Sonnentag, S. (2005). Organizational error management culture and its impact on performance: A two-study replication. *Journal of Applied Psychology, 90*, 1228–1240.

Vancouver, J. B., Thompson, C. M., Tischner, C., & Putka, D. J. (2002). Two studies examining the negative effect of self-efficacy on performance. *Journal of Applied Psychology, 87*, 506–516.

West, M. A., & Farr, J. L. (1990). Innovation at work. In M. A. West & J. L. Farr (Eds.), *Innovation and creativity at work* (pp. 3–13). Chichester, UK: Wiley.

Woehr, D. J., & Huffcutt A. I., (1994). Rater training for performance appraisal: A quantitative review. *Journal of Occupational and Organizational Psychology, 67*, 189–205.

Zhou, J., & George, J. M. (2001). When job dissatisfaction leads to creativity: Encouraging the expression of voice. *Academy of Management Journal, 44*, 682–696.

Zhou, J., & George, J. M. (2003). Awakening employee creativity: The role of leader emotional intelligence. *Leadership Quarterly, 14*, 545–568.

Zhou, J., & Oldham, G. R. (2001). Enhancing creative performance: Effects of expected developmental assessment strategies and creative personality. *Journal of Creative Behavior, 35*, 151–167.

Zhou, J., & Shalley, C. E. (2003). Research on employee creativity: A critical review and directions for future research. *Research in Personnel and Human Resources Management, 22*, 165–217.

Zhou, J., & Woodman, R. W. (2003). Managers' recognition of employee's creative ideas: A socio-cognitive model. In L. V. Shavinina (Ed.), *The international handbook on innovation* (pp. 631–640). Amsterdam: Elsevier Science.

4

Revisiting the "Error" in Studies of Cognitive Errors

Shabnam Mousavi and Gerd Gigerenzer

A few decades ago, cognitive scientists studied judgment errors to discover rules that govern our minds, just as visual errors were studied to unravel the laws of perception. This practice has generated a long list of so-called cognitive biases, with disappointingly little insight into how the human mind works. In this chapter, we present our diagnosis of this fruitless labor. An important difference between errors of perception and those of judgment is that visual perception is measured against properties that are objectively measurable, whereas judgments are traditionally evaluated against conventional norms for reasoning and rationality. Moreover, most visual perception errors have developed evolutionarily and are "good" in the sense of enhancing survival or adaptation to the natural or habitual environment, whereas judgmental errors are traditionally perceived as biases that encumber decision making. We maintain that humans can and do make both "bad" and "good" errors. Bad errors, such as spending more than what you earn, should be avoided, whereas good errors are valuable. An example of good errors is the grammatical mistakes that children make with irregular verbs, such as "I *breaked* my toy!" It is through continuous feedback to such errors that children master their native language (for a discussion of good errors, see Gigerenzer, 2005). Many innovations are results of playfulness, allowing for errors and learning from these. One celebrated failure in industry is the accidental development of a weak glue (instead of a strong one), which led to the invention of the Post-it™.* Good errors are side effects of reasonable inferences, explanations, or even serendipity.

Human judgments are usually considered erroneous when measured against logical and statistical norms of rationality. Hence, we address two essential questions of cognitive studies: What is rational judgment? How can one construct reasonable cognitive norms? To answer these questions, we scrutinize two phenomena famous for supposedly demonstrating

---

* We are thankful to Reza Kheirandish for providing us this illustrative example.

human logical and calculative errors: the Wason selection task and overconfidence. First, we show that the Wason selection task confuses logic with rationality, specifically social rationality. Second, we distinguish five types of overconfidence and the environmental structures in which they appear. These two steps bring to light our view of rational judgment and proper norms: Rational judgment must be evaluated against an ecological notion of rationality, which in turn requires constructing content-sensitive norms. In contrast to logical norms, which are content blind in assuming the truth of syntax, content-sensitive norms reflect the actual goals and specifics of the situation. Ecological rationality is about the match of decision-making strategies to the structure of information in the environment. This match is an ecological one but not a mirror image of the environment. Finally, while the literature on cognition claims that "debiasing" is hardly possible, we illustrate the contrary.

Finally, humans react to different representations of information differently. Thus, the way in which information is presented to decision makers can enhance or hinder sound judgment. Using examples from the medical field, in which patients and doctors have to make vital decisions under pressures of time, emotions, and money, we demonstrate how communication of information can be enhanced through transparent modes of presenting risk information.

Is a Violation of the Logical Truth Table an Error in Judgment?

One of the most studied reasoning problems, the Wason (1966) selection task (a Google search returned more than 150,000 entries), assumes logic as a norm for evaluating people's judgments and supposedly demonstrates logical errors.

Various Wason selection tasks share the same logical structure. In the four-card game, the following rule holds: *If P then Q*. The cards shown have information about four situations, with each card representing a situation. One side of a card tells whether *P* has happened, and the other side of the card tells whether *Q* has happened. The player needs to indicate only those cards that definitely need to be turned over to verify whether any of the cards violate the rule.

| *P* | not-*P* | *Q* | not-*Q* |

The answer that satisfies logical truth is to choose the P card (to see if there is a not-Q on the back) and the not-Q card (to see if there is a P on the back). This is because the rule is violated by any situation in which P happens and Q does not. But, this logic does not constitute a useful norm for all kinds of reasoning. For instance, let us apply it to the task of cheating detection. We will see that pragmatic goals affect judgment in ways that are not necessarily compatible with logical truth yet cannot be labeled as errors.

Consider the following conditional: "If an employee works on the weekend (P), then that person gets a day off during the week (Q)." In an experiment (Gigerenzer & Hug, 1992), two roles were assigned to participants: employee or employer. Both were given the same task: to detect whether the rule has been violated. For those who played the role of employee, their dominant response coincided with the logically correct answer, P and not-Q. However, for those who played the role of employer, the dominant response changed to not-P and Q. Why? The answer lies in the perspective of players, who define cheating according to their assigned roles. For the employee, the rule has been violated (i.e., cheating has occurred) if weekend work has been performed (P) but not rewarded with a day off (not-Q). To determine this, the employee needs to inspect P and not-Q, which coincides with the logically correct verification of the conditional. Therefore, in this particular case, the goal of verifying logical truth and that of detecting cheaters lead to the same pattern of information search. From the perspective of employers, however, cheating has happened when an employee takes a weekday off (Q) without earning it through weekend work (not-P). This explains why employers would select not-P and Q instead of trying to verify the logical truth that requires selecting P and not-Q. When logical truth is used as a benchmark, half of the players appear to be following it (the employees' action), whereas the other half do not. That is, using logical truth as a measure of erroneous choice is irrelevant. The achievement of social contract theory (Cosmides, 1989) was to replace the logical rule with the (evolutionary) motivation of cheating detection, thus predicting both observed outcomes. Gigerenzer and Hug (1992) took another step forward by decomposing the observed behavior into correspondence with cheating detection and obligations of a social contract (which defines roles and sets goals.). They showed that "it is *the pragmatics* of who reasons from what perspective to what end" that can sufficiently account for the observed behavior.

Proper norms are sensitive to pragmatics and semantics. We are against upholding logical and statistical rules as a universal yardstick for evaluating human judgment; such "content-blind" norms misrepresent the actual goals (pragmatics) of action. Notice, for example, that in the case of employee/employer role-playing, researchers who held logical truth to be the only yardstick for measuring human rationality evaluated some judgments as rational (e.g., when participants were the employee) and others as irrational (when participants were the employer). In contrast, a content-sensitive approach would predict and explain both patterns of judgments by the adaptive (not Kantian) goal of detecting whether the *other* party has cheated, which is a necessary part of social exchange, social contract, and forms of reciprocal altruism (Trivers, 2002).

An interesting result from the experiments reported in Gigerenzer and Hug (1992) is that some participating students remained true to their training in logic, ignoring the content and always looking for the logically correct answer. However, when interviewed, these subjects reported that what they chose according to logical norms "doesn't make sense."

Does this mean that we should abandon logic? No. Should we abandon logic as a universal, content-blind norm for rational thinking, as assumed by psychologists such as in Wason's (1966) four-card task and Tversky and Kahneman's (1983) conjunction fallacy problems? Yes. To establish a reasonable and useful theory of norms, we need to start with the actual goals of humans, not with the goal of maintaining logical truth. As shown again in the following section, many a puzzling "bias" resolves itself if the focus shifts to content-sensitive norms. We disagree with the view that "our minds are not built (for whatever reason) to work by the rules of probability" (Gould, 1992, p. 469). Rather, biases often appear when researchers ignore the semantics, assume the logical truth of syntax, and reduce the pragmatics of action to fit the observed behavior into convenient and conventional analytical frameworks. These are the systematic errors *committed by researchers*. This is not to say that ordinary people do not make bad errors, but that researchers should reason more carefully about good reasoning in the first place. This argument applies to the celebrated "overconfidence bias" as well.

Is "Overconfidence Bias" an Error?

"The most robust effect of overconfidence" (Odean, 1998, p. 1888) was elaborated in the framework of the research program generally referred to as *heuristics and biases* (Kahneman, Slovic, & Tversky, 1982). A continuing source of confusion within this program is the fact that different phenomena have been labeled with the same word. For instance, the labels "availability" and "representativeness heuristic" have been used for at least a dozen different phenomena, including mutually exclusive ones (Ayton & Fischer, 2004; Gigerenzer, 1996, 2006). The same applies to "overconfidence bias." In the following, we argue that there are at least five different phenomena labeled as overconfidence and its close cousin, (excessive) optimism. We then clarify the problematic nature of imposing improper standards as a benchmark for the evaluation of human judgment.

Overconfidence 1: Better Than Average

Svenson, Fischhoff, and MacGregor (1985) report "an optimism bias: a tendency to judge oneself as safer and more skillful than the average driver, with a smaller risk of getting involved and injured in an accident" (p. 119). Their study is a logical continuation of earlier studies (see also Svenson, 1978, and Johansson & Rumar, 1968). For example, 77% (in Svenson, 1981) of the subjects responded "yes" to the question of whether they drive more safely than the average driver. The authors explain: "It is no more possible for most people to be safer [drivers] than average than it is for most to have above average intelligence" (Svenson et al., 1985, p. 119). Yet, this claim is correct only if both safe driving and intelligence have distributions that are symmetric rather than skewed. In reality, the distribution of car accidents is asymmetric—unlike IQ distributions, which are standardized to be normally distributed and symmetric so that the median and mean are the same. The asymmetric (skewed) distribution of car accident data implies that more than 50% of drivers do in fact have fewer accidents than the mean. To check the size of the effect, we analyzed the data on approximately 7,800 drivers in the United States, reported by Finkelstein and Levin (2001). Among these American drivers, 80% indeed had fewer than the mean number of accidents.

This oversight of the difference between symmetric and asymmetric distributions can be counterbalanced by asking for percentiles or medians rather than averages or means and is sometimes done. Yet, asymmetry is not the only problem with the claim that people commit reasoning errors.

The second problem with the normative argument is that people are asked ambiguous questions for which correct answers cannot be determined. The question of whether one is an "above average" driver or an "above average" teacher leaves open to what "average" refers, and the statement will be interpreted in individual ways. For some, a good driver is one who causes no traffic accidents; for others, it is someone who can drive faster than others, hold a mobile phone and a cigarette while driving, or obey the law and drive defensively. The subjects who are asked to rate themselves on driving safety in comparison with the average driver will likely assume their understanding of a good driver. If people tend to be better than others at what they believe one should do, the ambiguous question is a second, independent reason to produce the "better-than-average" effect.

There is also a third reason why researchers need to be cautious with quick judgments about what is right and wrong. Consider the question, "Is your IQ above or below the mean of 100?" Such a question avoids the asymmetry argument and to some degree the second argument as well because it is fairly precise (although different IQ tests can lead to strikingly different results for the same person). Even here, the normative claim that "50% of all people have an IQ over 100" is not strictly correct. The "Flynn effect" implies that IQ climbs annually on average around 0.3 points, that is, there is a "year effect" associated with IQ measurements. Therefore, the statistical tests must be restandardized on a regular basis. Yet, the more years that pass after the last standardization, the more people will have above-average IQs. In these cases, the statement that more than 50% of people are better than average is, again, a fact about the world, not a judgment error.

The general point here is that to make statements about people's irrationality, researchers need to analyze the statistical structure of the environment carefully, pose clear questions, and take into account the reference points that people use for making their judgments. For the study on self-assessments of driving skills, this includes verifying the distribution of car accidents and what a person considers good driving. Only then can one decide whether a person's judgment is based on the kind of self-deception that the term *overconfidence* suggests or whether the problem lies mostly in habitual normative beliefs inside researchers' minds.

Overconfidence 2: (Too) Narrow Confidence Intervals

A second phenomenon also called overconfidence is that individuals tend to indicate too narrow confidence intervals. For instance, subjects of an experiment were asked to guess an interval for the next year's interest rate that included the true value with a probability of .90. The true value, however, was only included in 40 to 50% of the intervals (Block & Harper, 1991; Lichtenstein, Fischhoff, & Philips, 1982; Russo & Shoemaker, 1992). Because the intervals produced were, on average, too small, the conclusion was drawn that this is another case of "oversimplification." Note first that Overconfidence 1 and 2, in spite of sharing the same name, are distinct phenomena. Better than average (Overconfidence 1) does not imply narrow confidence intervals (Overconfidence 2) or vice versa. We do not know of a single study demonstrating that people who show the first phenomenon also show the second. Labeling different phenomena as the same without any evidence is one thing; calling a phenomenon an error is another.

Again, important insight would be gained by carefully specifying reasonable norms and models (rather than labels) of the cognitive processes that imply the observed phenomenon (see Juslin, Wennerholm, & Olsson, 1999; Winman, Hansson, & Juslin, 2004). Juslin, Winman, and Hansson (2007) presented a naïve sampling model, which offers a causal explanation for Overconfidence 2. This model is based on the fact that if everyday experiences (random samples) are transferred directly to a population, statistical theory implies that a systematic error will appear in estimating the variance and therefore in producing confidence intervals. Estimation of means, however, remains unaffected in this situation. Juslin et al.'s naïve sampling model enables precise predictions of produced confidence intervals and was successfully tested in a series of experiments. If the experimenter provides a probability (an average) and asks about an interval for this probability, the reported phenomenon of Overconfidence 2 appears. However, if an interval is provided and subjects are asked to provide the probability that the true value lies in the interval, the same phenomenon disappears. Juslin et al.'s scrutiny illustrates how this phenomenon (Overconfidence 2) can be elicited or eliminated from the same subject according to the setup and therefore cannot represent a stable personality characteristic or trait. Overconfidence 2 can be adequately explained by the fact that means are unbiased, but intervals (variance) are biased estimators.

Overconfidence 3: Mean Confidence Is Greater Than Percentage Correct

There is a third phenomenon also called overconfidence. It initially emerged from answers obtained to two-step questions such as "Which city lies further south: Rome or New York? How certain are you that your answer is correct?" Since the 1970s, many studies of this sort have shown that people express a level of certainty (confidence) in their own judgment that is higher than the average percentage of correct answers. The result was labeled "overconfidence" and put into the same category as the two previously mentioned phenomena, suggesting more evidence for the same irrational propensity. In one of these studies, Griffin and Tversky (1992) asserted, "The significance of overconfidence to the conduct of human affairs can hardly be overstated" (p. 414).

The probabilistic mental models theory (Gigerenzer, Hoffrage, & Kleinbölting, 1991) provided a model (rather than a label) for confidence judgments and predicted for the first time that Overconfidence 3 appears if the experimenter systematically selects the tasks and disappears if tasks are chosen randomly. These results were experimentally confirmed. As a rule, judgments were well adapted to reality but erred when participants were faced with selected and therefore nonrepresentative tasks. For example, the annual average temperature of a city is generally a good cue for its geographic location. When people are asked whether Rome or New York lies further south, Overconfidence 3 appears because knowledge about the annual average temperature of the two cities leads to an incorrect answer: Many do not know that this selected pair is an exception to the rule, and that Rome is actually higher in latitude than New York.

Gigerenzer et al. (1991) attributed the emergence and disappearance of overconfidence to selecting questions of this type versus using a representative set. This finding was first denied (Griffin & Tversky, 1992) but later supported in an extensive analysis of 130 studies, in which this form of overconfidence appeared only if tasks were selected systematically rather than randomly (Juslin et al., 2000).

Just as the naïve sampling model reveals the mechanisms of overconfidence in the case of seemingly narrow confidence intervals (Overconfidence 2), the probabilistic mental models theory reveals the phenomenon of overconfidence to be a mismatch between a mental model adapted to an environment and experimenters' biased task selection (Overconfidence 3). These theories explain and predict when Overconfidence 2 and 3 appear or disappear in the same person. The fact that the appearance or disappearance of Overconfidence 3 is subject to whether questions are sampled in a selective or representative way again demonstrates that the phenomenon is a poor candidate for a human trait or personality type, as the label suggests. Nonetheless, many researchers persist in asserting that overconfidence is a general trait or disposition.

Overconfidence 4: Overconfidence Equals Miscalibration

A fourth definition of overconfidence is miscalibration. Here, data are elicited from questions similar to those in the previous case (Overconfidence 3): "Which city lies further south: Rome or New York? How certain are you that your answer is correct?" What is analyzed, however, is not the discrepancy between the average confidence and the proportion correct, but rather the total calibration curve. For instance, a typical finding is that in all cases when people said they were 100% sure that their answer was correct, the proportion correct was only 80%; when people said that they were 90% confident, the proportion correct was only about 70%; and so on. Here, calibration is assumed to mean that the confidence categories match the proportion correct, and the mismatch is called miscalibration, or overconfidence. This again is a distinct phenomenon that can appear independently from Overconfidence 3. For example, although the difference between mean correct and percentage correct (Overconfidence 3) can be zero (e.g., when the curve for the proportion correct crosses the calibration line at 50%, as in a regression curve), miscalibration can be substantial.

For two decades, the fact that the actual curve of proportion correct differed from the calibration line was attributed to deficits of the human brain until it was noticed independently by Erev, Wallsten, and Budescu (1994) and Pfeifer (1994) that researchers had made an error: They had overlooked that a regression to the mean is at work. Confidence judgments tend to generate noisy data—that is, conditional variance is larger than zero, which is equivalent to assuming that the correlation between confidence and proportion correct is imperfect. An imperfect correlation implies that when the reported confidence ratings are high, the corresponding proportions correct will be smaller and hence resemble miscalibration and overconfidence. Typically, for general knowledge questions sampled randomly from a large domain, the regression line is symmetrical around the midpoint of the reported confidence scale (e.g., a midpoint of 50% when the confidence scale is from 0 to 100% and 75% when the confidence scale is from 50% to 100%; Juslin et al., 2000). That is, to unskilled eyes the mere presence of conditional variance can appear as systematic bias in the judgments, where there is only unsystematic noise—just as in Francis Galton's famous example that sons of tall fathers are likely to be smaller in height, and sons of small fathers are likely to be taller (see Stigler, 2002, for a detailed account). It is a normal consequence of regression, not a cognitive bias. In these environments, any intelligent system, human or computer, will produce patterns that mimic what has been called miscalibration or overconfidence.

If one estimates the confidence judgments from proportion correct (rather than vice versa), regression implies the mirror result: a pattern that looks as if there were underconfidence bias. When looking at all items that the participants got 100% correct, for instance, one will find that the average confidence was lower, such as 80%. This seems to indicate underconfidence. In contrast, when looking at all items for which participants were 100% confident, one finds that the average proportion correct was lower, such as 80%. This seems to indicate overconfidence. Erev et al. (1994) showed for three empirical data sets that regression toward the mean accounted for practically all the effects that would otherwise have been attributed to overconfidence or underconfidence, depending on how one plotted the data. Dawes and Mulford (1996, p. 210) reached the same conclusion for another empirical data set. In general, one could determine whether there is under-/overconfidence beyond regression by plotting the data both ways. This research shows that Overconfidence 4 appears largely due to researchers' misinterpreting regression to the mean as a cognitive illusion of their experimental subjects.

Overconfidence 5: Functional Overconfidence

The fifth and last phenomenon that is labeled overconfidence is of yet another nature. In this case, a functional and often profitable mechanism is at play. Many tasks, such as predicting the stock market, are so difficult that even experts do not perform better than laypeople, and sophisticated statistical strategies are not consistently better than intuition. For instance, in a comparison of a dozen optimization methods and the simple $1/N$ heuristic (equal allocation of money to all options) for investment allocation decisions, $1/N$ proved to be as good as or better than complex statistical methods in most cases (DeMiguel, Garlappi, & Uppal, 2009). In other studies, professional analysts performed worse than chance in picking stocks, while laypeople performed at chance level. Nevertheless, many customers would like to believe that experts can predict the stock market, or they tend to delegate responsibility for selecting strategies to experts. In such situations, experts who convey security and confidence stand a good chance of winning trust even if it is ill founded. In medicine, placebo effects are based on this kind of belief in the efficacy of treatments that have no real effect. Without unrealistic self-confidence, option advisers and astrologers would lose their customers—and some physicians their patients. Instances of functional overconfidence give us an opportunity to study the creation of the illusion of certainty and its role in determining human behavior (Gigerenzer, 2002). Functional overconfidence has little to do with the previously listed four types of overconfidence.

The story of the overconfidence bias may tell us more about the biased norms of many researchers in this field than the biased thinking of their experimental subjects. We pointed to two flaws: First, the practice of referring to different phenomena using the same label impedes understanding their nature; second, the norms against which people's judgments have been evaluated as flawed are often flawed themselves (see Gigerenzer, 1996, 2000). Moreover, deviations of human judgment measured against these standards are not predicted by formal psychological theories but are instead explained away by vague "irrational" notions such as overconfidence. Functional overconfidence, however, illustrates that there are parts of our external world where too much certainty is expected, and where judgments are thus directed by other incentives and goals than factual correctness.

Overconfidence research should be of particular interest to people who study the lack of progress made by psychological theories. Even after the normative problems had been pointed out in top journals such as *Psychological Review* (Gigerenzer et al., 1991; Juslin et al., 2000), many researchers nonetheless maintained that there is clear evidence for a general human tendency toward overconfidence, and one can still hear this message in social psychology, behavioral economics, and behavioral finance. Indeed, there is one clear instantiation of this bias: *the overconfidence of many researchers in a phenomenon called overconfidence.*

One of the dangers of focusing on toy problems such as the Wason selection task and many overconfidence tasks is that researchers lose sight of the real errors that people make in the world outside the laboratory. In the next section, we illustrate how an ecological rather than a logical approach to cognition can help people to learn how to reduce these real-world errors.

Helping People Avoid Judgment Errors

In this section, we present some findings from the study of decision-making processes in the health care arena on how doctors and patients make decisions. The health statistics that doctors obtain and communicate to their patients can be reported in different formats, such as relative risk (percentage change), absolute risk (how many in what total), or conditional probabilities, including sensitivity and false-positive rate. In what follows, we explain these modes of representation and recommend particular representations that reduce the potential of confusing health providers and decision makers.*

Absolute Risk Versus Relative Risk

"Mammography screening reduces the risk of dying from breast cancer by 20%" is an example of reporting results of a clinical study in terms of relative risk. In this form, the baseline information is concealed; we do not know "20% of how many." An absolute risk, in contrast, reveals the baseline risk. The absolute risk reduction is the absolute difference between the treatment and control groups. For instance, the same result can be reported in absolute risk terms as "Mammography screening reduces the risk of dying from breast cancer by about 1 in 1,000, from about 5 in 1,000 to about 4 in 1,000." That is, the 20% corresponds to 1 in 1,000. In a study with 160 gynecologists, one third did not know that the relative risk of 20% (sometimes reported as 25%) means 1 in 1,000 but instead believed that it means 20 or 200 in 1,000 (Gigerenzer et al., 2007). To communicate risk reduction associated with treatments, we therefore recommend using absolute risks, not relative risks.

Conditional Probability versus Natural Frequency

Prevalence, sensitivity, and false-positive rate are the usual pieces of information that physicians need when assessing positive predictive values, that is, the probability that a patient has a disease given a positive screening test result. Assume that in a particular region, the prevalence of the disease (probability of a woman having breast cancer) is 1%, sensitivity is 90% (probability of a woman with breast cancer testing positive), and the false-positive rate is 9% (i.e., probability of a woman without breast cancer testing positive). What then are the chances that a woman who has tested positive actually has breast cancer? When this information was provided

---

* This section heavily draws on the work of Gigerenzer, Gaissmaier, Kurz-Milke, Schwartz, and Woloshin, 2007.

to a group of gynecologists, their answers varied between a 1% and a 90% chance of cancer, with a majority of them overestimating the chances to be above 80%. However, when these physicians were given an alternative representation of the same information involving natural frequencies, their judgments improved dramatically, such that 87% of them correctly specified the chances of cancer for a woman with a positive test result to be about 1 in 10.

The first mode of presentation uses a *conditional probability* format, which is the probability of an Event A given an Event B (such as probability of testing positive if having cancer). Because of the difficulties it causes the majority of physicians in assessing the patient's chances of cancer, we consider it to be a nontransparent mode of risk communication. The alternative mode, called *natural frequency* format, communicates the same information in terms of raw counts for each group out of a certain total, as the following example shows: Of 1,000 women, we expect that 10 have breast cancer; of these 10 women with cancer, 9 test positive; and of 990 women without cancer, about 89 still test positive. Thus, we expect that only 9 out of 98 who test positive actually have cancer. Our recommendation is to use natural frequencies and avoid conditional probabilities.

In sum, whether physicians and patients make errors in estimating probabilities depends heavily on the way information is presented in the environment. We also believe that human choices can be enhanced—for instance, mistakes (or avoidable errors) can be reduced—by designing environments that facilitate the access to information (Gigerenzer & Hoffrage, 1995; Hertwig & Gigerenzer, 1999). As a guideline, we recommend that the questions given next always be asked about all types of risk information. The answers to these questions clarify the numbers and improve the reliability and correctness of the information communicated. These questions apply to any situation of decision making under risk; here, the explanations following each question are taken from the medical field.

1. "Risk of what?" Is it risk of getting a disease, developing a symptom, or dying from the disease?
2. "Time frame?" Risk changes over time and directly affects the evaluation of treatment options. So, it is important to present and receive information for specific time frames, such as "over the next 10 years," as opposed to "over a lifetime."
3. "How big?" The size of risk is best communicated in absolute terms, for example, 2 of 10,000 instead of a percentage with an unclear baseline. In addition, the associated risk attached to a certain condition (such as risk of cancer for smokers) can be compared to tangible fatal events (such as car accidents).

4. "Does it apply to me?" For the risk analysis available from a study to be relevant to you, the study group should include people like you. This means that you must share determining characteristics with this group. For example, if the age group of people in the study is different from yours and age is a significant determinant of the studied disease, the risk results are unlikely to apply to you. Recall that Overconfidence 2 was explained as the result of extending results from a random sample to a population. Keeping this in mind, you can avoid making the mistake of extending and using results that are found for a sample to which you do not belong.

An Ecological View of Error

In the study of cognitive errors, the tradition of focusing on logic as a universal yardstick for evaluating judgments (which produces a list of irrationalities) is an unfruitful practice because it measures deviations from content-blind norms. Content-blind norms accept logical, mathematical, or statistical truths at face value and implement them as such in models of decision making. However, generalizing from the world of models to the real world is often unjustified or inadequate. As shown in our close examinations of two popular "errors," namely, violations of logic in the Wason selection task and the overconfidence bias, content-sensitive norms are needed to evaluate behavior in the real world. Sensible norms can be conceived in an ecological framework, in which many biases disappear as soon as the actual structure of information in the environment is taken into account and the focus is turned to finding an *effective* match between the environment and available strategies. This ecological view of judgment and its resulting methods—for instance, utilizing favorable modes of information presentation—can enhance human performance.

The framework presented in this chapter does not entail exaggerated assumptions about cognitive abilities or oversimplification of the environment. In this ecological view of error, people of course make mistakes. As outlined, there are three types of errors: good errors, bad errors, and nonerrors. Some errors are useful and need be recognized as such (e.g., grammatical mistakes in the process of learning a language). Other errors are bad and should be avoided (e.g., misunderstanding medical test results). Many bad errors can be avoided by improving presentations of information, using forms that are easily absorbed and interpreted by human minds (e.g., use of absolute risks for communicating risk). Finally, some phenomena (e.g., overconfidence) that are mislabeled as errors can be easily explained if scientists include more than the conventional norms in their operational frameworks and allow for a wider scope of analysis.

Central to this chapter is the argument that the study of cognitive errors has not been successful in unraveling rules of human mind mainly because it focuses on abstract, incomplete, and sometimes irrelevant norms. One might ask, "Why does it matter if a scientific inquiry is focused on erroneous questions for a while? We can still learn *something*!" In light of what was presented in this chapter, we hope that you agree that it is time to rectify the questions and move forward. Allow us to present a strategy for doing so: to look beyond the most convenient analytical tools and rethink our habits of scientific practice. We hope that the current piece encourages less-biased practice of scientific inquiry into unraveling processes of the human mind.

References

Ayton, P., & Fischer, I. (2004). The hot hand fallacy and the gambler's fallacy: Two aces of subjective randomness? *Memory & Cognition, 32*, 1369–1378.

Block, R. A., & Harper, D. R. (1991). Overconfidence in estimation: Testing the anchoring and adjustment hypothesis. *Organizational Behavior and Human Decision Processes, 49*, 188–207.

Cosmides, L. (1989). The logic of social exchange: Has natural selection shaped how humans reason? Studies with the Wason selection task. *Cognition, 31*, 187–276.

Dawes, R. M., & Mulford, M. (1996). The false consensus effect and overconfidence: Flaws in judgment, or flaws in how we study judgment? *Organizational Behavior and Human Decision Processes, 65*, 201–211.

DeMiguel, V., Garlappi, L., & Uppal, R. (2009). Optimal versus naive diversification: How inefficient is the $1/N$ portfolio strategy? *Review of Financial Studies, 22*, 1915–1953.

Erev, I., Wallsten, T. S., & Budescu, D. V. (1994). Simultaneous over- and underconfidence: The role of error in judgment processes. *Psychological Review, 101*, 519–527.

Finkelstein, M. O., & Levin, B. (2001). *Statistics for lawyers* (2nd ed.). New York: Springer-Verlag.

Gigerenzer, G. (1996). On narrow norms and vague heuristics: A reply to Kahneman and Tversky (1996). *Psychological Review, 103*, 592–596.

Gigerenzer, G. (2000). *Adaptive thinking: Rationality in the real world*. New York: Oxford University Press.

Gigerenzer, G. (2002). *Calculated risks: How to know when numbers deceive you*. New York: Simon & Schuster. (UK edition: *Reckoning with risk: Learning to live with uncertainty*. London: Penguin Books, 2002)

Gigerenzer, G. (2005). I think, therefore I err. *Social Research, 72*, 1, 195–218.

Gigerenzer, G. (2006). Heuristics. In G. Gigerenzer & C. Engel (Eds.), *Heuristics and the law* (pp. 17–44). Cambridge, MA: MIT Press.

Gigerenzer, G., Gaissmaier, W., Kurz-Milke, E., Schwartz, L., & Woloshin, S. (2007). Helping doctors and patients make sense of health statistics. *Psychological Science in the Public Interest, 8*(2), 53–96.

Gigerenzer, G., Hoffrage, U., & Kleinbölting, H. (1991). Probabilistic mental models: A Brunswikian theory of confidence. *Psychological Review, 98*, 506–528.

Gigerenzer, G., & Hoffrage, U. (1995). How to improve Bayesian reasoning without instruction: Frequency formats. *Psychological Review, 102*, 684–704.

Gigerenzer, G., & Hug, K. (1992). Domain-specific reasoning: Social contracts, cheating, and perspective change. *Cognition, 43*, 127–171.

Gould, S. J. (1992). *Bully for brontosaurus: Further reflections in natural history*. London: Penguin Books.

Griffin, D. W., & Tversky, A. (1992). The weighing of evidence and the determinants of confidence. *Cognitive Psychology, 24*, 411–435.

Hertwig, R., & Gigerenzer, G. (1999). The "conjunction fallacy" revisited: How intelligent inferences look like reasoning errors. *Journal of Behavioral Decision Making, 12*, 275–305.

Johansson, G., & Rumar, K. (1968). Visible distances and safe approach speeds for night driving. *Ergonomics, 11*, 275–282.

Juslin, P., Wennerholm, P., & Olsson, H. (1999). Format dependence in subjective probability calibration. *Journal of Experimental Psychology: Learning, Memory, and Cognition, 25*, 1038–1052.

Juslin, P., Winman, A., & Olssen, H. (2000). Naive empiricism and dogmatism in confidence research: A critical examination of the hard-easy effect. *Psychological Review, 107*, 384–396.

Juslin, P., Winman, A., & Hansson, P. (2007). The naïve intuitive statistician: A naïve sampling model of intuitive confidence intervals. *Psychological Review, 114*, 678–703.

Kahneman, D., Slovic, P., & Tversky, A. (1982). *Judgment under uncertainty: Heuristics and biases*. Cambridge, UK: Cambridge University Press.

Lichtenstein, S., Fischhoff, B., & Phillips, L. D. (1982). Calibration of subjective probabilities: The state of the art up to 1980. In D. Kahneman, P. Slovic, & A. Tversky (Eds.), *Judgment under uncertainty: Heuristics and biases* (pp. 306–334). Cambridge, UK: Cambridge University Press.

Odean, T. (1998). Volume, volatility, price and profit when all traders are above average. *Journal of Finance, 53*, 1887–1934.

Pfeifer, P. E. (1994). Are we overconfident in the belief that probability forecasters are overconfident? *Organizational Behavior and Human Decision Processes, 58*, 203–213.

Russo, J. E., & Schoemaker, P. J. (1992). Managing overconfidence. *Sloan Management Review, 33*, 7–17.

Stigler, S. M. (2002). *Statistics on the table: The history of statistical concepts and methods*. Cambridge, MA: Harvard University Press.

Svenson, O. (1981). Are we all less risky and more skillful than our fellow drivers? *Acta Psyhologica, 47*, 143–148.

Svenson, O. (1978). Risks of road transportation in a psychological perspective. *Accident Analysis and Prevention, 10*, 267–280.

Svenson, O., Fischhoff, B., & MacGregor, D. (1985). Perceived driving safety and seatbelt usage. *Accident Analysis and Prevention, 17,* 119–133.

Trivers, R. (2002). *Natural selection and social theory* (Evolution and Cognition Series). New York: Oxford University Press.

Tversky, A., *Kahneman,* D. *(1983).* Extensional versus intuitive reasoning: The *conjunction* fallacy in probability judgment. *Psychological Review, 90,* 293–315.

Wason, P. C. (1966). Reasoning. In B. M. Foss (Ed.), *New horizons in psychology.* Hammondsworth, England: Penguin.

Winman, A., Hansson, P., & Juslin, P. (2004). Subjective probability intervals: How to reduce "overconfidence" by interval evaluation. *Journal of Experimental Psychology: Learning, Memory, and Cognition, 30,* 1167–1175.

5

Collective Failure: The Emergence, Consequences, and Management of Errors in Teams

Bradford S. Bell and Steve W. J. Kozlowski

On the evening of March 5, 2000, Southwest Airlines Flight 1455 overran the end of the runway while landing at Burbank–Glendale–Pasadena Airport in California. The plane broke through a metal blast fence at approximately 40 knots, skidded across Hollywood Way, and came to rest in front of a Chevron gasoline station. Of the 142 passengers on board, 2 received serious injuries, 42 received minor injuries, and the aircraft was written off. This was the first major accident in the 33-year history of Southwest, and the question was quickly raised regarding what went wrong. An investigation by the National Transportation Safety Board (NTSB) revealed that the probable cause of the accident was a fast, steep, and unstabilized approach (NTSB, 2002). The aircraft descended at an angle 3 or 4 degrees steeper than the typical flight path and touched down 2,150 feet down the runway traveling 44 knots over the target airspeed. The descent was so steep that it sounded automated "sink rate" warnings in the cockpit, yet both the captain and first officer ignored the warnings. The captain later reported that he knew he was not "in the slot" required for a normal landing and should have performed a go-around maneuver but said that he became "fixated on the runway." The first officer indicated in postaccident interviews that he was also aware that they were "out of the slot" but did not mention it because he believed the captain was taking corrective action. Southwest later fired both pilots and admitted that their actions in the incident were negligent.

In 1993, the National Aeronautics and Space Administration (NASA) started the *Mars Surveyor* program with the goal of conducting a series of missions to explore Mars. One of the missions added to the program in 1995 was the *Mars Climate Orbiter* (MCO), which was designed to carry instruments to study the surface, atmosphere, and weather of the planet.

The MCO was launched December 11, 1998, and traveled 416 million miles over the next 9½ months to reach Mars. However, shortly after firing its main engine to achieve elliptical orbit around Mars, radio contact was lost, and no further signal was ever received from the spacecraft. An investigation later revealed that a simple software error led to the spacecraft entering the Martian atmosphere at a lower-than-expected trajectory and its likely destruction due to atmospheric friction (Mars Climate Orbiter Mishap Investigation Board, 1999). Specifically, the navigation team at the Jet Propulsion Laboratory used metric units in its calculations, but the design and manufacturing team at Lockheed Martin provided important acceleration data in English units (Hotz, 1999). The result was that the effect of the thrusters was underestimated by a factor of 4.45—the required conversion factor from force in pounds to newtons—and the spacecraft drifted off course. A NASA official was later quoted as saying, "This was not a failure of Lockheed Martin. It was systematic failure to recognize and correct an error that should have been caught" (Hotz, 1999, p. A-1).

A fourth-year medical student on rotation in the pediatric intensive care unit (PICU) was invited to observe the operative repair of a congenital heart lesion (Wachter, 2005). As the surgical team prepped the patient for surgery, the student was surprised to see one of the team members insert a catheter without first performing a "sterile prep." However, the student was new to the PICU and assumed different procedures are used with pediatric patients, so she did not mention the incident to any of the team members. Three days later, the patient developed an infection, and while on rounds, the student presented the account of the catheter placement in the operating room. After rounds, the student was approached by two attendings. One commented that the information about the catheter should not have been presented during rounds because the patient's family might overhear. The second attending told the student that the information should have been presented in the operating room. Neither attending commended the student for bringing the incident to the attention of the team (Wachter, 2005).

Over the past two decades, there has been an ongoing shift from work organized around individual jobs to team-based work systems (Kozlowski & Bell, 2003). As the nature of work has grown more complex and dynamic, organizations have turned to teams to perform tasks that exceed the capabilities of a single individual. Because they have the potential to serve as self-correcting performance units (Hackman, 1993), teams are also being used in organizations that emphasize quality and high reliability. Yet, as the examples illustrate, teams are not infallible. In fact, the potential for teams to fail is real because they are often used in situations characterized by a high risk of errors (e.g., aviation, medicine, military) and in situations for which even minor errors can have severe consequences (Salas, Cooke, & Rosen, 2008). As Salas, Kosarzycki, Tannenbaum, and Carnegie (2005) stated, "On the one hand, it is clear that teams can accomplish great things, but it is equally apparent that subfunctioning teams have the potential to fail spectacularly" (p. 98). Fortunately, most errors that occur in teams are unspectacular or are remedied before they lead to disaster. More minor errors are nevertheless important because they can impede team effectiveness, can stimulate team learning, and represent warning signs to potentially larger and more consequential errors (Edmondson, 2004).

Although it is widely acknowledged that teams can and often do fail, the conditions that give rise to errors in teams, the implications of these errors for team effectiveness, and the ability of teams to prevent and manage errors are not well understood (Bauer & Mulder, 2007; Tjosvold, Yu, & Hui, 2004). This may be due, in part, to the fact that errors in teams rarely have simple causes or consequences. Team performance is influenced by factors occurring not only at the team level (e.g., communication, coordination) but also at levels above (e.g., culture, climate) and below (e.g., individual performance), which can make it difficult to determine the root cause of a team failure (Rosen et al., 2008). Also, although errors often have negative consequences for team performance, they represent an important stimulus of team learning and a foundation for continuous improvement and adaptive capabilities. Accordingly, to understand the impact of errors on the effectiveness of a team over time, it is important to look more closely at how teams manage errors when they occur.

The goal of the current chapter is to examine the emergence, consequences, and management of errors in teams. We begin by discussing the origin and emergence of errors in teams. We argue that errors in teams can originate at both the individual and collective levels and suggest this distinction is important because it has implications for how errors propagate within a team. We then consider the paradoxical effects of errors on team performance and team learning. This discussion highlights the importance of error management in teams so that errors can prompt learning while mitigating their negative consequences. Thus, we focus significant attention on the challenge of error prevention and error management in teams and highlight numerous factors that can influence these processes. We conclude the chapter with a discussion of important research gaps and outline an agenda for future work in this area.

Origin and Emergence of Errors in Teams

Hofmann and Frese (Chapter 1, this volume) define errors as actions that "unintentionally fail to achieve their goal if this failure was potentially avoidable." This definition highlights several widely accepted and fundamental properties of errors. First, an error occurs when there is a deviation from something else, namely, some external goal, standard, or desired behavior. Second, errors are unintended, which differentiates them from intentional deviations from standards or goals (i.e., violations). For instance, an error can occur when there is a failure to execute an intended action or plan or when the plan itself is inappropriate or inadequate to achieve the intended outcome (Rasmussen, 1986; Reason, 1990). Third, errors are potentially avoidable. That is, the error cannot be attributed to the intervention of some chance agency. This property distinguishes errors from the concept of risk, for which one knowingly accepts some probability of failure (Hofmann & Frese, Chapter 1, this volume).

Human error is unavoidable, particularly in today's complex and dynamic work environments. Recognizing this, organizations have adopted a variety of strategies to mitigate the effects of human error, including training, human factors engineering, and work design. One specific approach is to design work around teams because team members can serve as redundant systems to help catch one another's mistakes and provide support when needed (Salas, Wilson, Murphy, King, & Salisbury, 2008). However, researchers have also suggested that placing a task in a team context may have little impact on error rates for a number of reasons, including social distractions and social loafing (Hollenbeck, Ilgen, Tuttle, & Sego, 1995). Although there is evidence that teams sometimes make fewer mistakes than individuals (Baker, Day, & Salas, 2006; Foushee, Lauber, Baetge, & Acomb, 1986), it is also clear that errors are a common occurrence in teams, even in high-reliability settings (Edmondson, 2004; Thomas, 2004).

To understand the nature of errors at the collective level, it is important to consider the origin of errors in teams. Errors within teams can originate and manifest at both the individual and collective levels of analysis, and we suggest that the level at which an error originates has important implications for how it propagates through the team and the ability of the team to identify and manage the error. In the sections that follow, we first consider individual error within the context of teams and then extend our examination to the collective level.

Individual Error in Teams

Hofmann and Frese (Chapter 1, this volume) develop an integrative error taxonomy based on the fundamental premise that behavior in organizations is goal directed. Their taxonomy is based on action theory (Frese & Zapf, 1994), and its basic structure is similar to other prominent error taxonomies in the field, such as Reason's (1990) generic error-modeling system (GEMS) and Rasmussen's (1982) skill-knowledge-rule framework. In particular, they argue that execution of a goal requires a number of actions and subactions or tasks, as well as higher-order regulation processes that coordinate and monitor the cascading sequence of actions. This hierarchy of actions and regulatory processes can be decomposed into four different levels, with different types of errors occurring at each level. For example, the lowest level of regulation is the sensorimotor level, which involves the nonconscious execution of skill-based scripts or automatic movement sequences. Errors at this level arise from stereotyped and automatic movements, such as typing an incorrect key. In contrast, the conscious, or intellectual, level of regulation involves active cognitive processing and the development of goals and plans. Errors at this level, referred to as knowledge errors, occur when a person develops an inappropriate or inadequate goal or plan due to resource constraints (e.g., inability to access prior knowledge) or incomplete or incorrect knowledge. In between the nonconscious and conscious levels of regulation is the level of flexible action patterns, which involves well-known actions that are triggered by environmental cues and adapted to the situation (Hofmann & Frese, Chapter 1, this volume). Errors at this level typically occur when a person misclassifies a situation and applies a rule (i.e., plan) that is inappropriate or deficient given the circumstances. Finally, there is an overarching level of regulation, the heuristic level, which involves the general approaches, or metacognitive strategies, that individuals apply toward goal accomplishment. Errors at this level arise in the processes that individuals use to formulate goals, conduct information searches, develop plans, and monitor and process feedback.

Although this taxonomy focuses on errors that affect individual goal achievement, individual error can also lead to collective failure. Kozlowski, Gully, Nason, and Smith (1999) described how team member errors, if undetected and uncorrected, can "propagate throughout the network if they are sent to a central role or along a critical path" (p. 272). For example, on July 3, 1988, the *USS Vincennes* mistakenly shot down an Iranian Airbus, killing 290 civilians. Although multiple factors contributed to this disaster, identification errors committed by the radar console operators were paramount (Fisher & Kingma, 2001). The operators misidentified the Airbus A300 as an Iranian Air Force F-14 and mistakenly claimed that the aircraft was descending when it was actually climbing. These information-processing errors (i.e., mapping errors; see Table 1.1 in Chapter 1) by the individual operators permeated throughout the command-and-control team and ultimately facilitated the erroneous decision to shoot down the aircraft. As another example, on April 26, 1994, China Airlines Flight

140 crashed on its approach to land in Nagoya, Japan, killing 264 of the 271 individuals on board. An investigation revealed that while performing the approach the first officer inadvertently pressed the takeoff go-around (TOGA) button located on the forward edge of the throttles, which increases thrust to takeoff levels. Subsequent efforts by the crew to correct the situation resulted in a steep climb, at which point the plane stalled and crashed. Thus, the sensorimotor error committed by the first officer triggered a sequence of actions by the crew, many of which were inappropriate given the circumstances, which ultimately doomed the flight.

In both of the examples provided, individual error was a primary cause of the failure of the team. However, these two examples also highlight the different ways in which individual error can propagate in team settings. In the case of the *Vincennes*, the operators' errors led directly to the erroneous decision to shoot down the aircraft. Thus, the causal chain linking the individual error and team failure was short and direct. However, in the case of the China Airlines flight, the first officer's inadvertent press of the TOGA button did not directly cause the accident. Rather, the error triggered a flawed recovery operation by the crew, which led the aircraft to stall and crash. Thus, the first officer's error was a precipitating event that was distally linked to the accident through a subsequent chain of errors committed by the crew. Whether an error serves as a direct or indirect cause of team failure may have important error management implications. For instance, when an error has indirect consequences, a team may have more time and opportunity to identify and trap the mistake before it has an impact on other parts of the system.

Interestingly, both examples involve teams performing in complex task environments characterized by highly interdependent workflow arrangements in which members must diagnose, problem solve, or collaborate simultaneously (Van de Ven, Delbecq, & Koenig , 1976). Some authors have argued that team members should be better able to catch each other's mistakes when interdependence is high, but it may depend on the nature of that interdependence. Salas, Sims, and Burke (2005), for example, argue that mutual performance monitoring is critical to catching mistakes prior to or shortly after they occur and suggest that this process is easier in collaborative tasks than in coordination tasks. Similarly, Druskat and Pescosolido (2002) argue that heedful interrelating, or an awareness of team member interdependence, increases the capability of a team to reduce process errors and adapt to unexpected events. Consistent with these arguments, we suggest that team failure that is rooted in individual error may occur more often in situations characterized by low levels of interdependence. For example, when a team is composed of distributed experts who perform their work separately and then combine it to form the final product (i.e., pooled/additive work arrangements), team members will find it more difficult to monitor and evaluate fellow members'

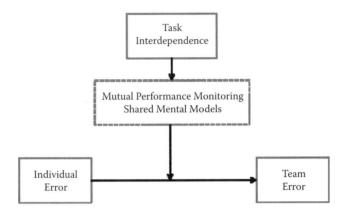

FIGURE 5.1
The moderating influence of task interdependence on the relationship between individual and team error.

work. In contrast, when there is a high level of workflow interdependence (i.e., reciprocal or intensive workflow arrangements) accompanied by a shared understanding of the task and team responsibilities, team members will have greater opportunity and capacity to identify and correct individual errors before they have an impact on team performance (Salas, Sims, et al., 2005) (Figure 5.1).

Collective Error

Although sometimes the product of individual error, it is more often the case that team failure is due to errors that occur at the collective level, most commonly a breakdown in internal team processes (Sims & Salas, 2007). Research has shown that a majority of aviation accidents are due to failures in coordination among the crew (Freeman & Simmon, 1991), and the medical community has identified communication failures as the leading cause of inadvertent patient harm (The Joint Commission, n.d.). Yet, teams often attribute errors to individual causes, which has important implications for error prevention and management. Leonard, Graham, and Bonacum (2004), for instance, note that in medicine there is a deeply embedded belief that quality of care is the result of being well trained and trying hard. The result is that errors are typically viewed as personal failures, rather than a failure of the team or larger organizational system, and are therefore often not reported or openly discussed.

Hofmann and Frese (Chapter 1, this volume) argue that both individuals and collectives engage in functionally equivalent action cycles (cf., DeShon, Kozlowski, Schmidt, Milner, & Wiechamnn, 2004; Kozlowski & Bell, 2008), and we should be able to observe similar types of collective errors at the different levels of action regulation (with the exception of sensorimotor errors, which are inherently individual level). Although errors at the individual and collective levels may be functionally similar (i.e., have similar effects or outputs), they differ structurally (Morgeson & Hofmann, 1999). That is, the processes that produce the functional outcomes differ across levels; at the individual level, they are intrapersonal, and at the collective level, they are interpersonal. For example, Gersick and Hackman (1990) propose that teams often develop habitual routines, which exist "when a group repeatedly exhibits a functionally similar pattern of behavior in a given stimulus situation without explicitly selecting it over alternative ways of behaving" (p. 69). Although habitual routines can yield functional benefits (e.g., save time and energy), when a group fails to recognize a novel situation or notice changes in a familiar situation, it risks engaging in a habitual routine when an alternative response is needed. Thus, habitual routines can lead to dysfunction because the team miscodes the situation and performs a correct behavior in the wrong situation. These dysfunctional habitual routines are functionally similar to individual habit errors that occur at the level of flexible action patterns. Yet, they are structurally different because they represent patterns of behavior or interactions (interpersonal) rather than individual action (intrapersonal) (Kozlowski & Klein, 2000). Also, habitual routines are maintained through socioemotional factors that exist within groups, such as social entrainment and group norms, whereas habit errors are perpetuated by cognitive biases (e.g., frequency gambling, similarity matching) (Reason, 1990).

The interpersonal processes that underlie collective error can occur at the dyadic level or the team level. Consider the vignette on Southwest Airlines Flight 1455 that opened this chapter. On the surface, it may appear that this accident was caused by the actions of the captain, who was the pilot flying on the approach. However, a closer examination reveals that the accident was rooted in errors that occurred at the dyadic level, in the interaction between the captain and first officer. Both the captain and first officer were aware that the aircraft was not "in the slot," but neither of the pilots mentioned the deviation from normal operations. When the automated warnings sounded in the cockpit, the captain stated, "that's all right," and there was no further discussion of the issue (NTSB, 2002). The lack of communication in the cockpit led to a low level of situation awareness, such that the captain became fixated on the runway, and the first officer incorrectly assumed the captain was taking corrective action. In their analysis of 300 civilian aviation incident reports, Jentsch, Barnett, Bowers, and Salas (1999) found that the loss of situation awareness by the flight crew represented the primary error committed in approximately half of the mishaps. Further, the authors found that when the captain was the pilot flying the plane, the first officer tried to correct the error in only

30% of the incidents. In the Southwest case, the loss of situation awareness resulted in neither the captain nor the first officer recommending a go-around maneuver, which is the airline's standard operating procedure when an aircraft is determined to be out of the slot. Since both interacting parties contributed to the failure, the accident represents a collective failure rather than the failure of a single individual (Hofmann & Frese, Chapter 1, this volume). Although this case focuses on a two-person crew, errors can originate at the dyadic level in larger teams as well. Consider, for example, teams characterized by a sequential workflow arrangement, in which work and activities flow undirectionally from one member to another (Bell & Kozlowski, 2002). Under these conditions, collective failure is likely to stem from errors that originate not at the collective level but rather at the dyadic level, such as a breakdown in coordination or communication as work transitions from one member to the next.

When teams are engaged in more collaborative and interdependent work, however, collective errors are likely to originate at the team level. Pearsall, Ellis, and Bell (2008), for example, demonstrated how errors in the transactive memory behaviors of a team can have an impact on team performance. Transactive memory is a group-level shared system for encoding, storing, and retrieving information that combines the knowledge possessed by each team member with a collective awareness of who knows what (Wegner, 1987). Transactive memory develops and operates in teams through the communication of expertise-specific information between team members. In their study of 69 command-and-control teams characterized by a high level of task interdependence, Pearsall et al. (2008) found that errors often occurred in these communication behaviors, primarily as members attempted to update the directory of expertise in the team or allocate information to their teammates. For instance, team members frequently shared inaccurate task information and sent important information to the wrong person in the team. The authors found that the errors had a negative effect on team performance by creating deficiencies in transactive memory cognitions and mental model accuracy. The errors in transactive memory behaviors represent an example of a collective mapping error because they involve information being processed and integrated incorrectly within the team (Hofmann & Frese, Chapter 1, this volume). Not only do the errors originate at the team level, but also they manifest in the collective knowledge structures (i.e., transactive memory cognitions, mental models) that exist within the team.

Although team errors can originate at both the dyadic and collective levels, the frequency of error occurrence may vary across the two levels. Researchers have noted that intrateam processes increase in complexity with more team members (Kozlowski & Bell, 2003). For example, research has found that larger teams are more likely to experience coordination problems that interfere with performance (Lantané, Williams, & Harkins,

1979). Thus, errors in teamwork (e.g., coordination, communication) may be more likely to occur in more intensive interactions than in unidirectional or bidirectional dyadic interactions. At the same time, however, intensive teamwork arrangements that involve multiple team members may provide greater opportunity for teams to detect, capture, and correct errors. As noted, errors in teams often arise due to a breakdown in teamwork, but teamwork processes, such as mutual performance monitoring and backing up behavior, allow teams to create redundant systems that prevent and manage errors (Burke, Salas, Wilson-Donnelly, & Priest, 2004; Foushee et al., 1986). In dyads, these redundant systems are limited, and errors may be more likely to slip through undetected. Future research should examine the frequency with which errors in teams originate at the individual, dyadic, and team levels and the corresponding implications for error detection and correction within teams.

The Effects of Errors on Team Learning

Errors, by definition, are actions that fail to achieve their goal, which implies that they should have negative implications for team performance. Indeed, the cases discussed demonstrate how errors can lead to team performance failures, sometimes with disastrous consequences. This suggests that teams should strive to prevent errors if they are to achieve optimal performance. However, research conducted since the late 1990s has revealed that errors are also an important stimulus of team learning and, as such, are critical to team performance improvement efforts (Edmondson, 2004; Wilson, Burke, Priest, & Salas, 2005). Wilson, Goodman, and Cronin (2007, p. 1044) define group learning as "a change in the group's repertoire of potential behavior" and argue that the learning process in groups consists of the *sharing, storage*, and *retrieval* of group knowledge, routines, or behavior. To better understand the role of errors in team learning, we consider the potential effects of errors on each of these three group learning processes.

Sharing

Wilson et al. (2007) define sharing as "the process by which new knowledge, routines, or behavior becomes distributed among group members and members understand that others in the group possess that learning" (p. 1044). As a group construct, they argue that group learning takes place when members possess some new knowledge, routine, or behavior, *and* there exists an understanding that others in the group have the same knowledge and it is a property of the group. Because knowledge acquisition occurs at the individual level, the process of sharing transfers new knowledge to other group members and serves to legitimate the knowledge for the group. The process of knowledge sharing is a key component of most cognitively oriented theories of team learning. For example, Ellis and Bell (2005) argue that collaboration, which includes the sharing of information, critical discussion, and insight within the team, is a critical condition for team learning. They suggest that team members must collaborate effectively with one another to share their knowledge, experience, or ideas with their teammates.

Errors may have several implications for knowledge sharing within teams. First, errors highlight potential pitfalls and error-prone areas of a task, which in turn increases understanding of a system and facilitates the development of more accurate and complete mental models (Frese et al., 1988). Thus, errors may influence the content or quality of the information that is shared within a team and facilitate the development of more comprehensive shared mental models. Second, errors tend to be novel and unexpected events and therefore often involve a surprise reaction, which can induce mindfulness and stimulate learning (Bell & Kozlowski, 2009). In groups or teams, the surprise of an error may lead members to reflect on their performance, engage in collective sense making, and share knowledge. Bauer and Mulder (2007), for example, studied the learning activities of nurses after the experience of an error episode. After committing an error, they found that nurses often engaged in socially oriented learning activities, such as exchange with more experienced persons, joint root cause analysis, and the search for a new solution. The nurses emphasized the exchange with colleagues and supervisors as well as discussions in team meetings as critical activities for learning from errors.

Although errors may spur social exchange among team members, for knowledge sharing to be effective several subprocesses must take place. Specifically, a group must focus attention on the information that is to be learned, develop a shared understanding of the specific learning, and reach a shared understanding of how the new knowledge will be used in the future (Wilson et al., 2007). Yet, a number of individual, team, and organizational barriers can make it difficult for teams to fully engage in these processes. For example, team members may resist focusing on failure because doing so presents a threat to their self-esteem or may avoid revealing or discussing errors with colleagues for fear of being perceived as incompetent and losing valued organizational rewards (e.g., promotions, bonuses) (Cannon & Edmondson, 2001). One strategy for overcoming these barriers is to use formal processes to initiate and structure the analysis and discussion of errors within a team. In their theory of dynamic team leadership, Kozlowski and colleagues (Kozlowski, Gully, McHugh, Salas, & Cannon-Bowers, 1996; Kozlowski, Watola, Jensen, Kim, & Botero,

2009) describe how team leaders should monitor team performance and then lead the team in collective reflection and process feedback to promote learning by identifying errors, diagnosing them, and developing ways to improve individual and team performance. Salas, Klein, et al. (2008) discuss the importance of debriefs or after-action reviews in the medical community. They state: "The debriefing process is used to allow individuals to discuss individual and team-level performance, identify errors made, and develop a plan to improve their next performance" (p. 519). Formal debriefs or audits should be used not only to discuss failures but also to examine "near misses," which can hold important lessons for preventing future errors (Wilson et al., 2005). Although formal processes can provide a forum for the sharing of knowledge gained from errors, effective knowledge sharing requires a supportive interpersonal climate. We discuss the influence of climate on error management in teams in more detail elsewhere in this chapter.

In this section, we examined how performance errors may stimulate knowledge sharing in teams. However, errors may also occur *during* the team learning process, with potentially more negative consequences. In the discussed study by Pearsall et al. (2008), for example, the most common error in the transactive memory behaviors of the teams involved team members incorrectly sharing knowledge with their teammates, and these directory-updating errors had a negative effect on team cognitions and performance. Future research should examine how errors that occur during the sharing, storage, and retrieval processes have an impact on team learning and performance.

Storage

For group learning to occur, new knowledge not only must be shared within the team but also must be stored and retained so that it can be exploited over time (Wilson et al., 2007). Groups have access to a wide range of storage repositories, including the memories of group members (i.e., transactive memory), information technology (e.g., databases, expertise systems), and structural repositories such as standard rules and procedures. Wilson et al. (2007) argue that the effectiveness of these different group repositories for storage and retrieval may depend on the type of knowledge stored. Specifically, they suggest that explicit knowledge can be stored in any of the three types of repositories, but tacit knowledge may be more easily stored and retrieved in human memory systems.

Unfortunately, little research has focused on group-level storage, which makes it difficult to determine precisely how errors may influence this feature of group learning. However, research conducted at the individual level of analysis suggests that errors may have an impact on both the content and process of group-level storage. As discussed, errors can enhance the coherence and breadth of individuals' knowledge structures (Bell & Kozlowski, 2009). At the collective level, errors may have similar effects on the depth and breadth of knowledge stored in a team's repositories. For example, Tucker, Nembhard, and Edmondson (2006) argue that teams can engage in two types of learning activities. The first is "learn-what," which describes learning activities aimed at identifying best practices. The second type is "learn-how," which refers to learning activities that are aimed at understanding how and why a particular practice works so that it can be adapted to different situations. They suggest that learn-how occurs when teams engage in trial-and-error experimentation and reflect on their actions, including the mistakes that they made. Similarly, the team compilation model proposed by Kozlowski et al. (1999) suggests that teams can engage in the exploration of different structures or ways of performing their task under different situational demands to build their repertoire. As they state: "Adaptive teams will have created a repertoire of networks for different situations as well as corresponding indicators that signify when a switch in configuration must occur" (p. 273).

In addition to influencing the content of information stored in memory, errors may prompt important storage processes in groups. Wilson et al. (2007) discuss how elaborative processing can improve storage through the enhancement of retrieval cues. For example, group discussion can serve as a form of elaboration or practice as the team considers how to respond to future problems or challenges. As discussed, research has found that errors can prompt social exchange as individuals seek information and engage in joint problem solving. In addition, research on learning has shown that when individuals engage in exploratory learning and make errors, they engage in more effortful processing of information and acquire knowledge at a higher level of regulation (Frese & Zapf, 1994). In the context of teams, errors may lead to effortful processing of new information as members engage in more elaborated discussions aimed at diagnosing the causes of the failure and planning for future scenarios. As Wilson et al. (2007) argue, this elaboration should strengthen the memory record and facilitate the persistence and recall of the knowledge over time.

Retrieval

The final component of group learning is retrieval, which means that groups can locate and access stored knowledge for later use (Wilson et al., 2007). Although retrieval is a necessary condition for group learning to occur, it also has received minimal attention in the literature on group learning. Thus, little is known about the retrieval process and how it is influenced by features of groups (e.g., social processes, stability of membership) and other contextual factors (e.g., time). Yet, based on what we currently know about the retrieval process in groups, it is likely that errors affect the ability of a group to find and access stored knowledge. In particular, errors may affect retrieval through two general mechanisms. First, Wilson et al. (2007) argued that the retrieval of group learning is closely intertwined with the group learning processes discussed in this chapter. This suggests that errors may have an impact on retrieval indirectly through their effects on knowledge sharing and storage. For example, knowledge sharing increases the number of group members able to recognize and respond to retrieval cues. Thus, by stimulating social exchange and sense making in groups, errors may facilitate the subsequent retrieval of new knowledge. In addition, the strength of a stored memory record influences the likelihood that the knowledge will be retained and successfully retrieved when needed. As argued, errors may lead teams to engage in elaborative processing, which may strengthen the knowledge record and allow the team to retrieve the information over time.

In addition to these indirect effects, errors may directly influence the likelihood that a team will engage in retrieval processes and be able to apply the retrieved knowledge to a new situation. Wilson and colleagues (2007) noted that for retrieval to occur a group or one of its members must first recognize the need to access stored knowledge. The retrieval mechanism is triggered when features of a stimulus situation cue particular learning stored in the knowledge repository of the group. However, as noted, teams often fail to recognize novelties in their task environment and as a result may retrieve outmoded routines (Gersick & Hackman, 1990). In addition, when teams have significant tenure, previous learning and established mental models may inhibit the retrieval of new, updated learning. Errors can serve as important feedback to a team, highlighting misunderstandings or shortcomings that frustrate effective action and triggering the retrieval of new or additional information (Cannon & Edmondson, 2001; Tjosvold et al., 2004). Yet, as Hofmann and Frese (Chapter 1, this volume) note, error detection is difficult because the "error signal" is often nested within a flow of feedback characterized by a high degree of uncertainty and equivocality. Moreover, an undetected error cannot stimulate learning. Thus, teams should engage in practice activities designed to build a broader repertoire of behaviors and develop greater adaptive capacity (Kozlowski et al., 1999). For example, exploration and trial-and-error experimentation expose teams to a wider range of task material. Although the material may not be relevant to the team's current task, it is valuable knowledge that a team can access and utilize when faced with a novel situation (Bell & Kozlowski, 2009). Finally, when teams diagnose error episodes and plan for future challenges (e.g., self-correction), they also generate retrieval cues they can use to recognize situations in which specific knowledge should be accessed (Kozlowski et al., 1996).

Error Management Within Teams

Hofmann and Frese (Chapter 1, this volume) argue that there are two primary strategies that can be used to deal with errors. The first is error prevention, in which the focus is on reducing the occurrence of errors. The second is error management, in which the focus is on disconnecting the error from error consequences. Although most complex organizations focus on preventing errors, there are several limitations to this approach. First, a pure error prevention approach overlooks the fact that errors are ubiquitous, and even the most well-designed systems can never eradicate all human error, particularly in the complex and dynamic organizations of today (Bauer & Mulder, 2007; Reason, 1990). Second, error prevention reduces the learning potential of errors and ignores the fact that some errors have long-term positive consequences for individual, team, and organizational effectiveness (Hofmann & Frese, Chapter 1, this volume).

However, as Hofmann and Frese note in Chapter 1, both error prevention and error management share the common goal of reducing the negative consequences of errors. Thus, they suggest that organizations adopt a dual-prong approach aimed at preventing error consequences. Specifically, organizations not only should focus on error prevention to try to prevent errors from occurring but also should add a second layer of defense in the form of error management, which involves intercepting and correcting errors that occur before they produce negative consequences. In the following section, we examine the process of error management, focusing attention on the goals of the process and the necessary conditions for error management to be successful. We then consider factors that influence error detection and error management in team contexts. In particular, we examine three categories of factors—team characteristics, team climate, and team interventions—that have an impact on a team's ability to detect, trap, and recover from failure.

The Error Management Process

Kontogiannis (1999) argues that there are three processes in error management. The first is error detection or realizing that an error is about to occur or suspecting that an error has occurred. He notes that error detection can occur in different stages of performance. First, errors can be detected in the outcome stage, such that a mismatch between expected effects and observed outcomes signals an error. In this case, error detection occurs "after the fact," after the action cycle has concluded. Failures in error detection at this stage often stem from the challenges involved in attending to actual outcomes and remembering expected outcomes. Second, errors can be detected during the execution stage when individuals recognize a mismatch between the actions being executed and the actions specified in their plans. Finally, errors can be detected during the planning stage, such that individuals recognize wrong intentions or mismatches between intentions and formulated plans. Error detection during the execution and planning stages is based on indicators of error that emerge during the action cycle and is influenced by the clarity, or strength, of these error signals (Hofmann & Frese, Chapter 1, this volume). After an error has been detected, it is then important to explain why the error occurred. Error explanation represents the second process of error management, and it is important for learning from errors, as discussed previously. In addition, error explanation facilitates the final process of error management, error correction, which involves the modification of an existing plan or development of a new one to compensate for the mistake.

The success of error management in reducing the negative consequences of errors is influenced greatly by the timing of error detection and recovery. Hofmann and Frese (Chapter 1, this volume) argue that errors that remain after the conclusion of the action cycle are more likely to have a negative impact on the performance of other entities within the organization. In contrast, errors that are detected during the action cycle can potentially be captured before these negative consequences are realized. The primary objective of error management, therefore, should be to trap errors during the action cycle to prevent or minimize their consequences on the system (Helmreich, Wilhelm, Klinect, & Merritt, 2001). It is also important to detect errors quickly because errors that remain latent in a system can grow and accrue over time, resulting in much more serious long-term consequences (Hofmann & Frese, Chapter 1, this volume). In addition, efficient error detection and recovery help to reduce error cascades that can occur as one error leads to other errors. For example, error management can help reduce the frustration and stress that often accompany errors, thereby preserving cognitive resources required for successful performance and minimizing debilitating motivation losses. Finally, error management contributes to secondary error prevention. As individuals diagnose errors and uncover their potential causes, they are in a better position to prevent the error from occurring in the future (Hofmann & Frese, Chapter 1, this volume; Kontogiannis, 1999).

Error management is a difficult and complex process, particularly in team settings. Thomas (2004), for example, examined error management among flight crews and found that nearly half of all errors remained undetected. Further, he found that less than a fourth of all errors were effectively managed by the flight crews. Research suggests that there are several necessary conditions for error management to be successful in team environments. One of the most important is that there must be a general expectation that errors will occur. If individuals do not recognize the possibility of errors, then little effort will be put forth to determine if an error has occurred (Hofmann & Frese, Chapter 1, this volume). Kontogiannnis (1999) argues that error suspicion and curiosity can overcome complacency and help individuals maintain vigilance for subtle changes in the environment. In teams, individuals must devote attention to monitoring not only their own performance but also that of their teammates. Research suggests that mutual performance monitoring is critical for the reduction of catastrophic errors (Salas, Burke, & Stagl, 2004; Wilson et al., 2005), and team leaders can play an important role in monitoring the progress of a team toward its goals and providing the feedback necessary for error detection and management (Salas, Burke, & Stagl, 2004).

Mutual performance monitoring requires not only the expectation that errors will occur but also the ability to recognize errors in others' performance. Hutchins (1994) argued that for errors to be detected by other persons, the detector must possess sufficient knowledge of the task being performed to be able to determine whether it has been performed correctly and must have knowledge of the possible goals associated with a task. Accordingly, researchers have suggested that error management in teams requires that members possess a shared understanding of the task and team responsibilities. As Salas, Rosen, Burke, Nicholson, and Howse (2007) state: "Team members must share mental models of the task, team, and equipment in order to be aware of their teammates' surroundings, and to correctly recognize abnormalities in their fellow team members' performance" (p. B81). Shared awareness not only alerts team members to latent errors but also facilitates team coordination efforts to contain and minimize errors that occur (Wilson et al., 2005). One factor that has been shown to contribute to shared awareness is the transparency of organizational and technical systems (Hofmann & Frese, Chapter 1, this volume; Kontogiannis, 1999). Transparency allows team members to develop a better conceptualization or mental model of the system. However, research suggests that a multitude of factors can shape the error management process in teams; we explore several of these in the next section.

Factors That Influence Error Management in Teams

Given the importance of error management to team learning and performance, a growing number of studies have focused attention on better understanding the factors that enhance or inhibit the capability of a team to detect and correct its mistakes successfully. In this section, we examine three broad categories of factors—team characteristics, team climate, and team interventions—and their influence on error management in team contexts.

Team Characteristics

Salas, Burke, and Stagl (2004) note that all teams are not created equal, and different types of teams face different challenges to prevent and manage errors. For instance, in commercial aviation, errors are often due to failures in coordination that arise from the fluid nature and short life span of flight crews. In medical teams, error management is often made more difficult by status differences and clear lines of hierarchy that determine who can question others' actions. Rather than focusing on error management in specific types of teams, it is more informative to consider the implications of different team characteristics for error management.

Indeed, Kozlowski and Bell (2003) argue that the value of team typology is in understanding the factors that constrain and influence effectiveness for different types of teams. They highlight six features that capture a majority of the unique characteristics that distinguish different team forms. One feature they discuss is team member diversity. As discussed earlier in this section, for team members to recognize the mistakes of others they must possess a shared understanding of task and team responsibilities. It may be more difficult for teams to develop this shared awareness when members are drawn from diverse backgrounds.

Rosen et al. (2008), for example, note that one of the barriers to error management in health care is the fact that medical teams are comprised of individuals from highly diverse backgrounds in terms of expertise, training, and experience. Van der Vegt and Bunderson (2005) suggest that one of the keys to overcoming the disruptive tendencies of expertise diversity in teams may be collective team identification. They argue that when there is a shared sense of identification with the team, members will be more motivated to interact and exchange information in the face of diversity. Indeed, they find that expertise diversity is negatively related to learning and performance for teams with low collective identification but positively related to learning and performance among teams with high levels of collective identification. However, they also find some evidence that even among teams with high levels of collective identification, very high levels of expertise diversity impede team learning and performance. Nonetheless, teams characterized by a high level of homogeneity face their own challenges to error management due to groupthink and a failure to engage in mutual learning (Wong, 2004). Ultimately, it appears that teams that possess moderate levels of diversity combined with high levels of collective identification may be best equipped to manage and learn from their errors.

A second key feature discussed by Kozlowski and Bell (2003) concerns the performance demands that teams face. Complex teams are characterized by performance demands that require coordinated individual performance in real time, the ability to adapt to dynamic goals and contingencies, and a capacity for continuous improvement (Kozlowski et al., 1999). In contrast, simple teams are characterized by minimal performance demands that allow for pooled or additive workflow arrangements. Although complex teams may be more susceptible to errors that occur due to a breakdown in teamwork (e.g., coordination/collaboration), the high level of workflow interdependence may facilitate mutual performance monitoring and create redundant systems for trapping errors. However, when performance demands are high, teams may have trouble detecting errors because in complex action cycles error signals are often unclear (Hofmann & Frese, Chapter 1, this volume). Salas, Burke, Fowlkes, and Wilson (2004) suggest that complexity may also moderate the effects of diversity on team performance. In particular, they suggested that cultural diversity may facilitate performance on simple tasks. However, on complex tasks that require significant coordination, cultural diversity may result in teams not performing to their full potential. It will be important for future research to further explore the relationship between team complexity and error management given the need for complex teams to use errors as a source of learning and continuous improvement.

Although having received less attention, researchers have examined several other team characteristics, such as collocation/spatial distribution and temporal characteristics, as potential factors that influence error management. Fletcher and Major (2006), for example, examined differences in error correction across teams utilizing one of three communication modalities: face to face, audio, and audio plus a shared workplace application. They did not find any difference in uncorrected errors between the face-to-face and audio-only conditions, but they did find teams left more uncorrected errors in the audio-only condition than in the shared workspace condition. This finding lends some support to the authors' argument that communication modalities low in information richness may make it more difficult for teams to monitor and correct their errors. Cooke, Gorman, Duran, and Taylor (2007) compared the performance of teams with experience working together in a command-and-control setting to inexperienced teams on an uninhabited aerial vehicle simulation. They found that experienced teams exhibited fewer errors on one of the training segments concerned with coordination, suggesting that individuals familiar with these kinds of coordination situations were able to transfer this teamwork knowledge between tasks. Edmondson, Bohmer, and Pisano (2001) also suggest that team stability can influence team learning. They argue that keeping team members together helps to develop transactive memory systems, which can enhance team learning. However, they also note that stable teams may fall into habitual routines and fail to respond to changing conditions. Overall, the effects of temporal issues, such as team life cycle, on error management warrant future research attention.

Team Climate

In addition to team characteristics, prior research suggests that the climate of a team can influence the error management process. Cannon and Edmondson (2001) argue that two capabilities must exist for organizations to learn from their failures: Members must be able and willing to take risks, and they must be able to acknowledge failure rather than cover it up. The extent to which members engage in these activities is influenced greatly by the climate that exists within the team. Edmondson (1999), for example, argues that team psychological safety, which was defined as "a shared belief that the team is safe for interpersonal risk taking" (p. 354), is important for creating a supportive interpersonal climate in which members can raise concerns or reveal mistakes without it being held against them. Similarly, Burke et al. (2004) argue that in high-reliability environments, team members must treat every potential opportunity as a learning event and must be willing to raise concerns. Mutual performance monitoring has little value if team members are afraid to speak up when they notice something out of the ordinary. As Burke et al. (2004, p. i101) state, "The ability to speak up in a non-threatening and respectful manner (deference to expertise) is a hallmark of learning organizations and the teams within them."

Yet, teams often face numerous barriers to creating a climate that supports the process of identifying and learning from failures. Edmondson (2004), for example, discusses how nurses and doctors face extreme pressures on their time, which can limit opportunities to engage in root cause analysis and instead force quick patches to problems. For teams to manage and learn from their mistakes effectively, time must be devoted to reflective activities, such as debriefings (Salas, Klein, et al., 2008). Helmreich et al. (2001) point out that the professional culture in aviation is one in which many pilots have unrealistic perceptions of invulnerability, and there is a strong sense of hierarchy, which can prevent crew members from acknowledging their own or others' mistakes.

Although the organizational and professional context in which a team operates can influence how members respond to errors, there exists some evidence that the group context may represent a more powerful influence on members' attitudes and behaviors. Edmondson et al. (2001), for example, studied 16 hospitals implementing a new technology for cardiac surgery. They found that organizational differences, such as resources and senior management support, were not associated with implementation success. Rather, successful implementers engaged in a qualitatively different team learning process than those that were unsuccessful. All of the successful implementers, and only one of the unsuccessful implementers, established a climate in which lower-status team members were comfortable speaking up with observations. Using data from 51 teams in a manufacturing organization, Cannon and Edmondson (2001) showed that beliefs about failure were shared within and varied between groups in the organization. These differences were found despite the fact that all of the teams were embedded in the same organizational context, which emphasized the importance of learning from mistakes. These results suggest that although the organizational context is important, factors in the more proximal team context may have a greater influence on how members view errors and whether they leverage failure as a learning opportunity.

Given the importance of team members' shared beliefs about failure, it is important to understand the antecedents of these climate perceptions. Prior research suggests that leaders are in a position to reframe failure as essential to learning, thereby shaping team members' attitudes toward errors. Cannon and Edmondson (2001), for example, found that leader coaching (e.g., initiation of discussions about improvement) encouraged team members to adopt productive attitudes and behaviors about failure. Edmondson (1996) found that detected unit error rates were strongly and positively associated with high scores on nurse manager direction setting and coaching, which suggests that certain leaders establish a climate of openness that facilitates the reporting and discussion of errors. In addition to shaping learning-oriented beliefs in teams, leaders need to coach subordinates on how to discuss failure constructively and need to set a clear direction for the team so members can recognize deviations when they occur (Cannon & Edmondson, 2001). Although a collective orientation toward learning often promotes higher levels of team performance, research by Bunderson and Sutcliffe (2003) suggests that an emphasis on learning may compromise performance when teams are already performing at a high level. Their results suggest that leaders may only want to encourage active experimentation and risk taking when established activity patterns have proved ineffective.

Team Interventions

A third and final category of factors that can influence error management in teams involves human resource management interventions specifically designed to enhance members' capacity to identify and rectify deficiencies in individual and team performance. Although observers have noted that the traditional human resource system is ill equipped to support and sustain team-based work units, research has examined the efficacy of several different human resource practices for helping teams manage their errors. Much of this work has focused on various training strategies that can be used to develop high-reliability teams. For example, simulation-based training can allow teams to practice trapping and recovering from errors and can provide feedback that is useful for preventing future errors (Salas, Cooke, et al., 2008). In addition, self-correction training has been used to help teams correct unsafe behaviors and avoid them in the future, and perceptual control training can facilitate cognitive skills, such as noticing, performance monitoring, and situational awareness, that can aid teams in error identification (Wilson et al., 2005). For a more detailed discussion of the role of training and development in facilitating error management in teams, refer to Chapter 6 by Weaver, Bedwell, and Salas in this volume.

A large body of research has also examined the role of technology and automation in helping teams to manage and avoid errors. Research evidence is mixed with respect to how automation influences error management in teams. Wright and Kaber (2005), for example, reviewed four different types of automation and concluded that certain types, such as automation of information acquisition, enhance teamwork skills and team effectiveness, while other types, such as decision automation, can compromise team performance. Further, the effects of automation can differ depending on task difficulty. For example, decision automation has been found to have negative effects on team effectiveness at high levels of task difficulty but positive effects under conditions of low task difficulty (Wright & Kaber, 2005). These results may be explained by the fact that technology does not always ease teamwork demands and in fact can often increase them. For example, Webster and Cao (2006) found that more verbal exchanges were required when surgical teams moved from conventional laparoscopy to a robotic system. In addition, the use of automation to reduce specific types of errors can be accompanied by increases in other types of errors. For example, the availability of automated decision aids can lead decision makers to use these aids as a replacement for vigilant information seeking and processing, a phenomenon known as *automation bias* (Mosier, Skitka, Dunbar, & McDonnell, 2001). Automation bias can result in a failure to respond to system irregularities when not prompted to do so by an automated device (omission error) and inappropriately following an automated recommendation without verifying it against other available information or despite contradictory information (commission error). In summary, automation may aid teams in avoiding errors, but such systems should not be viewed as a substitute for active error management in teams.

Future Research Directions

The goal of this chapter was to advance theory and research on collective failure by focusing on the emergence, consequences, and management of errors in teams. We recognized early on that this would be a challenging endeavor given the paucity of research on errors in teams. Although the literature on team effectiveness has grown dramatically in recent decades, relatively little of this work has focused on errors specifically and on their implications for team learning and performance (Tjosvold et al., 2004). The more extensive literature on human error and its effects on individual learning and performance provides valuable insight into the potential effects of errors in team contexts, but one also needs to exercise caution in generalizing findings across levels of analysis. Accordingly, we conclude the chapter with a discussion of several important gaps in research on errors in teams and highlight potential directions for future work in this area.

Understanding the Emergence of Errors in Teams

As we have discussed, errors in teams can occur at the individual, dyadic, or collective level, and the level at which an error originates may have important implications for how it propagates within a team to influence performance. However, team errors have often been studied by examining failures in team performance, which provides limited information about the root cause of an error. Salas, Cooke, et al. (2008), for example, suggest that in team environments it is important to distinguish between errors that arise during task work processes and those that occur during teamwork processes. Both types of errors may undermine team performance, but errors in task work processes suggest deficiencies in individual capacity and performance, whereas errors in teamwork processes point to problems surrounding team capacity, processes, and performance. Thus, information about the origin of different types of errors in teams is valuable for guiding root cause analyses and developing effective error management strategies. It will also be important for future research to study the emergence of errors within teams over time. For example, whether an error emerges early or late in the performance cycle of a team and the length of time an error remains latent in a system can influence the chances of a team successfully trapping the consequences of the error (Hofmann & Frese, Chapter 1, this volume). Also, whether an error reoccurs in future performance episodes is an important indicator of whether a team has successfully learned from its mistakes. Cross-sectional studies can only provide a limited understanding of the nature of errors in teams and the effectiveness of error management processes.

Mapping the Effects of Errors on Team Learning Processes

Prior work has provided convincing evidence that detected errors represent an important stimulus of team learning. Teams that openly reveal, discuss, and reflect on their mistakes expand their repertoire of potential behaviors and exhibit performance improvement over time. What is less clear, however, is how errors facilitate and shape the *process* of group learning. As Kozlowski and Ilgen (2006) noted, team learning has rarely been assessed directly as a construct and instead has been inferred from changes in team performance. Further, prior research has focused significant attention on particular group learning processes, such as knowledge sharing, but has neglected others, such as knowledge storage and retrieval (Wilson et al., 2007). The result is that we possess only a limited understanding of the processes involved in team learning, which makes it difficult to speculate about how errors may influence these processes. In addition, little work has examined how errors that occur *during* these processes have an impact on team learning and performance. Thus, it will be important for future research to focus attention on mapping the effects of errors onto critical team learning processes. We believe that the model of group learning specified by Wilson et al. (2007) may prove useful in such efforts as it provides an account of the information-processing activities that should take place for a group to learn (cf., Kozlowski and Bell, 2008). In this chapter, we have discussed how errors may influence these processes of group learning, but future research is needed to test our propositions empirically.

Identifying the Factors That Influence Error Management in Teams

Errors within teams are unavoidable. However, whether an error results in learning and performance improvement or has a detrimental effect on team performance will depend largely on how well the team manages the error. Research has provided significant insight into the error management process and the challenges that surround error detection and recovery (Hofmann & Frese, Chapter 1, this volume; Kontogiannis, 1999). We know that teams often struggle to manage their errors, resulting in a large proportion of errors remaining either undetected or uncorrected (e.g., Thomas, 2004). However, our ability to reverse this trend is hampered by the fact that we currently possess a limited understanding of the factors that enhance and inhibit error management in team contexts. For example, prior work has established team member diversity as having an important influence on error management, but the impact of other team characteristics, such as team life cycle, on error management have rarely been examined. This may be due, at least in part, to the fact that the vast majority of research that has examined errors in teams has been conducted in health care, aviation, and military settings. This is not surprising when one considers that teams perform much of the work in these settings, and errors in these contexts often have dire consequences. Yet, medical teams (e.g., nursing units, surgical teams), flight crews, and combat teams are all examples of action teams, characterized by members who possess specialized expertise, a high degree of integration with other work units, and work cycles that involve brief, repeated performance events (Sundstrom, De Meuse, & Futtrell, 1990). The overrepresentation of action teams in this literature limits our understanding of the implications of errors for other types of teams, such as product and service, project, and top management teams. More important, it makes it difficult to determine how different team characteristics, such as workflow interdependence and boundary permeability, may influence error management. For example, project teams require little external synchronization, which may provide greater opportunity for them to trap errors before they have an impact on other entities within the organization. Moving forward, it will be important for research in this area to focus on a wider variety of team types to understand better how the features that distinguish among different team forms influence the process of error management.

References

Baker, D. P., Day, R., & Salas, E. (2006). Teamwork as an essential component of high-reliability organizations. *Health Services Research, 41,* 1576–1598.

Bauer, J., & Mulder, R. H. (2007). Modelling learning from errors in daily work. *Learning in Health and Social Care, 6,* 121–133.

Bell, B. S., & Kozlowski, S. W. J. (2002). A typology of virtual teams: Implications for effective leadership. *Group and Organization Management, 27*(1), 14–49.

Bell, B. S., & Kozlowski, S. W. J. (2009). Toward a theory of learner-centered training design: An integrative framework of active learning. In S. W. J. Kozlowski & E. Salas (Eds.), *Learning, training, and development in organizations* (pp. 263–300). New York: Routledge.

Bunderson, J. S., & Sutcliffe, K. M. (2003). Management team learning orientation and business unit performance. *Journal of Applied Psychology, 88,* 552–560.

Burke, C. S., Salas, E., Wilson-Donnelly, K., & Priest, H. (2004). How to turn a team of experts into an expert medical team: Guidance from the aviation and military communities. *Quality and Safety in Health Care, 13,* i96–i104.

Cannon, M. D., & Edmondson, A. C. (2001). Confronting failure: Antecedents and consequences of shared beliefs about failure in organizational work groups. *Journal of Organizational Behavior, 22,* 161–177.

Cooke, N. J., Gorman, J. C., Duran, J. L., & Taylor, A. R. (2007). Team cognition in experienced command-and-control teams. *Journal of Experimental Psychology: Applied, 13,* 146–157.

DeShon, R. P., Kozlowski, S. W. J., Schmidt, A. M., Milner, K. A., & Wiechmann, D. (2004). A multiple-goal, multilevel model of feedback effects on the regulation of individual and team performance. *Journal of Applied Psychology, 89,* 1035–1056.

Druskat, V. U., & Pescosolido, A. (2002). The content of effective teamwork mental models in self-managing teams: Ownership, learning and heedful interrelating. *Human Relations, 55,* 283–314.

Edmondson, A. C. (1996). Learning from mistakes is easier said than done: Group and organizational influences on the detection and correction of human error. *Journal of Applied Behavioral Science, 32,* 5–28.

Edmondson, A. C. (1999). Psychological safety and learning behavior in work teams. *Administrative Science Quarterly, 44,* 350–383.

Edmondson, A. C. (2004). Learning from failure in health care: Frequent opportunities, pervasive barriers. *Quality and Safety in Health Care, 13,* ii3–ii9.

Edmondson, A. C., Bohmer, R. M., & Pisano, G. P. (2001). Disrupted routines: Team learning and new technology implementation in hospitals. *Administrative Science Quarterly, 46,* 685–716.

Ellis, A. P. J., & Bell, B. S. (2005). Capacity, collaboration, and commonality: A framework for understanding team learning. In L. L. Neider & C. A. Shriesheim (Eds.), *Understanding teams: A volume in research in management* (pp. 1–25). Greenwich, CT: Information Age.

Fisher, C. W., & Kingma, B. R. (2001). Criticality of data quality as exemplified in two disasters. *Information and Management, 39,* 109–116.

Fletcher, T. D., & Major, D. A. (2006). The effects of communication modality on performance and self-ratings of teamwork components. *Journal of Computer-Mediated Communication, 11,* 557–576.

Foushee, H. C., Lauber, J. K., Baetge, M. M., & Acomb, D. B. (1986). *Crew factors in flight operations III: The operational significance of exposure to short-haul air transport operations* (Technical Memorandum No. 88342). Moffett Field, CA: NASA-Ames Research Center.

Freeman, C., & Simmon, D. A. (1991). Taxonomy of crew resource management: information processing domain. In R. S. Jensen (Ed.), *Proceedings of 6th Annual International Symposium on Aviation Psychology* (pp. 391–397). Columbus, OH: The Ohio State University.

Frese, M., Albrecht, K., Altmann, A., Lang, J., Papstein, P. V., Peyerl, R., et al. (1988). The effects of an active development of the mental model in the training process: Experimental results in a word processing system. *Behaviour and Information Technology, 7,* 295–304.

Frese, M., & Zapf, D. (1994). Action as the core of work psychology: A German approach. In H. C. Triandis, M. D. Dunnette, & L. M. Hough (Eds.), *Handbook of industrial and organizational psychology* (2nd ed., Vol. 4, pp. 271–340). Palo Alto, CA: Consulting Psychologists Press.

Gersick, C. J. G., & Hackman, J. R. (1990). Habitual routines in task-performing groups. *Organizational Behavior and Human Decision Processes, 47,* 65–97.

Hackman, J. R. (1993). Teams, leaders, and organizations: New directions for crew-oriented flight training. In E. L. Wiener, B. G. Kanki, & R. L. Helmreich (Eds.), *Cockpit resource management* (pp. 47–69). Orlando, FL: Academic Press.

Helmreich, R. L., Wilhelm, J. A., Klinect, J. R., & Merritt, A. C. (2001). Culture, error, and crew resource management. In E. Salas, C. A. Bowers, & E. Edens (Eds.), *Improving teamwork in organizations: Applications of resource management training* (pp. 305–331). Mahwah, NJ: Erlbaum.

Hollenbeck, J. R., Ilgen, D. R., Tuttle, D. B., & Sego, D. J. (1995). Team performance on monitoring tasks: An examination of decision errors in contexts requiring sustained attention. *Journal of Applied Psychology, 80,* 685–696.

Hotz, R. L. (1999, October 1). Mars probe lost due to simple math error. *Los Angeles Times,* p. A-1.

Hutchins, E. (1994). *Cognition in the wild.* Cambridge: MIT Press.

Jentsch, F., Barnett, J., Bowers, C. A., & Salas, E. (1999). Who is flying this plane anyway? What mishaps tell us about crew member role assignment and air crew situation awareness. *Human Factors, 41,* 1–14.

The Joint Commission. (n.d.). *2007 National Patient Safety Goals.* Retrieved November 25, 2008, from http://www.jointcommission.org/GeneralPublic/NPSG/07_npsgs.htm

Kontogiannis, T. (1999). User strategies in recovering from errors in man-machine systems. *Safety Science, 32,* 49–68.

Kozlowski, S. W. J., & Bell, B. S. (2003). Work groups and teams in organizations. In W. C. Borman, D. R. Ilgen, & R. J. Klimoski (Eds.), *Handbook of psychology (Vol. 12): Industrial and organizational psychology* (pp. 333–375). New York: Wiley.

Kozlowski, S. W. J., & Bell, B. S. (2008). Team learning, development, and adaptation. In V. I. Sessa & M. London (Eds.), *Group learning* (pp. 15–44). Mahwah, NJ: LEA.

Kozlowski, S. W. J., Gully, S. M., McHugh, P. P., Salas, E., & Cannon-Bowers, J. A. (1996). A dynamic theory of leadership and team effectiveness: Developmental and task contingent leader roles. In G. R. Ferris (Ed.), *Research in personnel and human resource management* (Vol. 14, pp. 253–305). Greenwich, CT: JAI Press.

Kozlowski, S. W. J., Gully, S. M., Nason, E. R., & Smith, E. M. (1999). Developing adaptive teams: A theory of compilation and performance across levels and time. In D. R. Ilgen & E. D. Pulakos (Eds.), *The changing nature of work performance: Implications for staffing, personnel actions, and development* (pp. 240–292). San Francisco: Jossey-Bass.

Kozlowski, S. W. J., & Ilgen, D. R. (2006). Enhancing the effectiveness of work groups and teams. *Psychological Science in the Public Interest, 7,* 77–124.

Kozlowski, S. W. J., & Klein, K. J. (2000). A multilevel approach to theory and research in organizations: Contextual, temporal, and emergent processes. In K. J. Klein & S. W. J. Kozlowski (Eds.), *Multilevel theory, research, and methods in organizations: Foundations, extensions, and new directions* (pp. 3–90). San Francisco: Jossey-Bass.

Kozlowski, S. W. J., Watola, D., Jensen, J. M., Kim, B., & Botero, I. (2009). Developing adaptive teams: A theory of dynamic team leadership. In E. Salas, G. F. Goodwin, & C. S. Burke (Eds.), *Team effectiveness in complex organizations: Cross-disciplinary perspectives and approaches* (pp. 109–146). New York: Routledge Academic.

Lantané, B., Williams, K., & Harkins, S. (1979). Many hands make light the work: The causes and consequences of social loafing. *Journal of Personality and Social Psychology, 37,* 822–832.

Leonard, M., Graham, S., & Bonacum, D. (2004). The human factor: The critical importance of effective teamwork and communication in providing safe care. *Quality Safety and Health Care, 13,* i85–i90.

Mars Climate Orbiter Mishap Investigation Board. (1999, November 10). *Phase I report*. Retrieved November 17, 2008, from ftp://ftp.hq.nasa.gov/pub/pao/reports/1999/ MCO_report.pdf

Morgeson, F. P., & Hofmann, D. A. (1999). The structure and function of collective constructs: Implications for research and theory development. *Academy of Management Review, 24,* 249–265.

Mosier, K. L., Skitka, L. J., Dunbar, M., & McDonnell, L. (2001). Aircrews and automation bias: The advantages of teamwork? *The International Journal of Aviation Psychology, 11,* 1–14.

National Transportation Safety Board. (2002). *Aircraft accident brief: Southwest Airlines flight 1455, Boeing 737–300, N668SW, Burbank, California, March 5, 2000* (AAB-02–04). Washington, DC: Author.

Pearsall, M., Ellis, A. P. J., & Bell, B. S. (2008). *Slippage in the system: The effects of errors in transactive memory behavior on team performance.* Academy of Management Annual Meeting Proceedings. Briarcliff Manor, NY: Academy of Management.

Rasmussen, J. (1982). Human errors: A taxonomy for describing human malfunction in industrial installations. *Journal of Occupational Accidents, 4,* 311–333.

Rasmussen, J. (1986). *Information processing and human-machine interaction.* Amsterdam: North-Holland.

Reason, J. (1990). *Human error.* New York: Cambridge University Press.

Rosen, M. A., Salas, E., Wilson, K. A., King, H. B., Salisbury, M., Augenstein, J. S., et al. (2008). Measuring team performance in simulation-based training: Adopting best practices for healthcare. *Simulation in Healthcare, 3,* 33–41.

Salas, E., Burke, C. S., Fowlkes, J. E., & Wilson, K. A. (2004). Challenges and approaches to understanding leadership efficacy in multi-cultural teams. *Advances in Human Performance and Cognitive Engineering Research, 4,* 341–384.

Salas, E., Burke, C. S., & Stagl, K. C. (2004). Developing teams and team leaders: Strategies and principles. In D. Day, S. J. Zaccaro, & S. M. Halpin (Eds.), *Leader development for transforming organizations* (pp. 325–355). Mahwah, NJ: Erlbaum.

Salas, E., Cooke, N. J., & Rosen, M. A. (2008). On teams, teamwork, and team performance: Discoveries and developments. *Human Factors, 50,* 540–547.

Salas, E., Klein, C., King, H., Salisbury, M., Augenstein, J. S., Birnbach, D. J., et al. (2008). Debriefing medical teams: 12 evidence-based best practices and tips. *The Joint Commission Journal on Quality and Patient Safety, 34,* 518–527.

Salas, E., Kosarzycki, M. P., Tannenbaum, S. I., & Carnegie, D. (2005). Principles and advice for understanding and promoting effective teamwork in organizations. In R. J. Burke & C. Cooper (Eds.). *Leading in turbulent times* (pp. 95–120). Malden, MA: Blackwell.

Salas, E., Rosen, M. A., Burke, S., Nicholson, D., & Howse, W. R. (2007). Markers for enhancing team cognition in complex environments: The power of team performance diagnosis. *Aviation, Space, and Environmental Medicine, 78,* B77–B85.

Salas, E., Sims, D. E., & Burke, C. S. (2005). Is there a "big five" in teamwork? *Small Group Research, 36,* 555–599.

Salas, E., Wilson, K. A., Murphy, C. E., King, H., & Salisbury, M. (2008). Communicating, coordinating, and cooperating when lives depend on it: Tips for teamwork. *The Joint Commission Journal on Quality and Patient Safety, 34*, 333–341.

Sims, D. E., & Salas, E. (2007). When teams fail in organizations: What creates teamwork breakdowns? In J. Langan-Fox, C. L. Cooper, & R. J. Klimoski (Eds.), *Research companion to the dysfunctional workplace: Management challenges and symptoms* (pp. 302–318). Northampton, MA: Elgar.

Sundstrom, E., De Meuse, K. P., & Futrell, D. (1990). Work teams: Applications and effectiveness. *American Psychologist, 45*, 120–133.

Thomas, M. J. W. (2004). Predictors of threat and error management: Identification of core nontechnical skills and implications for training systems design. *The International Journal of Aviation Psychology, 14*, 207–231.

Tjosvold, D., Yu, Z., & Hui, C. (2004). Team learning from mistakes: The contribution of cooperative goals and problem-solving. *Journal of Management Studies, 41*, 1223–1245.

Tucker, A. L., Nembhard, I. M.., & Edmondson, A. C. (2006). *Implementing new practices: An empirical study of organizational learning in hospital intensive care units* (Working paper 06–049). Boston, MA: Harvard Business School.

Van De Ven, A. H., Delbecq, A. L., & Koenig, R. (1976). Determinants of coordination modes within organizations. *American Sociological Review, 41*, 322–328.

Van der Vegt, G. S., & Bunderson, J. S. (2005). Learning and performance in multidisciplinary teams: The importance of collective team identification. *Academy of Management Journal, 48*, 532–547.

Wachter, R. M. (2005, December). *Low on the totem pole.* Retrieved November 17, 2008, from http://www.webmm.ahrq.gov/case.aspx?caseID=110

Webster, J. L., & Cao, C. G. (2006). Lowering communication barriers in operating room technology. *Human Factors, 48*, 747–758.

Wegner, D. M. (1987). Transactive memory: A contemporary analysis of the group mind. In B. Mullen & G. R. Goethals (Eds.), *Theories of group behavior* (pp. 185–208). New York: Springer-Verlag.

Wilson, J. M., Goodman, P. S., & Cronin, M. A. (2007). Group learning. *Academy of Management Review, 32*, 1041–1059.

Wilson, K. A., Burke, C. S., Priest, H. A., & Salas, E. (2005). Promoting health care safety through training high reliability teams. *Quality and Safety in Health Care, 14*, 303–309.

Wong, S. (2004). Distal and local group learning: Performance trade-offs and tensions. *Organization Science, 15*, 645–656.

Wright, M. C., & Kaber, D. B. (2005). Effects of automation of information-processing functions on teamwork. *Human Factors, 47*, 50–66.

6

Team Training as an Instructional Mechanism to Enhance Reliability and Manage Errors

Sallie J. Weaver, Wendy L. Bedwell, and Eduardo Salas

Current models of organizational error underscore the inevitability of human error (Bogner, 1994, 2000, 2004; Moray, 1994; Reason, 1990). As noted by Davies (2001), "because humans design, manufacture, maintain, and operate systems, human error has a role in 100% of accidents" (p. 265). Furthermore, the increasing complexity of organizations and work itself makes predicting and preparing for all potential events nearly impossible. This likelihood for human error demands training that includes development of skills relevant to recognizing and recovering from these errors. As underscored by Hofmann and Frese (Chapter 1, this volume), a purely prevention-focused approach to errors is inadequate in complex, dynamic work environments.

Teams offer a unique opportunity to improve the overall reliability of an organization by increasing the ability to anticipate and recover from the unexpected (Hofmann, Jacobs, & Landy, 1995). In this sense, reliability has been defined as the organization's capacity to "repeatedly produce collective outcomes that meet or exceed minimal quality expectations" (Hannan & Freeman, 1984, p. 153). Effective teamwork is one mechanism for enhancing organizational reliability due to its inherent use of checks and balances (e.g., backup behavior, closed-loop communication, mutual performance monitoring, and the ability to leverage expertise). Certainty of human error, however, requires that we train teams to recognize conditions that might lead to error, effectively support each other to recover from and correct errors during workflow, and prevent such conditions in the future.

The purpose of this chapter is to highlight team training as an instructional mechanism for improving team reliability through error management and prevention. To this end, we first outline the principles of high reliability. Second, we outline the relationship between team-based work, errors, and the pursuit of reliability. Third, we describe the team training process and present a framework for conceptualizing team training designed to facilitate error management and reliability. In the final section, we present several potential avenues for future research.

High Reliability

The Principles of High Reliability

A certain set of organizations, high-reliability organizations (HROs), have mastered the ability to remain adaptive, anticipate the unexpected, and maintain reliability (Weick & Sutcliffe, 2001; Weick, Sutcliffe, & Obstfeld, 1999). They operate in complex, high-risk environments in which the impact of error can be catastrophic; yet, they are able to learn from, adapt to, and actually utilize this complexity to their advantage. Furthermore, these organizations are able to mitigate and contain the impact of the few errors that do occur due to their perpetuation of a culture of reliability and cooperative approach (Weick, 1987). Nuclear submarines (e.g., Bierly & Spender, 1995) and the U.S. naval aircraft carrier fleet (e.g., Rochlin, 1989) are examples cited in the literature.

Weick and Sutcliffe (2001) conceptualize HROs as exhibiting five specific value dimensions as part of their culture: (a) sensitivity to operations, (b) commitment to resilience, (c) deference to experience, (d) reluctance to simplify, and (e) a preoccupation with failure. We provide some brief examples of these dimensions; however, detailed definitions appear in Table 6.1. *Sensitivity* to operations refers to a continuous search for even minor deficiencies in normal operations, which can lead to errors. This sensitivity, however, is focused solely on quick response to rectify the situation as opposed to placing blame. Organizations demonstrate a *commitment to resilience* by quickly containing errors that occur despite existing safety nets and by creating workarounds to maintain system functioning. *Deference to expertise* refers to the consideration that the location of expertise changes depending on the situation—meaning that authority must shift to those who are best equipped to interpret and lead response efforts (i.e., those with the most expertise relevant to any given situation). *Reluctance to simplify* entails an organizational desire to create a full and nuanced picture of operations, not merely a broad overarching generalization. HROs maintain unprecedented levels of reliability by paying attention to details easily missed without concerted detection efforts. Finally, organizations demonstrating *preoccupation with failure* recognize the learning power of all errors and near misses. To this end, they encourage both

TABLE 6.1

The Five Cultural Values Exhibited by HROs

HRO Cultural Value	Definition[a]
Preoccupation with failure	"[HROs] treat any lapse as a symptom that something is wrong with the system, something that could have severe consequences if separate small errors happen to coincide at one awful moment. [They] encourage reporting of errors, they elaborate experiences of near miss for what can be learned, and they are wary of the potential liabilities of success, including complacency, the temptation to reduce margins of safety, and the drift into automatic processing." (p. 10)
Sensitivity to operations	"…An ongoing concern with the unexpected. Unexpected events usually originate in 'latent failures' which are loopholes in the system's defenses, barriers and safeguards who's potential existed for some time prior to the onset of the accident sequence, though usually without any obvious bad effect.… Normal operations may reveal deficiencies that are 'free lessons' that signal the development of unexpected events.… People in HROs know that you can't develop a big picture of operations if the symptoms of those operations are withheld." (p. 13)
Commitment to resilience	"Resilience is a combination of keeping errors small and of improvising workarounds that keep the system functioning. These avenues of resilience demand deep knowledge of the technology the system, one's coworkers, and one's self, and the raw materials. HROs put a premium on experts; personnel with deep experience, skills of recombination, and training. They mentally simulate worst case conditions and practice their own equivalent of fire drills." (pp. 14–15)
Deference to experience	"HROs cultivate diversity, not just because it helps them notice more in complex environments, but also because it helps them do more with the complexities they spot. Rigid hierarchies have their own special vulnerability to error. Errors at higher levels tend to pick up and combine with errors at lower levels, thereby more prone to escalation.… "HROs push decision making down—and around. Decisions are made on the front line, and authority migrates to the people with the most expertise, regardless of rank." (p.16)
Reluctance to simplify	"HROs take deliberate steps to create more complete and nuanced pictures. They simplify less and see more. Knowing that the world they face is complex, unstable, unknowable, and unpredictable, they position themselves to see as much as possible. They encourage boundary spanners who have diverse experience, skepticism toward received wisdom, and negotiation tactics that reconcile differences of opinion without destroying the nuances that diverse people detect." (pp. 11–12)

[a] From Weick, K. E., & Sutcliffe, K. M. *Managing the unexpected*. San Francisco: Jossey-Bass, 2001. (With permission of John Wiley & Sons.)

error and near-miss reporting. In addition, HROs (a) continuously and actively seek knowledge, (b) are focused on reward systems that simultaneously recognize the cost of errors and benefits of reliability, and (c) persist in communicating the entire picture to all organizational levels (Roberts & Bea, 2001a, 2001b).

High-Reliability Teams

HROs offer a unique opportunity to study methods aimed at minimizing and mitigating error, especially through the role of teams. High-reliability teams (HRTs) play a vital role in achieving high-reliability status (Baker, Day, & Salas, 2006; Batt, 2004; Wilson, Burke, Priest, & Salas, 2005). Such teams are usually hierarchically structured and work under conditions of high task interdependence with intense communication between team members; it is necessary to pool information from multiple sources while remaining in constant coordination (Salas, Cannon-Bowers, & Johnston, 1997). Wilson and colleagues (2005) pinpoint how specific team behaviors support organizations in recognizing the five HRO cultural values (see Table 6.2).

Although these are similar to behaviors exhibited by traditional teams, HRTs differ in their ability to demonstrate these behaviors reliably while operating at the edge of human capacity—under intense stress in a dynamic, complex environment (Roberts & Rousseau, 1989; Wilson et al., 2005).

However, there remains limited empirical evidence regarding the role of HRTs in HROs and the training needed to reach HRT status. In this chapter, we build on the work of Wilson, Salas, and colleagues to present a framework of team training to reduce errors. We argue that team training is one mechanism vital to achieving high-reliability status because training equips such teams with the knowledge, skills, and attitudes (KSAs) necessary to operate as expert teams (see Figure 6.1) and offer theoretically based propositions for future research. By pinpointing areas for future research, we aim to initiate and drive further investigation to uncover those aspects contributing to training effectiveness in reducing errors and saving lives.

To provide firm context for the framework, we begin by defining teams and teamwork. Crew resource management (CRM) training in aviation is then discussed to exemplify the role of team training in transitioning an entire industry toward high reliability.

Team-Based Work, Errors, and the Pursuit of Reliability

The context of the work environment is changing. Work is more complex, distributed, and dynamic. Organizations increasingly turn to teams to

TABLE 6.2
HRT Values

Value	Team Level	Source
Sensitivity to operations	*Closed-loop communication*: Team's ability to exchange information accurately and clearly and to acknowledge receipt of information. *Information exchange*: Team members' ability to speak clearly, concisely, and in an unambiguous manner with other team members. Information exchange skills should be transportable, meaning improvement in their ability to communicate increases across tasks. *Shared situation awareness*: Team's ability to develop shared mental models of the environment (internal and external) to apply correct task strategies and anticipate future situations.	Cannon-Bowers, Tannenbaum, Salas, & Volpe, 1995 Endsley, 1999
Commitment to resilience	*Backup behavior*: The capability of team members to give, seek, and receive task instructive feedback. Assisting team members to perform their tasks. Backup behavior can be achieved by providing a teammate with verbal feedback or coaching, helping a teammate behaviorally in carrying out actions, or assuming and completing a task for a teammate. *Performance monitoring*: Team's ability to monitor team members' performance and provide constructive feedback. *Shared mental models*: Team's ability to share compatible knowledge pertaining to individuals' roles in the team, the roles of fellow team members, their characteristics, and the collective requirements needed for effective team interaction.	Cannon-Bowers, Tannenbaum, Salas, & Volpe, 1995 Dickinson & McIntyre, 1997
Deference to expertise	*Assertiveness*: The willingness of team members to communicate ideas and observations in a manner that is persuasive to other team members. Allows team members to provide feedback, state and maintain opinions, address perceived ambiguity, initiate actions, and offer potential solutions. *Collective orientation*: Some team members have been found to be more collectively oriented than others, meaning that they exhibit more interdependent behaviors in task groups. *Expertise*: Knowing how to do something well is gained through experience.	Blickensderfer, Cannon-Bowers, & Salas, 1997 Smith-Jentsch, Salas, & Baker, 1996

(continued)

TABLE 6.2 (Continued)
HRT Values

Value	Team Level	Source
Reluctance to simplify	*Adaptability/flexibility*: Team's ability to gather information from the task environment and adjust its strategies by reallocating resources and using compensatory behaviors such as backup behavior.	Cannon-Bowers, Tannenbaum, Salas, & Volpe, 1995
	Planning: Planning both prior to and during a mission helps teams improve performance by setting goals, sharing relevant information, clarifying members' roles, prioritizing tasks, discussing expectations and environmental characteristics and constraints.	Stout, Cannon-Bowers, & Salas, 1996
Preoccupation with failure	*Error management*: Based on understanding the nature and extent of error, changing conditions found to induce error, and determining and training behaviors that decrease errors.	Cannon-Bowers, Tannenbaum, Salas, & Volpe, 1995
	Feedback: Team's ability to provide constructive feedback, seek feedback on its performance, and accept feedback from others.	
	Team self-correction: Team's ability to monitor and categorize its behavior to determine its effectiveness, which generates instructive feedback so that team members can review performance episodes and correct deficiencies.	

Source: Adapted from Wilson, K. A., Burke, C. S., Priest, H. A., & Salas, E. Promoting health care safety through training high reliability teams. *Quality and Safety in Healthcare*, 14, 303–309, 2005. With permission from BMJ Publishing Group Ltd.

Training as an Instructional Mechanism 149

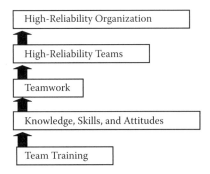

FIGURE 6.1
The path from team training to high reliability.

maintain effectiveness in this ever-evolving work environment, especially when the goals are difficult, the tasks are interdependent, and the consequences of errors are high (Salas, Kosarzycki, Tannenbaum, & Carnegie, 2004). For thorough coverage of issues related to teams and errors, see Bell and Kozlowski's Chapter 5 in this volume. We offer a brief overview to provide context for our discussion of team training.

What Constitutes a Team and Teamwork?

A *team* is defined as a distinguishable set of two or more people interacting toward a common goal with specific roles and boundaries on tasks that are interdependent and are completed within a larger organizational context (Kozlowski & Bell, 2003; Salas, Dickinson, Converse, & Tannenbaum, 1992). Teams differ from groups in that their tasks are interdependent—they require teamwork to be achieved; that is, team members must adjust their efforts based on the efforts of their team members to accomplish team goals (Brannick & Prince, 1997; Dickinson & McIntyre, 1997). Teamwork is comprised of the behaviors (e.g., closed-loop communication, mutual support), cognitions (e.g., shared mental models), and attitudes (e.g., collective efficacy, trust, cohesion) that combine to make adaptive interdependent performance possible (Cannon-Bowers, Salas, Tannenbaum, & Mathieu, 1995). These characteristics allow teams to achieve outcomes greater than those that could possibly be achieved by any one team member working alone. Teams of experts may come together and fail if they are not also experts in effective teamwork (Weaver, Wildman, & Salas, 2009).

Advantages of Team-Based Work

Working in teams creates a wider support system for employees. Teams can be self-correcting; that is, they can compensate for one another and reallocate tasks or functions as needed. Effective teams also engage in a cycle of "prebrief–performance–debrief" to diagnose performance, revise team goals, and anticipate future needs (Salas, Rosen, Burke, & Goodwin, 2009). Good teams provide backup behavior when others are fatigued, stressed, or overloaded—all common sources of error when working individually (Reason & Hobbs, 2003). Teams have been conceptualized as "workload sponges," meaning that, collectively, they can tolerate greater workload, stress, and pressure than individuals working independently (Salas, Wilson, Murphy, King, & Salisbury, 2008). Furthermore, good teams hold shared mental models, enabling them to anticipate next steps, and members can support one another to make sure these steps are accomplished. They also manage and optimize performance outcomes, which can translate into fewer errors, better decisions, and a greater chance of success.

Team-based organizations have higher productivity, service and product quality, and customer satisfaction; reduced employee turnover; and a flatter overall management structure (Glassop, 2002; Gross, 1997). In one specific health care example, a team-based approach utilized in the office of Veteran's Affairs led to a 30% improvement in productivity, increased ratings of customer satisfaction by 10%, and nearly $2 billion dollars in savings ("Five Case Studies," 2002). Research also suggests that teams contribute to organizational safety (a correlate of reliability and error management) and organizational responsiveness to business environment change ("Five Case Studies," 2002; Salas, Burke, Bowers, & Wilson, 2001; Wilson et al., 2005). An important caveat, however, is that teams can only have a positive impact on organizational outcomes when team members know how to work both effectively and efficiently—communicate, coordinate, and cooperate as an expert team, not just a team of experts (Salas, Wilson et al., 2008).

Why Is Team Training Suggested as an Error Reduction and Management Technique?

Team training provides a vehicle for refining a team of experts into an HRT, which is a team comprised of members who individually possess expert levels of both task work and teamwork KSAs (Salas, Rosen, Burke, Goodwin, & Fiore, 2006). Due to high teamwork capabilities, HRTs are able to adapt in highly complex and dynamic environments utilizing high levels of communication and coordination. These teams mitigate errors through several mechanisms, such as mutual performance monitoring, backup behavior, closed-loop communication, and collective efficacy (Weaver et al., 2009). In this sense, teamwork is both a mitigation technique and a final catch for errors that slipped through previous system layers.

Several meta-analyses indicate that team training has a positively impact on important outcomes. Specifically, Salas, DiazGranados, Klein and colleagues (2008) analyzed 45 studies of team training (93 effect sizes) and found that overall team training had a positive effect on team functioning (ρ = .34). In terms of specific outcomes, team training had a significant impact on cognitive outcomes (ρ = .42), affective outcomes (ρ = .35), process outcomes (ρ = .44), and team performance outcomes (ρ = .39). Furthermore, team training accounted for 12% to 19% of the variance in team performance. In addition, training content was a significant moderator such that training content focused on teamwork rather than task work and had a greater impact on these outcomes. Similar results were found in a prior meta-analysis with seven studies using 28 effect sizes (Salas, Nichols, & Driskell, 2007) suggesting that training improved team performance (r = .29). Specifically, team coordination and adaptation training (a derivative of CRM) exerted the largest effects on team performance (r = .61), followed by guided team self-correction training (r =. 45).

From a theoretical perspective, Burke, Wilson, and Salas (2005) present a team-based strategy model for enhancing organizational reliability, providing guidelines that refer explicitly to team training. Team training provides a mechanism through which team members learn the behaviors, attitudes, and cognitions relevant to expert teams and practice utilizing tools and processes aimed at error management and mitigation. By providing opportunities to practice using these tools and processes, team training can enhance reliability by allowing teams to practice error management strategies. The framework developed by Wilson and colleagues (2005) links specific team-level KSAs to five value dimensions of HROs. Building on this framework, we argue that team training is one ideal methodology for reducing and mitigating errors. CRM training is one such example in the aviation community, through their successful efforts to enhance reliability over the past 20 years.

CRM and Aviation

CRM training was designed to increase team reliability and reduce errors by increasing teamwork behaviors and teaching team members to maximally utilize all available resources maximally (Salas et al., 2001; Salas, Prince, et al., 1999; Salas, Rhodenizer, & Bowers, 2000). After several catastrophic accidents, the Federal Aviation Administration (FAA) and National Transportation Safety Board (NTSB) pinpointed teamwork and communication issues as contributing causes and thus began advocating CRM to streamline operations and optimize interpersonal behaviors among both air and ground crews (NTSB, 1979). Since 1990, carriers have been required to provide CRM training for all flight crews and integrate CRM concepts into technical training for pilots and ground crews (Helmreich, Merrit, & Wilhem, 1999).

CRM programs target behaviors such as communication, decision making and leadership, adaptability and flexibility, assertiveness, and situational awareness (Alonso et al., 2006; Powell & Hill, 2006; Seamster & Kaempf 2001). Current CRM training concentrates on identifying and recovering from inevitable errors (Salas, Wilson, Burke, & Bowers, 2002). CRM aims to stimulate reliability by fostering a more realistic awareness of personal limitations and abilities (Helmreich & Merrit, 1999). For example, CRM often includes examples of recognizing and mitigating errors, with practice scenarios in which trainees have the opportunity to practice recovering from errors in a nonconsequential environment (Helmreich et al., 1999).

Evaluations of CRM training indicated that it can improve team performance by 6% to 20% (Salas, Fowlkes, Stout, Milanovich, & Prince, 1999; Salas et al., 2001). After adoption by the airline industry, for example, the total number of fatalities due to U.S. carrier accidents has significantly declined from 285 in 1988 to just 1 in 2007 (NTSB, n.d.). Other high-risk industries (i.e., health care) are integrating team training concepts to manage errors and achieve high reliability (e.g., Baker et al., 2006; Musson & Helmreich, 2004; Pronovost, et al., 2006). Yet, multilevel CRM evaluation in the context of error management remains warranted.

Team Training as a Mechanism to Reduce and Manage Errors

Although seemingly intuitive, for team training to have a positive impact on reliability and error management, teamwork processes must be clearly linked to the tasks and processes underlying performance. Essentially, teamwork must clearly underlie processes related to reliable performance. We present a framework for developing and evaluating team training to increase reliability and enhance error management (see Figure 6.2). The framework serves as an organizational heuristic, integrating evidence derived from training and learning research (e.g., Aguinis & Kraiger, 2009; Salas & Cannon-Bowers, 2003) with principles of high reliability to map the process of identifying, designing, and evaluating team training to enhance reliability. Each component of the framework is discussed in detail next.

Team Coordination Audit

Traditionally, training needs analysis is the first step in training program design (Goldstein & Ford, 2002); however, teamwork is not fully captured

Training as an Instructional Mechanism 153

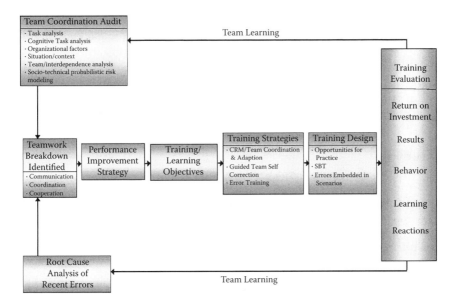

FIGURE 6.2
A framework for training teams to reduce errors.

in the traditional definitions of such needs analysis. Therefore, a team coordination audit (TCA) is an essential first step in developing team training to enhance reliability and manage errors. This audit focuses on detailing the communication, coordination, and cooperation demands that pose potential risk for errors and the team member KSAs needed to handle these risk points effectively. The TCA concept is based on Salas, Wilson, et al.'s (2008) conceptualization of teamwork as an essential component of the shared cognition necessary for reliability. Shared cognition specifically refers to the team's collective knowledge, awareness of the current situation, and decision making (Cooke, Salas, Cannon-Bowers, & Stout, 2000).

A TCA should include precise delineation and analysis of (a) those processes and efforts necessary to maintain accurate and efficient information exchange among team members (communication), (b) the behavioral and cognitive strategies necessary to integrate individual efforts precisely (coordination), and (c) the affective mechanisms that motivate team members to work together (cooperation). In some respects, the TCA process could be considered a subtask in a job/KSA analysis in a traditional training needs analysis. Table 6.3 offers a summary of several key questions to be investigated in the context of a TCA.

A key feature of a TCA is the consideration of both internal and external teamwork requirements. Communication, coordination, and cooperation requirements must be analyzed from both intra- and interteam

TABLE 6.3

Key Questions to Be Answered During a Team Coordination Audit

Communication

Information Exchange
- Did team members seek information from all available resources?
- Did team members pass information within a timely manner before being asked?
- Did team members provide "big picture" situation updates?

Phraseology
- Did team members use proper terminology and communication procedures?
- Did team members communicate concisely?
- Did team members pass complete information?
- Did team members communicate audibly and ungarbled?

Closed-Loop Communication
- Did team members acknowledge requests from others?
- Did team members acknowledge receipt of information?
- Did team members verify that information sent was interpreted as intended?

Coordination

Knowledge Requirements
- Did team members have a common understanding of the mission, task, team, and resources stress, based on their belief in their collective abilities?
- Did team members share common expectations of the task and team member roles and responsibilities?
- Did team members share a clear and common purpose?
- Did team members implicitly coordinate in an effective manner?

Mutual Performance Monitoring
- Did team members observe the behaviors and actions of other team members?
- Did team members recognize mistakes made by others?
- Were team members aware of their own and others' surroundings?

Backup Behavior
- Did team members correct other team members' errors?
- Did team members provide and request assistance when needed?
- Did team members recognize when one performed exceptionally well?

Adaptability
- Did team members reallocate workload dynamically?
- Did team members compensate for others?
- Did team members adjust strategies to situation demands?

(continued)

TABLE 6.3 (Continued)
Key Questions to Be Answered During a Team Coordination Audit

Cooperation

Team Orientation
- Did team members put group goals ahead of individual goals?
- Were team members collectively motivated, and did they show an ability to coordinate?
- Did team members evaluate each other and use inputs from other team members?
- Did team members exhibit "give-and-take" behaviors?

Collective Efficacy
- Did team members exhibit confidence in fellow team members?
- Did team members exhibit trust in others and themselves to accomplish their goals?
- Did team members follow team objectives without opting for independence?
- Did team members show more and quicker adjustment of strategies across the team when under stress, based on their belief in their collective abilities?

Mutual Trust
- Did team members confront each other in an effective manner?
- Did team members depend on others to complete their own tasks without "checking up" on them?
- Did team members exchange information freely across team members?

Team Cohesion
- Did team members remain united in pursuit of mission goals?
- Did team members exhibit strong bonds and desires to remain a part of the team?
- Did team members resolve conflict effectively?
- Did team members exhibit less stress when performing tasks as a team rather than as individuals?

Source: Adapted from Wilson, A., Salas, E., Priest, H. A., & Andrews, D. Errors in the heat of battle: Taking a closer look at shared cognition breakdowns through teamwork. *Human Factors, 49*(2), 243–256, 2007. With permission from Sage Publications.

perspectives. For example, a surgical team depends on the preoperative team to communicate particular events and patient prep information prior to a surgery. Breakdowns in this interteam communication can lead to catastrophic errors for the patient and team members alike.

A TCA may also incorporate aspects of sociotechnical probabilistic risk assessment (ST-PRA; Marx & Slonim, 2003). Whereas more traditional models focus on modeling risk in mechanical systems, ST-PRA focuses on modeling risk associated with human error and at-risk behaviors. In addition, ST-PRA offers a mechanism to account for effects of behavioral norms on reliable performance (Marx & Westphal, 2008). A detailed discussion of the full ST-PRA methodology is beyond the scope of this chapter (see Mohaghegh, Kazemi, & Mosleh, 2009); the basic principles of ST-PRA usually include the use of fault tree models to outline task/event processes by linking each step of the task/event through a series of AND gates and OR gates. The probability of each event (or frequency of occurrence) is estimated and then combined to calculate the actual estimated risk of particular culminating events. Thus, breakdowns in team processes could be listed as events, estimates provided for their frequency/probability, and an actual risk probability calculated to quantify impacts of teamwork-related errors. Similar procedures for audits in aviation have been effective (e.g., International Civil Aviation Organization, 2002).

Root Cause Analysis of Recent Events

Root cause analysis (RCA) refers to a family of problem-solving methods aimed at clearly identifying the causal factors and the cause–effect relationships leading to failures in quality control and errors (Rooney & Vanden Heuvel, 2004). RCA is a means of identifying the nonobvious factors related to errors and focuses on discovering their controllable causes to implement corrective action to prevent and better manage future events. There is no singular method for conducting RCA; however, common methods are similar to the charting stages of ST-PRA and usually involve the creation of a cause–effect chart or factor tree (for detailed coverage of RCA techniques, see Vanden Heuvel et al., 2008).

RCA has been used successfully to uncover breakdowns in teamwork in a variety of industries by mapping relationships and key information/resource transfers. For example, communication, a key component in teamwork, has been cited as a root cause in nearly 70% of medical errors (Joint Commission, 2008). Thus, it is vital in determining whether team training will meaningfully affect organizational and team reliability and management of errors.

Is Teamwork a Key Point of Risk?

The results of the TCA and RCA should provide the basis for determining whether potential (or past) breakdowns in teamwork are (a) meaningful points of risk to reliable performance and (b) potential barriers/facilitators of error management. Several teamwork KSAs have been mapped to the core principles of high reliability (Wilson et al., 2005). For example, closed-loop communication and other forms of information exchange help promote shared situational awareness. Closed-loop communication involves the clear exchange of concise information, with the receiver acknowledging receipt of the information and acquiring confirmation from the information provider that the receiver has accurate understanding (McIntyre & Salas, 1995). Information exchange is crucial for reliable performance.

Considering health care as an example, emergency medical services (EMS) technicians must provide vital signs and other important information to emergency room (ER) staff. ER doctors, in turn, need to provide nurses with information regarding the amount and type of medication and required procedures, while nurses must keep doctors abreast of the patient's vital signs throughout the execution of emergency procedures. The flow of information helps ensure shared situation awareness among team members, which has a positive impact on reliability and error management by improving coordination efforts, aiding decision making, and alerting members to latent errors (Cannon-Bowers, Tannenbaum, Salas, & Volpe, 1995; Endsley, 1999).

Other important KSAs deal with the concept of *deference to expertise*. This includes collective orientation, assertiveness, and expertise as required KSAs. Status or rank is often heavily weighted when considering the value of information/advice. However, it may not always be the most senior member who has the most relevant information. HRTs consider and leverage the knowledge of all team members (Driskell & Salas, 1992a, 1992b). A team with collective orientation seeks input from all members, allows members to be assertive, and recognizes value in expertise, regardless of rank (Wilson et al., 2005). In an aircrew, the pilot is highest ranking; however, all members have expertise (e.g., navigator). Each member is situationally aware from a slightly different perspective. Within a collectively oriented team, members would feel empowered to express concerns, ultimately leading to error avoidance.

Performance Improvement Strategy

Once a clear link between teamwork breakdowns and reliability is established, a performance improvement strategy must be determined. While team training is one of the most common techniques applied to enhance team performance, it is not a panacea. The TCA and RCA may point to issues of job design, workflow, or other factors. If team training is selected, reliability and error management principles must be embedded within the organizational culture for transfer maximization to ensure that behaviors occur beyond the training environment.

Training/Learning Objectives

Training objectives indicate specific, observable changes in learner KSAs that should be evident in their end-of-training performance (Goldstein & Ford, 2002). Clear objectives provide a mechanism for determining the content, strategy, and evaluation methods for team training. Traditionally, three main components are included in effective learning objectives: (a) performance (i.e., the "terminal" behavior), (b) criteria (i.e., standards for the behavior, and (c) conditions (Phillips & Phillips, 2008). Training objectives developed for team training to enhance reliability and error management should focus on competencies determined to underlie reliable performance in the TCA and RCA.

Mapping Instructional Strategies to Identified Learning Objectives

Just as various tasks require different KSAs, specific KSAs are best taught through various instructional methods. Wilson and colleagues (2005) outlined several instructional strategies suitable for transforming teams into HRTs (refer to Table 6.4): cross-training, perceptual contrast training, team coordination training, team self-correction training, scenario-based training, and guided error training. These are reviewed next.

Cross-Training

Two key characteristics of HRTs are *sensitivity to operations* and *commitment to resilience* (Wilson et al., 2005). KSAs related to these team behaviors are closed-looped communication, information exchange, shared situational awareness, backup behavior, monitoring, and shared mental models. Teams with complex tasks in dynamic environments need members who can anticipate needs of fellow team members, provide necessary support (i.e., backup behavior), and clearly communicate (i.e., information exchange) to ensure that the team is "on the same page" (i.e., shared situational awareness and shared mental models). For this to occur, members must know differential roles/responsibilities of other team members. Cross-training allows members to develop a shared understanding of these role requirements, as well as the interdependence of member tasks (e.g., Marks, Sabella, Burke, & Zaccaro, 2002). Without this understanding, members may not see value in providing backup behavior, for example.

Generally, three levels of cross-training are considered: positional clarification, positional modeling, and positional rotation (Blickensderfer, Cannon-Bowers, & Salas, 1998). The main distinction between these levels is the amount of interactive participation on the part of the trainee. Positional clarification is simply the exchange of information. Positional modeling is just as the name implies—modeling of the typical required duties. Finally, positional rotation actually involves the trainee experiencing other team members' positional requirements by actively engaging in tasks and duties assigned to his or her teammates. The level of training will have an impact on the level of KSA development. For example, team members who experienced positional rotation will be able to complete the tasks of other team members who are overloaded or unable to perform their duties, such as the floater nurse, who can step in for several different nurses on multiple hospital units (Wilson et al., 2005). The increased involvement (such as with positional rotation) in addition to the understanding can lead to greater

TABLE 6.4

Instructional Strategies to Target KSAs Required for Error Reduction

Training Strategy	Definition	Source
Cross-training	Team members are able to gain a clear understanding and shared representation (shared mental model) of how the team functions as well as how each individual's tasks and responsibilities are interrelated to those of other team members.	Volpe, Cannon-Bowers, Salas, & Spector, 1996
Perceptual contrast training	Team members are actively involved in learning to create a deeper and more acute understanding of the instructional material. Team members may be required to actively compare the defining characteristics of contrasting cases, which will give them a keener understanding of the concepts.	Schwartz & Bransford, 1998
Team coordination training	Team members improve skills such as team coordination, communication (both explicit and implicit), and backup behavior. Practice opportunities are also provided for additional competencies that lead to effective coordination.	Bowers, Blickensderfer, & Morgan, 1998; Entin & Serfaty, 1999
Team self-correction training	Team members are taught to assess the effectiveness of their own behavior and that of others. Team members are also taught how to provide constructive feedback and correct deficient behaviors.	Blickensderfer, Cannon-Bowers, & Salas, 1997; Smith-Jentsch, Zeisig, Acton, & McPherson, 1998
Scenario-based training	Team members are given an opportunity to experience critical learning events (such as common instances when errors and unsafe behaviors are occurring) embedded in the training scenarios, allowing them to learn from a meaningful framework.	Rosen & Salas, 2008; Salas & Cannon-Bowers, 2000; Stout, Cannon-Bowers, Salas, & Milanovich, 1999
Guided error training	Team members are able to experience errors, react to errors, and see the consequences of errors through guided practice. Errors are used as a function of information that provides feedback to trainees so that they may develop better learning and knowledge transfer strategies to be used in the real task environment.	Oser, Cannon-Bowers, Salas, & Dwyer, 1999; Carlson, Lundy, & Schneider, 1992

Source: Adapted from Wilson, K. A., Burke, C. S., Priest, H. A., & Salas, E. Promoting health care safety through training high reliability teams. *Quality and Safety in Healthcare*, 14, 303–309, 2005. With permission from BMJ Publishing Group Ltd.

sensitivity to operations and commitment to resilience, necessary team behaviors for promoting reliability and error management.

Perceptual Contrast Training

These same KSAs that Wilson and colleagues (2005) characterized as necessary for sensitivity to operations and commitment to resilience, particularly the cognitive skills such as monitoring, can be successfully taught with another instructional strategy. Perceptual contrast training involves the presentation of contrasting scenario/event alternatives (Bransford, Franks, Vye, & Sherwood, 1979). Trainees then are encouraged to recognize the differences between these contrasting alternatives and conceptualize the positives and negatives of each. Several scenarios should be presented that are job relevant and demonstrate both successful and unsuccessful actions. Shared situational awareness and monitoring requires members to notice what could be subtle cue differences (Bransford et al., 1979). The ability to "notice" these differences in complex environments with multiple inputs and little time for making life-altering decisions differentiates expert HRTs from others. "Noticing" is a dimension of many of the previously noted cognitive skills and thus is critical for creating reliable work environments.

Team Coordination and Adaptation Training

Much research has focused on the important role that coordination plays in teamwork (Oser, MacCallum, Salas, & Morgan, 1989). Others have noted that effective coordination strategies are particularly important for HRTs (Serfaty & Entin, 1998). As noted, CRM, one type of coordination training, has been particularly successful, leading to improved reliability records and fewer accidents attributed to poor teamwork (Salas et al., 2001). CRM focuses on teaching trainees to acknowledge that human error will occur and how to focus their efforts on managing errors through effective teamwork (Serfaty & Entin, 1998). A critical component of CRM is that all team members are trained. This helps inform all individuals involved in the effort of the potential for failure. In complex situations, errors can mean the difference between life and death. By focusing on teamwork competencies that can help reduce the likelihood of errors and the development of strategies for dealing with errors, team coordination strategies such as CRM can help lead to a greater commitment to resilience and sensitivity to operations, which ensures development of shared mental models, a crucial skill for HRTs.

Team Self-Correction Training

HRT members constantly monitor the situation to identify potentially problematic behaviors before they result in errors (Wilson et al., 2005). Once identified, these behaviors must be corrected and avoided in the future. Team self-correction training teaches team members to identify the potentially damaging errors as well as provide timely, constructive feedback to help correct behaviors that may lead to error (Smith-Jentsch, Cannon-Bowers, Tannenbaum, & Salas, 2008; Stout, Cannon-Bowers, & Salas, 1996). Throughout training, a facilitator actively guides the team through the process by focusing discussion, promoting a climate of learning, eliciting comments from all members, and demonstrating how to provide effective feedback (Smith-Jentsch, Payne, & Johnston, 1996; Tannenbaum, Smith-Jentsch, & Behson, 1998). The goal of this intensive training is to teach teams to correct the issues themselves, without outside intervention (Blickensderfer, Cannon-Bowers, & Salas, 1997). Active participation by all team members prepares the team to utilize this process on its own, which can lead to increased expertise and a reduction of errors.

Scenario-Based Training

Scenario-based training involves the use of highly structured scripts of real-word situations in an operationally relevant environment with strategically placed trigger events (Fowlkes, Dwyer, Oser, & Salas, 1998; Oser, Cannon-Bowers, Salas, & Dwyer, 1999; Prince, Oser, Salas, & Woodruff, 1993). Trigger events are designed to elicit specific, targeted behaviors clearly defined prior to training. Feedback is provided to trainees based on their performance (e.g., Rosen, Salas, Silvestri, Wu, & Lazzara, 2008).

The scenarios and trigger events create opportunities for team members to engage in effective error management behaviors in complex environments. Practice opportunities allow for enactment of multiple solutions in a safe environment with little consequence for error (Wilson et al., 2005), allowing members to learn crucial skills of adaptability, flexibility, and planning. This will better equip teams to deal with nonroutine events and errors resulting in failure.

Guided Error Training

Researchers have suggested that the inclusion of errors in training provides meaningful feedback that leads to greater learning and transfer as compared to error avoidance training (e.g., Frese & Altman, 1989; Keith, Chapter 2, this volume). One particular strategy is guided error training. This involves actively leading trainees into errors and supporting error correction efforts in an effort to enable trainees to learn corrective strategies for dealing with error outcomes (Karl, O'Leary-Kelly, & Martocchio, 1993; Lorenzet, Salas, & Tannenbaum, 2005). This particular method has been found to lead to improved performance as well as increased self-efficacy (Lorenzet et al., 2005). Allowing trainees to see what errors look like, what outcomes can result from errors, and the appropriate corrective action to take is invaluable experience that can lead to expertise in increased efficacy with error management as well as the ability to self-correct when errors do indeed occur. Both skills are important for creating and maintaining a culture of reliability and error management orientation.

Leadership Training

Leadership is an integral component of effective team functioning. Whether leadership functions such as goal setting, performance monitoring and alignment, and enhancing the development of key emergent states such as trust and psychological safety are performed by a single leader or distributed among several team members, they must be performed effectively for the team to function. Models of leadership in the team context underscored that effective leadership focuses on facilitating team coherence, integration, and interconnectivity (Zaccaro, Heinen, & Shuffler, 2009). Therefore, in this context, leadership training may be focused not only on an individual leader but also on facilitating these skills among team members.

Meta-analytic and review results suggested that leadership training, particularly strategies that incorporate behavioral modeling, have been effective in terms of increased knowledge and performance (Burke & Day, 1986). However, investigation is needed regarding the impact of leadership training on team reliability and error management. In one example, transformational leadership training specifically focused on medical error prevention demonstrated significant reductions in self-reported errors/injuries (e.g., accidental needle pricks, bruises, strains) among medical staff (Mullen & Kelloway, 2009).

Training Evaluation

Inherent in our framework is the notion of training evaluation. *Evaluation* refers to the systematic process of gathering both descriptive and judgmental information of a training program (Goldstein & Ford, 2002). However, training evaluation is not common practice in many organizations. Those who do conduct evaluations often focus on trainee reactions rather than issues of knowledge acquisition and performance improvements (i.e., transfer). The goal of any organizationally sponsored training program is to develop job-related KSAs necessary for successful performance. To determine training effectiveness, measures of training success (criteria) must also be based on job-related KSAs, specifically those targeted in training. Essentially, criteria should be developed that allow for the determination of whether the training increased targeted, job-related KSAs immediately after training and again once the trainee was back on the job to ensure that transfer occurred (Goldstein & Ford, 2002).

There are a number of issues related to training criteria that should be considered, such as distinctions between behavior, performance, and effectiveness; criterion relevancy; and the debate between multiple or composite criteria. To evaluate training most accurately, it is necessary to measure multiple dimensions of performance that were targeted in training. There are many training evaluation theories. Each has merits, focusing on different elements of training. For brevity, neither the theories nor training criteria variables are discussed. The point is to underscore the importance of creating an organizing framework for how training evaluation will be conceptualized as it can have an impact on the transfer (i.e., employees see that training is important if it is tied to evaluations).

Transfer of Training

Research has shown that the instructional techniques described are suitable for training teamwork-related KSAs (e.g., CRM; Serfaty & Entin, 1998). We, as well as others (e.g., Wilson et al., 2005), propose that these training strategies are necessary, but not sufficient, for achieving HRT status, which is characteristic of reliable teams. Additional variables have an impact on the relationship between training strategy and successful acquisition of desired KSAs. Training is of little value to an organization if trainees do not transfer newly attained KSAs to their organizational setting (Kozlowski & Salas, 1997). Training transfer refers to the application of prior training to job-specific tasks, thus influencing performance (Holding, 1991). There are many models of training transfer (e.g., Baldwin & Ford, 1988). For our purposes, we submit that transfer of training is *required* if selected instructional strategies are to have any meaningful impact on reliability and actual error management. A more thorough treatment of the issues related to transfer and learning through errors can be found in Chapter 2 of this volume.

Other Team Training Issues

Two categories of variables should also be considered in developing team training for errors: training needs/constraints (target audience, available resources, desired immersion level) and transfer climate (supervisory support, opportunity to perform, change conditions).

Training Requirements/Constraints

Training requirements and constraints include the target audience, available resources, and desired level of immersion. These variables help determine a number of issues, such as which training strategies are most effective (i.e., most suited for the audience or level of immersion) and which are most feasible given the resource constraints. Depending on the TCA results, a combination of training strategies may be the most effective methodology to address targeted KSAs. Training delivery will depend on the target audience. If the audience is distributed, depending on the availability of resources, it may not be feasible to have face-to-face training; therefore, instructional strategies that utilize electronic delivery mechanisms may be most appropriate and effective. We are not suggesting that the training requirements/constraints should become the main focus of what instructional strategies should be used; this should still be largely based on the content of the training, in this case, the desired KSAs. However, the target audience, the available resources, and the desired level of immersion should be considered as important factors in determining the most appropriate instructional strategies.

Transfer Climate

Multiple variables can be considered part of a transfer climate, including supervisory support, opportunity to perform, and conditions of change. Trainees must perceive a supportive transfer climate. Research has suggested that pretraining contextual cues (i.e., managerial support) have a significant impact on outcome expectancies (Broad & Newstrom, 1992). Researchers have tied management support to beliefs in the utility of training (Cohen, 1990) and intentions to transfer (Baldwin & Magjuka, 1991).

Perhaps more important, employees distinguish between permission to attend training and actual support of training and training content (Baldwin & Magjuka, 1991). This implies that actual training acceptance and display of managerial support increase employee perceptions, which leads to greater outcome expectancies, not the acknowledgment that those employees should attend training. Forms of managerial support include removal of situational constraints (Latham & Crandall, 1991; Peters, O'Connor, & Eulberg, 1985). Latham and Crandall (1991) noted eight different situational constraints: (a) lack of job-related information; (b) lack of or faulty tools and equipment; (c) lack of or low-quality materials and supplies; (d) required services and help from others; (e) lack of time; (f) limiting physical aspects of the work environment; (g) lack of job-relevant authority/autonomy; and (h) budgetary constraints.

Finally, while there are numerous variables that can be considered as conditions of change, we leverage the definition proposed by Latham and Crandall (1991) that conditions of change are any variables that affect trainee outcome expectation that is primarily determined by persons or events external to the immediate supervisor or job role. Since there are many views on what exactly is the most important of these variables (cf. Latham, 1988; Salinger, 1973), the main issue we wish to address is that the organizational systems in place must be congruent and supportive for training to transfer (Latham & Crandall, 1991).

Where Do We Go From Here?

While the framework just described provides a foundation for team training designed to enhance reliability and error management, it also raises several questions for future research. The following points of consideration are offered as food for thought to drive both theoretical and practical discussions of the relationships between teamwork, team training, errors, and reliability.

> Consideration 1: Determine which teamwork competencies and processes are key "barriers" that keep potential adverse events from progressing into full-blown errors and are thus most effective in optimizing reliability and error management.
>
> Theoretical and practical models of error demonstrate that errors are not spontaneous incidents; they usually begin with some adverse event and, if left unchecked, can align with other adverse conditions to result in full-blown error. As detailed in this chapter, teamwork competencies are barrier mechanisms useful in managing adverse events. For example, closed-loop communication can keep an incorrect dosage of medicine from being administered to a patient. As such, teamwork could effectively manage an event (e.g., nurse thought he heard "500-mg dosage," but when he verbally checked, the doctor was able to clarify that the dosage should be "50 mg") and keep it from progressing into administration of an incorrect dosage to a patient. However, there is a broad range of teamwork competencies that theoretically should have an impact on reliability and error management. Clear empirical evidence linking specific teamwork competencies to reliability, however, must be established.
>
> Consideration 2: Develop new mechanisms for quantifying and diagnosing errors beyond simple error rate calculations.

We must rethink how errors are measured to measure validly and understand the impact of team training on reliability and error management. Calculating organizational errors in terms of traditional errors rates is problematic. Underreporting and low base rates contribute to the severely limited diagnosticity of such metrics. In addition, little organizational learning can occur from analyses that rely completely on error rates. They provide no indication of near misses or insight on behaviors that kept near misses from escalating to full error status. This issue plays a role in the fact that existing literature has yet to provide much solid evidence regarding the impact of team training on actual error reduction (e.g., Leape & Berwick, 2005; Salas et al., 2001). However, several studies in the health care field suggest innovative methods for calculating errors from a more comprehensive, systems viewpoint. For example, Pronovost and colleagues (2006) present a comprehensive unit-based safety program (CUSP) model for measuring culture and helping organizations learn from mistakes that are not captured in traditional error rate calculations, such as near misses. Mann and colleagues (2006) also present three unique metrics designed as quality improvement tools in obstetrics. Specifically, their "adverse outcome index" captures the percentage of opportunities (e.g., deliveries) in which one or more adverse events occurred. The metrics contained in the index were assigned a numeric score based on ratings of these events by subject matter experts (SMEs). These numeric scores are then used to calculate a "weighted adverse outcome score" for each patient. A third "severity index" is also calculated for each instance when two or more adverse events occurred. Creating such indices will increase much-needed variability in our measures of errors and increase our ability to learn from such events. Key remaining questions include the degree to which various measurement methods are appropriate across a range of industries and the level of specificity required.

Consideration 3: Determine which combinations of team training strategies are most effective in enhancing reliability and error management.

The team training strategies discussed vary in the degree to which they tap specific team-level behaviors related to errors given that, by design, they were developed to concentrate on particular team behaviors. For example, team coordination training is designed to tap skills related to cooperation, coordination, and communication (i.e., backup behavior, assertiveness, and generation of shared mental models), whereas guided team self-correction training is designed to specifically target competencies related to feedback and team learning (cf. Smith-Jentsch, Cannon-Bowers, Tannenbaum, & Salas, 2008). Although this notion is relatively well supported, questions regarding the incremental validity of strategies targeting specific teamwork skills have not been thoroughly addressed. For example, from a definitional and theoretical perspective, cross-training, perceptual contrast training, and team coordination training all tap competencies related to backup behavior, generation of shared mental models, and mutual performance monitoring. However, little empirical evidence compares the effectiveness of these methods for training such behaviors (exceptions include Salas, DiazGranados, Klein et al., 2008). Both practitioners and theorists alike would benefit from direct comparisons of such strategies for reducing errors.

Consideration 4: Understand how aspects of the transfer climate (e.g., supervisory support, opportunity to perform, and conditions of change) moderate the effects of the various instructional strategies on team reliability and error management.

Helmreich and Merrit (1999) underscored that employees will report errors and embrace them as opportunities for learning because (a) they think it is their responsibility, (b) coworkers encourage the behavior, (c) they believe management will take immediate action to address the issue, and (d) management provides continued feedback on error-related issues related to job performance. The transfer climate literature has pinpointed several factors that affect transfer of training, including supervisor, peer, and subordinate support (Smith-Jentsch, Salas, & Brannick, 2001; Tracey, Tannenbaum, & Kavanagh, 1995); opportunities to perform trained KSAs (Ford, Quiñones, Sego, & Sorra, 1992; Quiñones, Ford, Sego, & Smith, 1995); rewards (or punishments) for use of trained competencies (Smith-Jentsch, Blickensderfer, Salas, & Cannon-Bowers, 2000); and resources (Rouiller & Goldstein, 1993). However, little work has focused on team training for reliability and error management. Fruitful future research efforts could be directed toward investigation of organizational and team factors that moderate transfer of team training KSAs to the actual job.

Consideration 5: Create a more solid understanding of how the feedback loop between team training evaluation and reliability/errors facilitates team learning.

High-reliability theory posits that teams learn through detailed analysis of errors, near misses, and successes, specifically in relation to dimensions of preoccupation with failure and commitment to resilience. Future literature is needed, however, to investigate empirically the impact of incorporating such information into this early analysis on training effectiveness.

In addition, we suggest that TCAs and RCAs should become a preemptive methodology as opposed to a reactive method for continuing to enhance reliability. In essence, organizations and teams should conduct these analyses on a regular schedule, regardless of whether a catalyst event occurs. The key to this link, however, is error reporting. In many organizations, the underreporting of errors is assumed to be a significant problem, and the reporting of near misses is almost nonexistent. For example, it is estimated that 50% of medical errors go unreported every year (Leape, 1994). Furthermore, error-reporting systems are often self-report systems that rely on voluntary reports by frontline employees and have rarely been designed to capture near misses (e.g., Leape, 2002), despite the inherent learning value of near misses. Without explicit encouragement of error reporting (i.e., preoccupation with failure), the potential for underreporting is great, thereby undermining the relationship between errors and organizational outcomes. For example, it is not uncommon for organizations to report minimum numbers of errors and demonstrate no detectable relationship with organizational outcomes. Only reported errors can be correlated with organizational outcomes; therefore, low reporting may hinder detection of an actual relationship. We argue that a culture of reliability encouraging error reporting without blame and developed through team training will underlie the relationship between reported errors, organizational outcomes, and team performance.

Conclusion

In this chapter, we have attempted to bridge the gap between several streams of research related to attempts to enhance reliability and error management, suggesting further exploration of team training as a mechanism for creating organizational reliability and optimal error management. Our framework attempts to shed light on the relationship between team training and error reduction/management. To that end, we suggested several points for further investigation of these relationships. We hope to stimulate evolution of a broader understanding regarding initiation and maintenance of reliability and effective team error mitigation/management.

Acknowledgment

This work was supported by funding from the Department of Defense (Award Number W81XWH-05-1-0372). All opinions expressed in this chapter are those of the authors and do not necessarily reflect the official opinion or position of the University of Central Florida or the Department of Defense.

References

Aguinis, H., & Kraiger, K. (2009). Benefits of training and development for individuals and teams, organizations, and society. *Annual Review of Psychology, 60*, 451–474.

Alonso, A., Baker, D. P., Holtzman, A., Day, R., King, H., Toomey, L., et al. (2006). Reducing medical error in the military health system: How can team training help? *Human Resource Management Review, 16*, 396–415.

Baker, D. P., Day, R., & Salas, E. (2006). Teamwork as an essential component of high-reliability organizations. *Health Services Research, 41*, 1576–1598.

Baldwin, T. T., & Ford, J. K. (1988). Transfer of training: A review and direction for future research. *Personnel Psychology, 41*, 63–105.

Baldwin, T. T., & Magjuka, R. J. (1991). Organizational training and signals of importance: Linking pretraining perceptions to intentions to transfer. *Human Resource Development Quarterly, 2*, 25–36.

Batt, R. (2004). Who benefits from teams? Comparing workers, supervisors, and managers. *Industrial Relations, 43*(1), 183–212.

Beaubien, J. M., & Baker, D. P. (2004). The use of simulation for training teamwork skills in healthcare: How low can you go? *Quality and Safety in Healthcare, 13*, i51–i56.

Bierly, P., & Spender, J. C. (1995). Culture and high reliability organizations: The case of the nuclear submarine. *Journal of Management, 21*, 639–656.

Blickensderfer, E. L., Cannon-Bowers, J. A., & Salas, E. (1997). Theoretical bases for team self-correction: Fostering shared mental models. In M. Beyerlein, D. Johnson, & S. Beyerlein (Eds.), *Advances in interdisciplinary studies in work teams series* (Vol. 4, pp. 249–279). Greenwich, CT: JAI Press.

Blickensderfer, E., Cannon-Bowers, J. A., & Salas, E. (1998). Cross-training and team performance. In J. A. Cannon-Bowers & E. Salas (Eds.), *Making decisions under stress: Implications for individual and team training* (pp. 299–311). Washington, DC: American Psychological Association.

Bogner, M. S. (1994). *Human error in medicine*. Hillsdale, NJ: Erlbaum.

Bogner, M. S. (2000). A systems approach to medical error. In C. Vincent & B. DeMol (Eds.), *Safety in medicine* (pp. 83–100). Amsterdam: Pergamon.

Bogner, M. S. (2004). *Misadventures in health care: Inside stories*. Mahwah, NJ: Erlbaum.

Bowers, C. A., Blickensderfer, E. L., & Morgan, B. B., Jr. (1998). Air traffic control specialist team coordination. In M. W. Smolensky & E. S. Stein (Eds.), *Human factors in air traffic control* (pp. 215–236). San Diego, CA: Academic Press.

Brannick, M. T., & Prince, C. (1997). An overview of team performance measurement. In M. T. Brannick, E. Salas, & C. Prince (Eds.), *Team performance assessment and measurement* (pp. 3–16). Mahwah, NJ: Erlbaum.

Bransford, J. D., Franks, J. J., Vye, N. J., & Sherwood, R. D. (1979). New approaches to instruction: Because wisdom can't be told. In S. Vosniadou & A. Ortony (Eds.), *Similarity and analogical reasoning* (pp. 470–497). Cambridge, UK: Cambridge University Press.

Broad, M., & Newstrom, J. W. (1992). *Transfer of training: Action packed strategies to ensure high payoff from training investments.* Reading, MA: Addison-Wesley.

Burke, C. S., Wilson, K. A., & Salas, E. (2005). The use of a team-based strategy for organizational transformation: Guidance for moving toward a high reliability organization. *Theoretical Issues in Ergonomic Science, 6*, 509–530.

Burke, M. J., & Day, R. R. (1986). A cumulative study of the effectiveness of managerial training. *Journal of Applied Psychology, 71*, 232.

Cannon-Bowers, J. A., Salas, E., Tannenbaum, S. I., & Mathieu, J. E. (1995). Toward theoretically based principles of training effectiveness: A model and initial empirical investigation. *Military Psychology, 7*(3), 141–164.

Cannon-Bowers, J. A., Tannenbaum, S. I., Salas, E., & Volpe, C. E. (1995). Defining team competencies and establishing team training requirements. In R. Guzzo, E. Salas, & Associates (Eds.), *Team effectiveness and decision making in organizations* (pp. 333–380). San Francisco: Jossey-Bass.

Carlson, R. A., Lundy, D. H., & Schneider, W. (1992). Strategy guidance and memory aiding in learning problem-solving skill. *Human Factors, 34*, 129–145.

Cohen, D. J. (1990). What motivates trainees. *Training and Development Journal, 36*, 91–93.

Cooke, N. J., Salas, E., Cannon-Bowers, J. A., & Stout, R. J. (2000). Measuring team knowledge. *Human Factors, 42*(1), 151–175.

Davies, J. M. (2001). Medical applications of crew resource management. In E. Salas, C. Bowers, & E. Edens (Eds.), *Improving teamwork in organizations: Applications of resource management training.* (pp. 265–282). Mahwah, NJ: Erlbaum Associates.

Dickinson, T. L., & McIntyre, R. M. (1997). A conceptual framework for teamwork measurement. In M. T. Brannick, E. Salas, & C. Prince (Eds.). *Team performance assessment and measurement* (pp. 19–43). Mahwah, NJ: Erlbaum.

Driskell, J. E., & Salas, E. (1992a). Can you study real teams in contrived settings? The value of small group research to understanding teams. In R. W. Swezey & E. Salas (Eds.), *Teams: Their training and performance* (pp. 101–124). Norwood, NJ: Ablex.

Driskell, J. E., & Salas, E. (1992b). Collective behavior and team performance. *Human Factors, 34*, 277–288.

Endsley, M. R. (1999). Situation awareness in aviation systems. In D. J. Garland & J. A. Wise (Eds.), *Handbook of aviation human factors. Human factors in transportation* (pp. 257–276) Mahwah, NJ: Erlbaum.

Entin, E. E., & Serfaty, D. (1999). Adaptive team coordination. *Human Factors, 41*, 312–325.

Five case studies on successful teams. (2002). *HR Focus, 79*(4), 18–20.
Ford, J. K., Quiñones, M. A., Sego, D. J., & Sorra, J. S. (1992). Factors affecting the opportunity to perform trained tasks on the job. *Personnel Psychology, 45,* 511–527.
Fowlkes, J. E., Dwyer, D. J., Oser, R. L., & Salas, E. (1998). Event-based approach to training. *The International Journal of Aviation Psychology, 8,* 209–221.
Frese, M., & Altman, A. (1989). The treatment of errors in learning and training. In L. Bainbridge & S. A. Quintanilla (Eds.), *Developing skills with information technology* (pp. 65–86). Chichester, UK: Wiley.
Glassop, L. I. (2002). The organizational benefits of teams. *Human Resources, 2,* 225–249.
Goldstein, I. L., & Ford, J. K. (2002). *Training in organizations: Needs assessment, development, and evaluation* (4th ed.). Belmont, CA: Wadsworth.
Gross, S. E. (1997). When jobs become team roles, what do you pay for? *Compensation and Benefits Review, 29*(1), 48–51.
Hannan, M. T., & Freeman, J. (1984). Structural inertia and organizational change. *American Sociological Review, 49,* 149–216.
Helmreich, R. L., & Merrit, A. C. (1999). *Culture at work in aviation and medicine.* Brookfield, VT: Ashgate.
Helmreich, R. L., Merrit, A. C., & Wilhem, J. A. (1999). The evolution of crew resource management training in commercial aviation. *International Journal of Aviation Psychology, 9*(1), 13–32.
Hofmann, D. A., Jacobs, R., & Landy, F. (1995). High-reliability process industries: Individual, micro, and macro organizational influences on safety performance. *Journal of Safety Research, 26*(3), 131–149.
Holding, D. H. (1991). Transfer of training. In J. E. Morrison (Ed.). *Training for performance: Principles of applied human learning* (pp. 93–125). Chichester, UK: Wiley.
International Civil Aviation Organization. (2002). *Line operations safety audit* (Doc No. 9803). Retrieved January 25, 2009, from http://www.icao.int/ANB/humanfactors/LUX2005/Info-Note-5-Doc9803alltext.en.pdf
Joint Commission on Accreditation of Healthcare Organizations. (2008, February 27). *Root causes for sentinel events, 2006.* Retrieved March 31, 2008, from http://www.jointcommission.org/SentinelEvents/Statistics/
Karl, K. A., O'Leary-Kelly, A. M., & Martocchio, J. J. (1993). The impact of feedback and self-efficacy on performance in training. *Journal of Organizational Behavior, 14,* 379–394.
Kozlowski, S. W. J., & Bell, B. S. (2003). Work groups and teams in organizations. In W. C. Borman, D. R. Ilgen, & R. J. Kilmoski (Eds.), *Handbook of psychology: Industrial and organizational psychology* (Vol. 12, pp. 333–375). London: Wiley.
Kozlowski, S. W. J., & Salas, E. (1997). A multilevel organizational systems approach for the implementation and transfer of training. In J. K. Ford (Ed.), *Improving training effectiveness in work organizations* (pp. 247–287). Hillsdale, NJ: Erlbaum.
Latham, G. P. (1988). Human resource training and development. *Annual Review of Psychology, 39,* 545–582.

Latham, G. P., & Crandall, S. (1991). Organizational and social factors. In J. Morrison (Ed.), *Training for performance: Principles of applied human learning* (pp. 259–295). Chichester, UK: Wiley.

Leape, L. L. (1994). Error in medicine. *Journal of the American Medical Association, 272,* 1851–1857.

Leape, L. L. (2002). Reporting of adverse events. *New England Journal of Medicine, 347,* 1633–1638.

Leape, L., & Berwick, D. (2005). Five years after "To err is human": What have we learned? *Journal of the American Medical Association, 293,* 2384–2390.

Lorenzet, S. J., Salas, E., & Tannenbaum, S. I. (2005). Benefitting from mistakes: The impact of guided errors on learning, performance, and self efficacy. *Human Resource Development Quarterly, 16,* 301–322.

Mann, S., Pratt, S., Gluck, P., Nielsen, P., Risser, D., Greenberg, P., et al. (2006). Assessing quality in obstetrical care: Development of standardized measures. *Joint Commission Journal on Quality and Patient Safety, 32,* 497–505.

Marks, M. A., Sabella, M. J., Burke, C. S., & Zaccaro, S. J. (2002). The impact of cross-training on team effectiveness. *Journal of Applied Psychology, 87,* 3–13.

Marx, D. A., & Slonim, A. D. (2003). Assessing patient safety risk before the injury occurs: An introduction to socio-technical probabilistic risk modeling in healthcare. *Quality and Safety in Health Care, 12*(Suppl. 2), ii-33–ii-38.

Marx, D. A., & Westphal, J. E. (2008). Socio-technical probabilistic risk assessment: Its application to aviation maintenance. *The International Journal of Aviation Psychology, 18*(1), 51–60.

McIntyre, R. M., & Salas, E. (1995). Measuring and managing for team performance: Emerging principles from complex environments. In R. Guzzo & E. Salas (Eds.), *Team effectiveness and decision making in organizations* (pp. 149–203). San Francisco: Jossey-Bass.

Mohaghegh, Z., Kazemi, R., & Mosleh, A. (2009). Incorporating organizational factors into probabilistic risk assessment (PRA) of complex socio-technical systems: A hybrid technique formalization. *Reliability Engineering and System Safety, 94,* 1000–1018.

Moray, N. (1994). Error reduction as a systems problem. In M. S. Bogner (Ed.), *Human error in medicine* (pp. 67–91). Hillsdale, NJ: Erlbaum.

Mullen, J. E., & Kelloway, E. K. (2009). Safety leadership: A longitudinal study of the effects of transformational leadership on safety outcomes. *Journal of Occupational Psychology, 82,* 253–272.

Musson, D., & Helmreich, R. (2004). Team training and resource management in healthcare: Current issues and future directions. *Harvard Health Policy Review, 5*(1), 25–35.

National Transportation Safety Board. Bureau of Accident Investigation (1979). *Aircraft accident report: United Airlines, Inc., Douglas DC-8-54, N8082U, Portland, Oregon, December 28, 1978* (NTSB-AAR-79-7). Washington, DC: National Transportation Safety Board.

National Transportation Safety Board. (n.d.). *Table 5. Accidents, fatalities, and rates, 1988 through 2007, for U.S. air carriers operating under 14 CFR 121, scheduled and nonscheduled Service (Airlines).* Retrieved June 16, 2008, from http://www.ntsb.gov/aviation/Table5.htm

Oser, R. L., Cannon-Bowers, J. A., Salas, E., & Dwyer, D. J. (1999). Enhancing human performance in technology-rich environments: Guidelines for scenario-based training. In E. Salas (Ed.), *Human/technology interaction in complex systems* (Vol. 9, pp. 175–202). Greenwich, CT: JAI Press.

Oser, R. L., MacCallum, G. A., Salas, E., & Morgan, B. B., Jr. (1989). *Toward a definition of teamwork: An analysis of critical team behaviors* (Technical Report 89–004). Orlando, FL: Naval Training Systems Center.

Peters, L. H., O'Connor, E. J., & Eulberg, J. R. (1985). Situational constraints: Sources, consequences, and future considerations. In K. Rowland & G. Ferris (Eds.), *Research in personnel and human resource management* (pp. 79–114). Greenwich, CT: JAI Press.

Phillips, J. J., & Phillips, P. P. (2008). *Beyond learning objectives: Expert strategies for developing learning objectives that get results*. Alexandria, VA: ASTD Press.

Powell, S. M., & Hill, R. K. (2006). My co-pilot is a nurse: Using crew resource management in the OR. *AORN Journal, 83*(1), 178–202.

Prince, C., Oser, R. L., Salas, E., & Woodruff, W. (1993). Increasing hits and reducing misses in CRM/ LOS scenarios: Guidelines for simulator exercise development. *The International Journal of Aviation Psychology, 3*(1), 69–82.

Pronovost, P. J., Berenholtz, S. M., Goeschel, C. A., Needham, D. M., Sexton, J. B., Thompson, D. A., et al. (2006). Creating high reliability in health care organizations. *Health Services Research, 41*(4, Part 2), 1599–1617.

Quinones, M. A., Ford, J. K., Sego, D. J., & Smith, E. M. (1995). The effects of individuals and transfer environment characteristics on the opportunities to perform trained tasks. *Training Research Journal, 1*, 29–48.

Reason, J. T. (1990). *Human error*. New York: Cambridge University Press.

Reason, J. T., & Hobbs, A. (2003). *Managing maintenance error: A practical guide*. Burlington, VT: Ashgate.

Roberts, K. H., & Bea, R. G. (2001a). Must accidents happen: Lessons from high reliability organizations. *Academy of Management Executive, 15*, 70–79.

Roberts, K. H., & Bea, R. G. (2001b). When systems fail. *Organizational Dynamics, 29*, 179–191.

Roberts, K. H., & Rousseau, D. M. (1989). Research in nearly failure-free, high-reliability organizations: Having the bubble. *IEEE Transactions on Engineering Management, 36*(2), 132–139.

Rochlin, G. L. (1989). Informal organizational networking as a crisis-avoidance strategy: U.S. naval flight operations as a case study. *Industrial Crisis Quarterly, 3*, 159–176.

Rooney, J. J., & Vanden Heuvel, L. N. (2004). Root cause analysis for beginners. *Quality Progress, 37*(7), 45–53.

Rosen, M., Salas, E., Silvestri, S., Wu, T. S., Lazzara, E. H. (2008). A measurement tool for simulation-based training in emergency training: The simulation module for assessment of resident targeted event responses (SMARTER) approach. *Journal of the Society for Simulation and Healthcare, 3*(3), 170–179.

Rosen, M. A., & Salas, E. (2008). Beyond the bells and whistles: When simulation-based team training works best. *Forum, 26*(4), 6–7.

Rouiller, J. Z., & Goldstein, I. L. (1993). The relationship between organizational transfer climate and positive transfer of training. *Human Resource Development Quarterly, 4*, 377–390.

Salas, E., Burke, C. S., Bowers, C. A., & Wilson, K. A. (2001). Team training in the skies: Does crew resource management (CRM) training work? *Human Factors, 43*, 641–674.

Salas, E., & Cannon-Bowers, J. A. (2000). Designing training systems systematically. In E. A. Locke (Ed.), *The Blackwell handbook of principles of organizational behavior* (pp. 43–59). Malden, MA: Blackwell.

Salas, E., & Cannon-Bowers, J. A. (2003). The science of training: A decade of progress. *Annual Review of Psychology, 52*, 471–499.

Salas, E., Cannon-Bowers, J. A., & Johnston, J. H. (1997). How can you turn a team of experts into an expert team? Emerging training strategies. In C. Zsambok & G. Klein (Eds.), *Naturalistic decision making* (pp. 359–370). Hillsdale, NJ: LEA.

Salas, E., DiazGranados, D., Klein, C., Burke, C. S., Stagl, K. C., Goodwin, G. F., & Halpin, S. M. (2008). Does team training improve team performance? A meta-analysis. *Human Factors, 50*(6), 903–933.

Salas, E., DiazGranados, D., Weaver, S. J., & King, H. (2008). Does team training work? Principles for healthcare. *Academic Emergency Medicine, 15*, 1002–1009.

Salas, E., Dickinson, T. L., Converse, S. A., & Tannenbaum, S. I. (1992). Toward an understanding of team performance and training. In R. J. Swezey & E. Salas (Eds.), *Teams: Their training and performance* (pp. 3–29). Norwood, NJ: Ablex.

Salas, E., Fowlkes, J. E., Stout, R. J., Milanovich, D. M., & Prince, C. (1999). Does CRM training improve teamwork skills in the cockpit? Two evaluation studies. *Human Factors, 41*, 326–343.

Salas, E., Kosarzycki, M. P., Tannenbaum, S. I., & Carnegie, D. (2004). Principles and advice for understanding and promoting effective teamwork in organizations. In R. J. Burke & C. Cooper (Eds.), *Leading in turbulent times* (pp. 95–120). Malden, MA: Blackwell.

Salas, E., Nichols, D. R., & Driskell, J. E. (2007). Testing three team training strategies in intact teams: A meta-analysis. *Small Group Research, 38*, 471–488.

Salas, E., Prince, C., Bowers, C., Stout, R., Oser, R. L., & Cannon-Bowers, J. A. (1999). A methodology for enhancing crew resource management training. *Human Factors, 41*, 161–172.

Salas, E., Rhodenizer, L., & Bowers, C. A. (2000). The design and delivery of CRM training: Exploiting available resources. *Human Factors, 42*, 490–511.

Salas, E., Rosen, M., Burke, C. S., Goodwin, G. F., & Fiore, S. (2006). The making of a dream team: When expert teams do best. In K. A. Ericsson, N. Charness, R. Hoffman, & P. Fletovich (Eds.), *The Cambridge handbook of expertise and expert performance* (pp. 439–453). New York: Cambridge University Press.

Salas, E., Rosen, M. A., Burke, C. S., & Goodwin, G. F. (2009). The wisdom of collectives in organizations: An update of the teamwork competencies. In E. Salas, G. F. Goodwin, & C. S. Burke (Eds.), *Team effectiveness in complex organizations* (pp. 39–79). New York: Routledge.

Salas, E., Wilson, K. A., Burke, C. S., & Bowers, C. A. (2002). Myths about crew resource management (CRM) training: Myths to avoid. *Ergonomics in Design*, 20–24.

Salas, E., Wilson, K. A., Murphy, C., King, H., & Salisbury, M. (2008). Communicating, coordinating, and cooperating when the life of others depends on it: Tips for teamwork. *Joint Commission Journal on Quality and Patient Safety, 34*, 333–341.

Salinger, R. D. (1973). *Disincentives to effective employee training and development.* Washington, DC: Bureau of Training: U.S. Civil Service Commission.

Schwartz, D. L., & Bransford, J. D. (1998). A time for telling. *Cognition and Instruction, 16,* 475–522.

Seamster, T. L., & Kaempf, G. L. (2001). Identifying resource skills management skills for airline pilots. In E. Salas, C. A. Bowers, & E. Edens (Eds.), *Improving teamwork in organizations: Applications of resource management training* (pp. 9–30). Mahwah, NJ: Erlbaum.

Serfaty, D., & Entin, E. E. (1998). Team coordination training. In J. A. Cannon-Bowers & E. Salas (Eds.), *Making decisions under stress: Implications for individual and team training* (pp. 299–311). Washington, DC: American Psychological Association.

Smith-Jentsch, K. A., Blickensderfer, E., Salas, E., & Cannon-Bowers, J. A. (2000). Helping team members help themselves: Propositions for facilitating guided team self correction (pp. 55–72). In M. M. Beyerlein, D. A. Johnson, & S. T. Beyerlein (Eds.), *Advances in interdisciplinary studies of work teams* (Vol. 6, pp. 55–72). Greenwich, CT: JAI Press.

Smith-Jentsch, K. A., Cannon-Bowers, J. A., Tannenbaum, S. I., & Salas, E. (2008). Guided team self correction. *Small Group Research, 39,* 303–327.

Smith-Jentsch, K. A., Payne, S., & Johnston, J. H. (1996). *Guided team self-correction: A methodology for enhancing experiential team training.* Paper presented at the 11th annual conference of the Society for Industrial and Organizational Psychology, When, how, and why does practice make perfect? K. Smith-Jentsch, Chair), San Diego, CA.

Smith-Jentsch, K., Salas, E., & Baker, D. P. (1996). Training team performance-related assertiveness. *Personnel Psychology, 49,* 909–936.

Smith-Jentsch, K. A., Salas, E., & Brannick, M. T. (2001). To transfer or not to transfer? Investigating the combined effects of trainee characteristics, team leader support, and team climate. *Journal of Applied Psychology, 86,* 279–292.

Smith-Jentsch, K. A., Zeisig, R. L., Acton, B., & McPherson, J. A. (1998). Team dimensional training: A strategy for guided team self-correction. In J. A. Cannon-Bowers & E. Salas (Eds.), *Making decisions under stress: Implications for individual and team training* (pp. 271–291). Washington, DC: APA.

Stout, R. J., Cannon-Bowers, J. A., & Salas, E. (1996). The role of shared mental models in developing team situational awareness: Implications for training. *Training Research Journal, 2,* 85–116.

Stout, R. J., Cannon-Bowers, J. A., Salas, E., & Milanovich, D. M. (1999). Planning, shared mental models, and coordinated performance: An empirical link is established. *Human Factors, 41,* 61–71.

Tannenbaum, S. I., Smith-Jentsch, K. A., & Behson, S. J. (1998). Training team leaders to facilitate team learning and performance. In J. Cannon-Bowers & E. Salas (Eds.), *Making decisions under stress. Implications for individual and team training* (pp. 247–270). Washington, DC: American Psychological Association.

Tracey, B. J., Tannenbaum, S. I., & Kavanagh, M. J. (1995). Applying trained skills on the job: The importance of the work environment. *Journal of Applied Psychology, 80,* 239–252.

Vanden Heuvel, L. N., Lorenzo, D. K., Hanson, W. E., Jackson, L. O., Rooney, J. R., & Walker, D. A. (2008). *Root cause analysis handbook: A guide to efficient and effective incident investigation* (3rd ed.) Brookfield, CT: Rothstein.

Volpe, C. E., Cannon-Bowers, J. A., Salas, E., & Spector, P. E. (1996). The impact of cross-training on team functioning: An empirical investigation. *Human Factors, 38*, 87–100.

Weaver, S. J., Wildman, J. L., & Salas, E. (2009). How to build expert teams: Best practices. In R. J. Burke & C. L. Cooper (Eds.), *The peak performing organization* (pp. 129–156). New York: Routledge Academic.

Weick, K. E. (1987). Organizational culture as a source of high reliability. *California Management Review, 29*, 112–127.

Weick, K. E., & Sutcliffe, K. M. (2001). *Managing the unexpected*. San Francisco: Jossey-Bass.

Weick, K. E., Sutcliffe, K. M., & Obstfeld, D. (1999). Organizing for high reliability: Processes of collective mindfulness. *Research in Organizational Behavior, 21*, 81–123.

Wilson, K. A., Burke, C. S., Priest, H. A., & Salas, E. (2005). Promoting health care safety through training high-reliability teams. *Quality and Safety in Healthcare, 14*, 303–309.

Wilson, K. A., Salas, E., Priest, H. A., & Andrews, D. (2007). Errors in the heat of battle: Taking a closer look at shared cognition breakdowns through teamwork. *Human Factors, 49*(2), 243–256.

Zaccaro, S. J., Heinen, B., & Shuffler, M. (2009). Team leadership and team effectiveness. In E. Salas, G. Goodwin, & C. S. Burke (Eds.), *Team effectiveness in complex organizations* (pp. 83–112). New York: Taylor & Francis.

7

Learning Domains: The Importance of Work Context in Organizational Learning From Error

Lucy H. MacPhail and Amy C. Edmondson

A growing body of research has increased scholarly and managerial awareness of the enormous potential for organizations to learn from errors (Edmondson, 1996; Keith & Frese, 2008; Van Dyck, Frese, Baer, & Sonnentag, 2005). Errors come in a variety of sizes and types, with a corresponding variety of potential lessons. Learning these lessons, however, is not easy. Organizations face psychological, social, technical, and practical barriers that often conspire to prevent them from extracting many or all of the lessons their errors provide (Argyris, 1990; Cannon & Edmondson, 2001). However, some groups and organizations do better than others in overcoming social and psychological barriers to learning from error (Edmondson, 1996).

This chapter investigates the range of work contexts in which errors occur in organizations and the implications of this variation for organizational learning from error. By *organizational learning from error*, we refer to organizational activities that both build understanding of what went wrong to cause an error and identify ways to prevent the same or similar errors from occurring in the future. We suggest that different kinds of work give rise to different conditions of error, and these distinctions influence which organizational approach and actions are best to maximize potential learning from an error. For example, an error committed and recognized quickly by an individual carrying out a well-understood task on an assembly line presents different learning implications than an error whose cause is not clear that occurs when multiple groups interact during an unusual patient care situation in a hospital. Given the variety of work situations in organizations, researchers and managers need to identify the essential contextual features that affect both the occurrence of errors and the best strategies for learning from them. In practical terms, determining who should be involved in discussions to analyze and correct errors can be a source of stress and contention in organizations. This chapter takes a first step toward addressing these issues by proposing a framework categorizing organizational work into four conceptually distinct domains, across which the nature and implications of error differ substantially. We use our framework to suggest different learning approaches for each error domain. Our aim is to build understanding of the appropriate classroom for an error—to shed light on how the nature of error differs across work contexts and how to focus an organizational learning process accordingly to maximize its benefits.

Consider a case that emerged in our research. In a major teaching hospital, we uncovered substantial disagreement among leaders about how best to process and learn from adverse medical events—errors and undesired outcomes that occur in the process of treating patients. Learning from errors was an organizational priority for the hospital, which had created a multilevel quality control system comprised of departmental and interdepartmental review committees. Contention arose over the types of errors that should be reviewed and remedied locally within a department versus those that required interdisciplinary input and broader organizational exposure.

Some physicians and quality managers found value in comprehensive review of *all* serious, unexpected adverse events that occurred in the hospital. Proponents of an inclusive review process argued that many errors involving a single department carried lessons that translated beyond that clinical area, and thus perspectives from multiple disciplines enriched discussions of root cause and corrective action.

Other clinicians believed that involving multiple disciplinary areas in the review of events was often unhelpful, wasting time without offering useful learning. They advocated a more contained, expert-based review process. One physician expressed frustration that certain adverse events in her department were inappropriately forced to senior-level multidisciplinary review groups when they could be more efficiently managed internally, attributing the problem to interdepartmental politics—in particular, the satisfaction of other groups in exposing the shortcomings of her high-performing department. Another physician did not believe that the unique challenges of treating patients in his department were sufficiently understood by clinicians outside his discipline when discussing errors, leading them to make inaccurate inferences about the causes of errors in his area. A department chair preferred to address errors involving workflow between his department and another outside formal review channels, finding that direct off-line communications with that department's chair resulted in faster and more productive solutions.*

---

* This case is described in an unpublished dissertation (MacPhail, 2010), available from the authors.

Which view is right? Should errors be reviewed locally within a department or broadly by interdisciplinary groups spanning multiple departments? Should the group convening to review an error include senior leadership, and if so, in what way? How should corrective steps to address the error be determined and then applied? In this chapter, we argue that the answer depends on the error—specifically, on the work context of the error. The best organizational learning model—whether local and contained, organizational and inclusive, or somewhere in between—will vary with core dimensions of this context, and organizational strategies for learning from error must vary accordingly.

We use the term *learning domain* to refer to essential features of the work context in which errors occur that may influence optimal organizational approaches to learning from them. We identify four archetypal domains: task execution, judgment, interpersonal coordination, and system interactions. The nature of work in these four domains ranges from well-specified, well-understood tasks and work processes that give rise to repeating routines contained within a defined group or department, involving minimal uncertainty (e.g., assembly lines, call centers); to tasks within a defined group involving an unfamiliar process or situation, requiring improvised decision making (e.g., new need or attribute of a customer); to those in which tasks are routine within one or more groups but completing the task requires individuals from these groups to actively coordinate their actions with those of other groups (e.g., medical care for a patient with multiple diseases); to those involving several groups confronting novel processes and novel interactions, in which "right" answers are often not known, and new knowledge must be developed to achieve desired outcomes (e.g., implementation of a new technology requiring coordination among multiple departments). Work in these domains differs in levels of predictability and complexity, and our proposed strategies for organizational learning from errors in each domain differ accordingly.

Errors in Organizations

Background

An *error* is an action that unintentionally deviates from a plan or accepted standard of performance and fails to achieve its desired goal, for which the failure was potentially preventable. Consistent with Hofmann and Frese (Chapter 1, this volume), we define an error in terms of action, not consequences of an action. An error as an action is distinct from its impact on organizational performance or on the customer or end user;

that impact may be severe or trivial but does not influence the nature of the error itself. While errors may sometimes result in desired outcomes (accidental discoveries, for example), we focus exclusively on errors whose outcomes—however harmful or inconsequential—are undesired. Thus, all failures are not errors, and all errors do not result in harm. More than simply "something that goes wrong" in an organization, an error refers to human behavior that is theoretically preventable—judgments, decisions, or actions that contribute to an unexpected, unintended result. An undesired outcome that is not influenced by human behavior is an *unavoidable adverse event* and is not attributable to error, and behavior that intentionally deviates from a standard is either *misconduct* or *experimentation*, depending on the context. Errors, by contrast, carry potentially negative consequences, either causing harmful impact or threatening to do so under other conditions if repeated.

Existing classification approaches have differentiated errors on numerous dimensions, including severity (such as harm or "near miss"; Morimoto, Gandhi, Seger, Hsieh & Bates, 2004); nature of outcome (e.g., medication overdose, intraoperative complication; Chang, Schyve, Croteau, O'Leary & Loeb, 2005; others); and cause (e.g., fatigue, environmental factors; Rubin, George, Chinn, & Richardson, 2003; Wiegmann & Shappell, 2003). Many existing error typologies are focused on identifying and rectifying the technical sources of error, thereby creating useful distinctions that can inform prevention strategies in organizations. For the most part, however, they are silent on the question of learning process—how an error should be reviewed and how lessons from it should be extracted and applied to optimize learning from the experience. Despite a body of work on analyzing failures and problem-solving activities, such as conducting root cause analysis, strategies for responding to varying kinds of organizational errors are poorly differentiated, leaving us well poised to examine errors but ill equipped to learn from them systematically.

Although some research has examined conditions that enable error reporting, showing, for example, that an interpersonal climate of psychological safety allows people to speak up about errors, this work has not delved deeply into what organizations do with errors once identified (Edmondson, 1996, 2003). The few recent studies that have investigated this question suggest that the answer is often, unfortunately, not much. Exploratory research on this question in health care organizations (Anderson et al., 2010; Ramanujam, Sirio, Keyser, & Thompson, 2009) found that reported errors often are not analyzed carefully or met with corrective actions to prevent them from reoccurring. To our knowledge, prior academic work has not addressed the question of whether and, if so, how organizational learning strategies in the aftermath of error detection should vary based on the nature of the error.

Discussing Error

Another stream of research has investigated interpersonal fear as a barrier to effective error management. When the task environment is characterized by reciprocal interdependence, active, real-time coordination is needed for effective performance, and when individuals in such environments are reluctant to raise concerns or communicate negative information to others whose performance relies on them, errors in coordination are more likely (e.g., Edmondson, 1996). Research in hospitals and other organizational settings has thus shown that psychological safety fosters error reporting and creates the conditions in which people are more able to interact mindfully and openly (e.g., Edmondson, 1999, 2003).

A central finding of this work is that psychological safety—a key dimension of interpersonal climate related to error—varies significantly at the group level of analysis (Edmondson, 1999). Psychological safety allows people to have open, data-driven conversations about errors, the processes that led to them, and the opportunities for improvement. Its significant variance across groups means that, even within a single organization, attitudes and behaviors related to error can vary widely across groups or departments. Error management, therefore, can be seen as a local phenomenon, and structuring a collective learning process at the team or group level is vital to effective organizational learning from error (Edmondson, 2002). Although human beings are endowed with both desire and ability for learning, neither small groups nor large organizations learn automatically. Interpersonal risk inhibits some of the necessary behaviors, but organizational routines tend to endure and have a permanence of their own, independent of the actors who engage in them (Gersick & Hackman, 1990; Levitt & March, 1988).

Organizational Learning From Error

Building on this prior work, this chapter suggests that a one-size-fits-all approach to learning from errors is unlikely to ensure that the lessons from an error are effectively identified and applied to work processes, training, and other organizational practices to prevent future errors. Although most organizations will undertake more thorough and inclusive reviews of errors as the severity of consequences increases, this intuitive response does not ensure effective learning or prevention of future errors, and it is overly reactive—errors with minor or narrowly avoided consequences when they initially occur may be overlooked until they eventually cause serious harm. Short of examining errors with major adverse outcomes, clarity about which errors need fuller reviews has remained elusive in most organizations. In the major teaching hospital previously discussed, for example, many clinical leaders recognized that all errors were not conducive to multidisciplinary review, and all disciplines did not benefit from involvement in discussions of all errors. Sorting errors accordingly, however, was done informally and inconsistently and often not based on criteria rooted in a theory of optimal learning. For example, this included requiring multidisciplinary review of an error involving a slip by one individual performing a practiced skill because it resulted in serious harm or, conversely, examining an error involving organization-level breakdowns locally within one or two departments because it did not result in harm and thus was not perceived as sufficiently meaningful to warrant multidisciplinary attention.

In this chapter, we offer a theory-driven approach to help maximize the value of structures and activities intended to promote learning from errors. While one may make a compelling case that almost all errors contain learning that may be extracted and transferred beyond the groups immediately involved in the problem, the unavoidable time constraints that organizations confront—and the abundance of errors that even the safest and greatest organizations experience—render comprehensive, organization-wide review of all errors an unrealistic and unwise goal.

Learning Domains

A learning domain is the situational context in which an error occurs, which we argue in this chapter is a critical determinant of the optimal organizational approach to learning from that error. We begin by identifying four domains in complex organizations created by the intersection of two important dimensions of work. In a further section, we propose different learning strategies for each domain. The first dimension differentiating work environments is *process uncertainty*, which pertains to the level of knowledge available for specifying work processes that lead to desired outcomes. The second is *actor interdependence*, or the extent to which tasks are carried out by individuals working relatively independently or require multiple individuals or groups to coordinate their knowledge or actions to perform effectively. The four error domains that result from the intersection of these two dimensions are not distinguished by the outcome of the error or its specific type (e.g., serious or minor consequences, commission or omission) but rather by the nature of the work in which an error occurs: the organizational members involved in the task that produced the error and the extent to which process knowledge in this work context is well established, certain, and complete. Similarly, learning strategies associated with each domain depend not on the severity or impact of the error but rather on the extent to which knowledge developed from reviewing

Learning Domains

"what went wrong" translates beyond the immediate particulars of the error to other tasks, work groups, or systems in the organization.

Our focus is on collective work because it is increasingly the nature of goal-oriented action in complex, knowledge-intensive organizations. Collective work thus frequently requires the successful execution of multiple process steps that involve interdependent individuals, with the steps often spanning departments or units. Collective work occurs at the individual level, within a group, between two groups or dyadically, and at the organization level (the composite action of multiple individuals and groups). Thus we consider both individual and intragroup task performance as well as intergroup and organizational activities that contribute to collective work.

Learning Domain 1: Task Execution

The first learning domain, task execution, is defined by the intersection of well-developed or familiar process knowledge and individual action, as depicted in Figure 7.1. Errors in task execution are contained within a single function or group and occur during the performance of work that is routine or well understood. Tasks with low uncertainty and low complexity still pose the potential for error, typically due to limitations of the human body (e.g., fatigue, distraction) or lack of resources (e.g., labor, equipment) required for successful performance. Technical execution errors may occur with an involuntary "slip" while performing a routine skill or exchange, such as the accidental addition of a decimal place on the prescribed dosage for a medication, or by distraction due to contextual problems such as an inadequate supply of ventilators on a hospital unit. Such errors occur within well-defined and reasonably bounded work groups or units, and relationships between cause and effect in carrying out the work are usually well understood. Some may give rise to substantial harm; others may cause trivial or no immediate impact.

		Actor Interdependence	
		Low	High
Process Uncertainty	Low	Task Execution	Interpersonal Coordination
	High	Judgment	System interactions

FIGURE 7.1
Learning domains.

For example, an error in the task execution domain may occur during a routine surgery: An experienced surgeon accidentally damages an organ (greater harm), or a new nurse who has just joined the surgery department is unable to locate a sponge and delays the procedure (lesser harm). Knowledge of how to perform the surgery error free might be virtually complete—this surgery has been done successfully hundreds of times at this hospital, in this department, perhaps even by this surgeon—but one or more narrowly scoped, easily identified unintended deviations can interfere with the intended outcome. The error is generally confined to the individual physician or nurse interacting with the demands of the task, and the relevant information or standards in question are largely specific to and contained within the surgery department or surgical specialty group (e.g., how to perform a procedure effectively or where sponges for this operating room are stored), and the involvement of outside groups in the error is incidental.

In another health care example, Edmondson (1996) describes an individual error in a routine work context: A postoperative cardiac surgery patient was given intravenous lidocaine rather than the intended drug, heparin, a clot-preventing blood thinner routinely administered after heart surgery. Lidocaine, an anesthetic and heart rhythm stabilizer, fortunately did not harm the patient in this case; however, the absence of heparin could have been fatal. Here, the perfusionist, a medical technician in the operating room, hung the wrong bag in the intravenous drip. The heparin and lidocaine bags looked alike and were stored nearby. The routine, well-understood procedure of delivering the drug produced an undesired outcome because of an unintended behavior by the medical technician, the accidental selection of the wrong intravenous drip bag. Note, as described by Edmondson (1996), such an error in routine task execution is often followed by additional errors of omission (also occurring during routine tasks) when subsequent actors assume that the first task in a sequence has been done correctly—in this case, other caregivers in the hospital unit failing to check the medications dripping at the bedside—and thereby fail to catch and correct the original error.

Learning Domain 2: Judgment

Actors facing novel situations are vulnerable to errors in the context of judgment tasks, our second learning domain. This domain similarly pertains to individual action and is contained to work within a defined group or function, but unlike the setting for execution errors, the judgment domain involves individuals confronting unfamiliar processes that require some degree of customization, improvisation or new decision making in the moment. The task itself may be subject to variability, such as the application of existing knowledge to a new situation or customer, or the organization may be implementing a new policy or practice. Process novelty in this domain may also characterize situations in which an individual seeks to explore a new approach to a practiced routine, such as through experimentation. The errors in this domain may involve a poorly reasoned selection among alternatives (e.g., an individual makes the wrong decision when faced with an unfamiliar customer request or a new consideration in an otherwise ordinary situation); poorly developed alternatives (e.g., an individual does not recognize a viable solution to a problem); or incomplete knowledge when introducing a new process (e.g., an individual attempts to adhere to a newly prescribed protocol but does not understand or is untrained in one step of the process). Judgment errors also refer to experimentation with a new action, resulting in negative consequences that could have been anticipated in advance. Sources of process unfamiliarity that may have an impact on a work group or department include a change in process, individual inexperience with an existing process, or variability in customer needs.

An error in the judgment domain that we found in our research occurred when a nurse, caring for a postoperative patient, calculated a morphine dose based on standard concentrations of the potent drug (Edmondson, Roberto, & Tucker, 2001). However, a new, more highly concentrated variety of the drug recently had been introduced to the hospital, identifiable by a different cartridge. The nurse recognized that the cartridge was unfamiliar, but he did not alter his process for delivering the drug—an error in decision making that led him to calculate a potentially lethal dose. Fortunately, in this case, the error was quickly recognized, and the patient was not greatly harmed, but it serves as an apt illustration of the potential for errors when individuals confront novel situations in otherwise routine work, especially in knowledge-intensive service delivery contexts such as health care.

Another example is demonstrated by a stylized simulation experience called the electric maze, described by Lee, Edmondson, Thomke, and Worline (2004), in which individuals must discover a viable path through an unfamiliar maze using trial and error. When individuals experiment to find their way through the maze, team members must step forward to test each new location, and a location that produces a beeping sound when stepped on requires the individual to return to the beginning of the maze to try again. Those initial forays that are unsuccessful, as outcomes of pure trial and error, cannot be deemed errors. However, when an individual steps on a location that already has been shown to be unviable (found to beep), this qualifies as an error created by an individual selecting the wrong alternative while performing an unfamiliar task.

Learning Domain 3: Interpersonal Coordination

Errors in our third domain occur at the boundaries or intersections between groups, disciplines, or functions. This domain, interpersonal coordination, and our fourth domain, discussed in the next section, are distinct from the task execution and judgment domains by characterizing work that is inherently interdependent in nature. When a task requires individuals from different units, disciplines, or backgrounds to coordinate their actions to produce a shared outcome, errors can occur at those points in the work process that demand alignment or clarity of responsibilities and goals between groups. Note that interpersonal coordination errors occur despite familiar, well-developed process knowledge within the different groups. In the coordination domain, work is routine within each group, as in the task execution domain, and process knowledge of specific components of the overall task is mature. Although the need for coordination between groups may be well understood, the coordination itself often cannot be programmed—that is, coordination across boundaries must be actively negotiated in each case because the ways in which group routines come together present opportunities for errors to occur. Aspects of the work assumed by members of one group to be common knowledge, whether specialized skills or task objectives, may not be shared by members of another group, resulting in miscommunication or inconsistencies at points of intersection. Often, interpersonal coordination errors also arise at points in the task at which differentiation of roles and responsibilities among intersecting groups is unclear, introducing the potential for actions to be overlooked, duplicated, or poorly aligned. Unlike the previous two domains, the need to integrate knowledge or action steps across groups in this domain creates vulnerabilities, rendering tasks perceived to be familiar within each group more complex. Work in this domain is at least partially customized to the situation or customer and is vulnerable to errors created by faulty integration of the knowledge or processes of those in separate groups, who must adapt their behavior to these variable needs.

Errors in the interpersonal coordination domain are the simplest form of what have been labeled "system" errors in literature on errors in high-risk industries such as health care and aviation (Roberts, 1990b; Weick, Sutcliffe, & Obstfeld, 1999). Reason (1990) described system errors using "Swiss cheese" as a metaphor—multiple layers of holes that unfortunately line up, allowing separate errors to pass through without correction, from the outer layers of active sources (e.g., actors or operators directly involved in the error) to the underlying preconditions for error (e.g., training, supervision, organizational resources), each layer of which must be peeled back to understand the latent factors that contributed to the final undesired outcome. We use the term *coordination* to define our third domain to specify a type of system error in which process knowledge is high within groups but low between them, allowing for incomplete, inconsistent, or misunderstood exchanges at their intersection. In our domain framework, this simple form of system error is differentiated from errors involving unfamiliar processes spanning the organization, a complex form of system error that we describe in further detail in the discussion of our fourth and final domain.

An error of interpersonal coordination may involve a newly convened interdisciplinary action team (Edmondson, 2003) or a dedicated multidepartment group formed for a specific project or purpose, or it may simply arise in the course of everyday work that appears typical from the perspective of individuals in each department but actually is more complex due to the combination of groups involved or circumstances of the tasks. For example, consider a case we identified in our research of a patient who received different instructions from two of his doctors on the best daily dosage for his prescription medication to manage his diabetes. Both physicians, members of different departments in the same large medical group, were performing what they perceived to be routine work in their disciplinary areas—diagnosing an understood problem and prescribing a frequently used treatment—but their expert opinions on the right dosage for this patient differed, resulting in an inconsistent, inefficient care delivery process.*

Another example of an error in the interpersonal coordination domain occurred at a major teaching hospital in the Midwest; a nurse with considerable knowledge of a patient being treated had concerns about the medication being given to that patient but was not heard, or at least not listened to, in the confusion of bedside rounds. The nurse did not forcefully articulate concerns, while the physician and others in the team failed to ask for input from the nurse or from each other, leaving the team's knowledge about the patient and the situation incomplete and giving rise to a consequential medication error. More generally, research on medical error has documented that handoffs—between shifts, between clinical roles, or between departments or organizations—are highly vulnerable to error (Peters & Peters, 2007).

Learning Domain 4: System Interactions

Errors in our fourth and final domain, system interactions, span intraorganizational boundaries and arise in novel territory, such as new processes, unfamiliar activities, or unexpected situations. System interaction errors occur when multiple elements—groups, tasks, knowledge, external conditions—converge in unpredicted or unprecedented ways, resulting

---

* This example was identified as part of a research study on coordination in the health care setting, the findings of which may be found in MacPhail, Neuwirth, and Bellows (2009).

in errors that are organizational in nature (Edmondson, Roberto, Bohmer, Ferlins, & Feldman, 2005). In contrast to interpersonal coordination errors, system interaction errors involve complex systems rather than just people; errors are likely in this domain due to the unfamiliar or unscripted nature of the work and the need for improvisation both within and between organizational groups. While preventable in theory, since they can be traced to human decisions and actions, errors in this domain are challenging—and sometimes impossible—to anticipate; indeed, the fact that an error has even occurred (rather than simply an unavoidable bad event) may be unclear if the confluence of elements that produced the error are unique, such that preexisting standards of performance appear not to apply.

An error in the system interaction domain is well illustrated by the highly publicized tragic death of Betsy Lehman from a massive drug overdose she received while receiving cancer treatment at the Dana-Farber Cancer Institute in 1994. One of the most prestigious cancer hospitals in the world, the Dana-Farber was a recognized leader in advanced cancer care, and the combination of complex disease and innovation often meant that work at the hospital was novel and uncertain. Lehman's death thus was initially attributed to the unpredictable nature of cancer and the risky treatments used to combat it, but retrospective analysis of the event revealed that in fact Lehman had been healthy enough to return home to her family had she not received the amount of a potent anticancer drug intended to be infused over a 4-day period once every day for 4 days in a row instead—a lethal amount that caused her heart to fail. The fatal error was traced to an ambiguous medication administration protocol written by one of Lehman's physicians that was then misinterpreted by multiple nurses. Because, in this leading-edge center, innovative treatment regimens were often new or unfamiliar to staff, as was the case here, multiple interdependent members of Lehman's health care team failed to identify that the physician's order had been misread. Moreover, as a research center, the clinician researchers at Dana-Farber spanned boundaries frequently—between units, departments, and hospitals—introducing further complexity and many opportunities for miscommunication. In a statement to a patient safety conference several years later, Lehman's mother aptly described the series of aligned mistakes that preceded her daughter's death as a "team-blind error" (reprinted in Millenson, 2002). The collection of actions among interdependent actors performing in a context of high task uncertainty resulted in a fatal, avoidable outcome.*

While previous work on errors in high-risk industries, notably that of Perrow (1984), proposed that systems with multiple interacting elements

---

* The Betsy Lehman case was described by Richard Bohmer and Ann Winslow in a Harvard Business School case study, The Dana-Farber Cancer Institute (Case 699–025), Harvard Business School Press, 1999.

linked to each other in unforgiving ways are vulnerable to "normal accidents" that are virtually inevitable, recent research on high-reliability organizations (HROs) has argued that complex system errors such as these can be prevented consistently through individual and collective vigilance (Roberts, 1990a; Weick & Roberts, 1993). Despite the extreme level of risk such complex systems face, high-reliability theorists maintain that organizations can be structured and managed such that the potential for errors is vastly minimized. Our intention is not to contribute to the debate on whether error-free performance is feasible in complex systems (cf. Perrow, 1984; Weick & Sutcliffe, 2001) but rather to observe that system interactions in organizations provide not only risk but also rich opportunities for learning. As a result of their often-uncharted nature, complex system interaction errors contain new information about organizational vulnerabilities that are broadly relevant across varied disciplines or defined work areas.

Crossing Domains

Organizational action may also cross learning domains in the context of a single work process; the subactions of a multiple-step work process involve different degrees of uncertainty and actor interdependence. For example, diagnosing an elderly patient who arrives at a hospital emergency room with chest pain and an acute headache may involve high uncertainty: active cognition by the emergency department physician performing assessment who does not yet know the etiology of the pain (judgment domain) and integration of the cardiology, neurology, and gerontology departments in problem solving when the source requires further investigation (system interactions). Treating that patient once diagnosed is more likely to involve task execution, with or without coordination between groups (low uncertainty as an established or predetermined program for treating the diagnosis is enacted).

Learning Strategies

In our error domain framework (Table 7.1), the approach of an organization for learning from an error should be contingent on the work context of the error. Errors involving varying levels of work process familiarity and actor interdependence contain different types of learning for organizations, such that their lessons must be extracted and diffused using

TABLE 7.1

Characteristics of the Four Learning Domains

	Task Execution	Judgment	Interpersonal Coordination	System Interactions
Process uncertainty	Low	High	Low	High
Actor interdependence	Low	Low	High	High
Key features of domain	Well-specified work processes	Individual action in an at least partially novel situation	Coordination at the intersection of two or more well-specified work processes	Multiple actors engaged in unspecified or novel interactions
Common sources of error	Process deviation	Cognition/ decision making	Confusion across role boundaries	Unpredictable interactions
Approach to analyzing and learning from error	Focused, local analysis of the work process and its vulnerabilities	Participation of a focused team of those in similar field of expertise in analysis	Participation of all groups or roles involved in the work where the error occurs	Broad organizational participation to analyze and track system vulnerabilities
Corrective solutions	Standardization and error proofing; training in correct process	Training in skills for decision making under uncertainty	Push for greater standardization of processes and interactions	Push for organization-wide vigilance, heedful interrelating; awareness of system vulnerabilities

different strategies. Familiar tasks present ripe opportunities for targeted fixes and work process fine-tuning, while errors involving novel activities are less straightforward to diagnose and often demand creative problem solving to identify solutions; work localized to an individual or defined work group can be effectively understood and improved within the contained space in which error arises, while breakdowns in achieving intended goals that occur at the linkages between individuals or work areas require broader organizational member involvement in analysis and correction.

Learning in the Task Execution Domain

In the task execution domain, strategies for learning from errors can be local and targeted within the defined group that performs the function or task. The goal of the learning process is to develop understanding of what went wrong, and to identify strategies for preventing recurrence of the same kind of error, by streamlining tasks, adding safeguards, or ensuring that individuals executing those routines have the necessary skills and supports to perform effectively in the future. In this domain, one goal of the learning process is to identify vulnerabilities in the existing work processes, such as aspects of the task that are particularly challenging to execute reliably or supply issues with necessary inputs not always available when or in the condition specified by the work processes. Identified vulnerabilities to error provide information about where and how to make improvements to reduce the chances of error.

Learning activities may involve root cause analysis led by a team leader knowledgeable about the work that produced the error, similar to problem-solving approaches prescribed by high-performance work models such as the Toyota production system (Spear & Bowen, 1999). Given the specialized and contained nature of the work, involvement of outside groups in investigating an error in task execution is unlikely to generate useful information and, in many cases, may be counterproductive—wasting time and encouraging unhelpful and even potentially misguided contributions to the learning process by individuals unfamiliar with the task domain. Although the idea that an outsider can provide a new perspective on a situation is often considered a truism, we argue that the risks of wasted time and unnecessary confusion introduced when individuals from other groups are involved in examining an error in task execution outweigh the potential advantages of a fresh perspective to add critical insights others might miss. This is not to say that an organization may not wish to review its work processes periodically and consider whether significant changes are warranted by changes in strategy or in the market, but simply to argue that errors in the task execution domain will best be learned through focused analysis conducted by those with experience in the specific processes at hand.

Learning in the Judgment Domain

Similar to task execution errors, errors in the second domain, judgment, may require focused discipline-based analysis to diagnose the causes of error and develop solutions that minimize identified risks. Unlike errors in the task execution domain, however, lessons gleaned from analysis of judgment or problem-solving errors may yield learning that translates beyond those immediately involved in the same type of task and working in the same group or department. We conceptualize errors in judgment as occurring when individuals face uncertainty that requires them to assess a situation and explicitly or implicitly choose among options. By definition, in our framework, these errors occur when processes are not highly scripted or certain and generally involve individuals.

If the organization is to learn from errors that occur in these relatively uncertain contexts, we propose that the learning process should begin with a group of individuals who understand the specific judgment situation and are able to provide a careful description and analysis of the error. This group may be reasonably contained to specialists within the area in which the area occurred, or distinct from the task execution domain, analysis may require the involvement of outside disciplines. For example, in seeking to learn from the judgment errors committed when the morphine dose was calculated based on the wrong standards, the hospital quickly convened a cross-functional team to review the error; the team included representation from every group that "touched" the error. This team, with participation from pharmacy, nursing, medicine, and information technology (IT), discovered factors contributing to the error that ranged from inadequate nurse training, to weak supervision of new nurses, to poorly designed medication labels from the IT group.

If review of an error in the judgment domain offers lessons that are applicable in the organization beyond the immediate context, this new knowledge and error prevention ideas developed through local analysis can be communicated after the review process to others in the organization who might face similar situations, despite not being involved in the current error. In this case, those who confront similar judgment challenges in other areas of the organization may benefit from ideas developed by the initial review process for minimizing future errors.

Learning in the Interpersonal Coordination Domain

Learning from errors that occur during interpersonal coordination requires participation of individuals from the involved groups for both diagnosis and problem solving. Because errors in this domain are caused by ambiguities created by the intersection of process knowledge that ties multiple routines or information sets across groups together to guide interdependent work, goals of learning processes in this domain often involve improved intergroup communication protocols—better delineation of workflow and responsibilities between interdependent groups and reinforcement of handoffs between them. These gaps and ambiguities also complicate the process of diagnosing causes of errors in this domain and identifying the precise point (or, in many cases, points) at which an error has occurred. The diffusion of responsibility and accountability across group boundaries in this work context may introduce unproductive, potentially detrimental political influences on the diagnostic process, such as shifting blame to less-powerful groups. In particular, the fundamental attribution error may lead each individual to blame the practices or intentions of individuals from the other group.

A psychologically safe environment for discussing error is crucial in each domain, but particularly in this one; given the need for individuals who do not work closely together routinely to have interpersonally threatening discussions about what went wrong, this will help to mitigate defensive posturing and allow groups to prosper from interdisciplinary dialogue. Every involved group must participate in problem-solving activities designed to correct the error to ensure that each group understands the work of others, and all available knowledge relevant to the task—some of which may be unique to individual groups—is shared and integrated.

Learning in the System Interactions Domain

Finally, learning from errors in the fourth domain, system interactions, requires the use of groups that are both interdisciplinary and positioned to authorize organizational changes. The goal of learning from errors in this domain is to transform existing routines and design novel systems that harness new process knowledge now available to the organization as a product of experiencing error. Doing so requires that participants in the learning process contribute varied specialized skills and disciplinary perspectives to analyzing system interaction errors and identifying opportunities in each of their respective work areas for eradicating identified risks. Likewise, the group must also be sufficiently senior to command broad organizational attention, enact changes with wide-reaching implications for organizational practice, and access resources necessary for change implementation.

Discussion

In this chapter, we present a framework for organizational learning from errors that emphasizes the importance of work context as a determinant of optimal learning strategy. We differentiate the context of an error on two primary dimensions, process uncertainty and actor interdependence, and our proposed strategies for learning from errors differ accordingly. Four archetypal learning domains are described. Errors in task execution, or unintended deviations from a prescribed process by individuals performing familiar activities, are likely to produce specialized insights relevant mainly to the department or work group in which they occur and thus can be met with targeted disciplinary learning strategies. Learning from errors in judgment, which are similarly contained to individuals performing within a defined work area but involve unfamiliar activities such as newly implemented processes or novel situations, may benefit from the involvement of one or more outside groups in root cause analysis and spreading pertinent learning, depending on the nature of the error. Errors in interpersonal coordination, or unintended deviations that occur at the intersection of groups or functions when processes familiar to individuals working within their discrete work areas must be integrated across group boundaries, require collaboration among the groups involved in the error for diagnosis and problem solving. Errors involving system interactions, system-level errors that arise in novel terrain—new processes, unfamiliar activities, or unexpected situations—and cross multiple groups or departments, are particularly rich in learning potential but also demand more intensive organizational responses to be addressed effectively. Learning from system interaction errors requires broad interdisciplinary involvement to examine the manifold converging factors at the root of the error, spread knowledge gleaned from the event, and enact changes in organizational practice.

Contribution

The unifying theme across these domains is that different errors require different management of the posterror learning process. For organizations to learn from errors, review processes and corrective responses must be adapted to the nature of an error such that useful lessons are identified, applied, and transferred effectively and efficiently. By differentiating errors on primary dimensions of the context, we aim to contribute to theory on organizational learning from errors by emphasizing the importance of flexible or contingent learning processes: approaches that vary with the circumstances of error. Well-established theory on organizational errors has recognized the importance of situational factors in both definition and analysis of error. For example, Reason (1990) defines skill-, rule-, and knowledge-based errors in terms of varying levels of predictability, and Perrow (1984) provides a framework of error in which unfamiliar or unexpected production sequences ("complex interactions") and high dependence among system subparts ("tight coupling") lead to unavoidable accidents in complex systems. These treatments, however, have focused on work context as a differentiator and predictor of error and have left the learning process after an error occurs largely unexplored. Our goal is to extend well-developed theory on the influence of situational or contextual task characteristics on the nature and likelihood of errors by training focus on the implications of the work context of error for organizational learning. We also focus on knowledge-intensive, relational work, which is not fully described by errors theory that is grounded in analysis of high-risk production environments (e.g., nuclear power).

Previous explorations of learning from error in organizations have focused on the need for rigorous error detection and analysis and have examined the operational and social-psychological complexities of achieving these goals successfully. But, the existing literature provides little guidance on how the learning process should be structured, leaving organizations left to muddle through the details—often either by force-fitting all errors into the same process or by devising mechanisms for sorting errors that are vulnerable to intergroup politics or do not allocate managerial time optimally for learning. We propose that the four quadrants of our error domains model pose distinct opportunities and challenges for organizational learning, and thus a uniform approach to managing these errors either will limit the learning that is achieved or will incur organizational costs (employee labor costs, burnout) that can be avoided with adaptive learning processes.

Our primary aim is to suggest that organizations resist the obvious temptation to engage broad organizational participation in analysis based on the severity of an error rather than on the nature of the work context in which the error occurred. For example, we suggest that analysis and learning from an individual deviation in executing a well-specified work process need not involve broad organizational representation, no matter how dire the consequences. Conversely, an error with minimal consequences that involved multiple groups and the intersection of unfamiliar processes is worthy of cross-functional participation and thoughtful analysis from multiple perspectives.

Limitations

Limitations of this chapter should be noted. First, we focus only on the work context in which an error occurs, specifically the extent to which process knowledge is mature and actors are interdependent, and do not discuss other factors that organizations should consider when determining the best strategy for learning from an error. Our intention is not to scope out the universe of factors that influence the way an organization learns from a particular error but rather to emphasize the importance of two key dimensions of the work context in which an error occurs—process uncertainty and actor interdependence—in shaping the optimal learning process. Other influences, such as industry regulation and professional norms, may change or constrain organizational decision making regarding the best approach for analyzing and correcting an error. Exploration of these factors through field research in different organizational settings would enrich understanding of the processes by which organizations are able to learn from errors most effectively under varying institutional conditions. Second, we develop our learning domain framework exclusively from our reading of the organizational literature on learning from errors and observations from our own research. Empirical research is needed to test, specify, and refine the domains we introduce in this chapter. Third, we focus only on errors and do not discuss problems and failures, which also contain the potential for triggering organizational learning (cf. Cannon & Edmondson, 2005; Tucker & Edmondson, 2003). Finally, we do not differentiate among varying forms of task knowledge and actor interdependence within our four domains. Knowledge that is codified in protocols versus developed informally through repeated performance of the same task and dependencies that are based mainly on resources (e.g., sharing funds or equipment) versus relationships (e.g., making decisions collectively) may need to be addressed differently. Within each of the error domains we specify, additional theoretical development and organizational research should examine the extent to which errors involving not only varying levels but also varying types of knowledge and interdependence require varying learning strategies.

Conclusion

As work in organizations becomes increasingly complex and knowledge intensive, organizational mastery of strategies for extracting and applying new knowledge from errors, problems, and other failures is crucial to survival, separating those organizations that are able to improve performance continuously from those that are not. Approaches to learning from errors that tailor learning processes to the unique nature of the error that has occurred are more likely to produce valuable information while protecting scarce organizational resources (i.e., time), and theory-driven decision making regarding how best to manage varying errors will support organizational responses to errors that are driven by learning goals rather than other potentially damaging influences on the learning process, such as intergroup politics. Our learning domain framework distinguishes among optimal organizational processes for learning from errors according to key features of the work context in which the error arose. By proposing targeted strategies for extracting lessons that we embedded in unintended deviations from desired performance, we aim to move closer to reliability in learning from errors, a goal that continues to elude most organizations.

References

Anderson, J., Ramanujam, R., Hensel, D., & Sirio, C. (2010). Reporting trends in a regional medication error data-sharing system. *Healthcare Management Science, 13*(1), 74–83.

Argyris, C. (1990). *Overcoming organizational defenses: Facilitating organizational learning.* Needham, MA: Allyn and Bacon.

Bohmer, R., & Winslow, A. *The Dana-Farber Cancer Institute* (Harvard Business School Case 699-025). Allston, MA: Harvard Business School Press, 1999.

Cannon, M., & Edmondson, A. C. (2001). Confronting failure: Antecedents and consequences of shared beliefs about failure in organizational work groups. *Journal of Organizational Behavior 22,* 161–177.

Cannon, M. D., & Edmondson, A. C. (2005). Failing to learn and learning to fail (intelligently): How great organizations put failure to work to innovate and improve. *Long Range Planning, 38,* 229–319.

Chang, A., Schyve, P., Croteau, R., O'Leary, D., & Loeb, J. (2005). The JCAHO patient safety event taxonomy: A standardized terminology and classification schema for near misses and adverse events. *International Journal for Quality in Health Care, 17*(2), 95–105.

Edmondson, A. (1996). Learning from mistakes is easier said than done: Group and organizational influences on the detection and correction of human error. *Journal of Applied Behavioral Science, 32,* 5–32.

Edmondson, A., Roberto, M., & Tucker, A. (2001). *Children's Hospital and Clinics (A)* (Harvard Business School Case 302-050). Allston, MA: Harvard Business School Press.

Edmondson, A. C. (1999). Psychological safety and learning behavior in work teams. *Administrative Science Quarterly, 44*(2), 350–383.

Edmondson, A. C. (2002). The local and variegated nature of learning in organizations: A group-level perspective. *Organization Science, 13*(2), 128–146.

Edmondson, A. C. (2003). Speaking up in the operating room: How team leaders promote learning in interdisciplinary action teams. *Journal of Management Studies, 40,* 1419–1452.

Edmondson, A. C., Roberto M. R., Bohmer, R. M. J., Ferlins, E. M., and Feldman, L. R. (2005). The recovery window: Organizational learning following ambiguous threats. In M. Farjoun & W. Starbuck (Eds.), *Organization at the limits: NASA and the* Columbia *disaster* (pp. 220–245). London: Blackwell.

Gersick, C. J., & Hackman, J. R. (1990). Habitual routines in task-performing groups. *Organizational Behavior and Human Decision Processes, 47,* 65–97.

Keith, N., & Frese, M. (2008). Effectiveness of error management training: A meta analysis. *Journal of Applied Psychology, 93*(1), 59–69.

Lee, F., Edmondson, A. C., Thomke, S., & Worline, M. (2004). The mixed effects of inconsistency on experimentation in organizations. *Organization Science, 15*(3), 310–326.

Levitt, B., & March, J. G. (1998). Organizational learning. *Annual Review of Sociology, 14,* 319–340.

MacPhail, L. (2010). *Work process failure and organizational learning in health care delivery settings*. Unpublished dissertation, Harvard University, Cambridge, MA.

MacPhail, L., Neuwirth, E., & Bellows, J. (2009). Coordination of diabetes care in four delivery models using an electronic health record. *Medical Care, 47*, 993–999.

Millenson, M. (2002). Pushing the profession: How the news media turned patient safety into a priority. *Quality and Safety in Health Care, 11*, 57–63.

Morimoto, T., Ghandhi, T. K., Seger, A. C., Hsieh, T. C., & Bates, D. W. (2004). Adverse drug events and medication errors: Detection and classification methods. *Quality and Safety in Health Care, 13*, 306–314.

Perrow, C. (1984). *Normal accidents: Living with high-risk technologies*. New York: Basic Books.

Peters, G. A., & Peters, B. J. (2007). *Medical error and patient safety: Human factors in medicine*. New York: CRC Press.

Ramanujam, R., Sirio, C. A., Keyser, D. J., & Thompson, D. (2009). *Reporting to report or reporting to learn? Examining the use of incident reporting data in hospitals*. Working Paper, Owen Graduate School of Management. Nashville, TN: Vanderbilt University.

Reason, J. (1990). *Human error*. New York: Cambridge University Press.

Roberts, K. (1990a). Managing high reliability organizations. *California Management Review, 32*(4), 101–113.

Roberts, K. (1990b). Some characteristics of one type of high reliability organization. *Organizational Science, 1*(2), 160–176.

Rubin, G., George, A., Chinn, D. J., & Richardson, C. (2003). Errors in general practice: Development of an error classification and pilot study of a method for detecting errors. *Quality and Safety in Health Care, 12*, 443–447.

Spear, S., & Bowen, H. (1999, September–October). Decoding the DNA of the Toyota production system. *Harvard Business Review*, pp. 97–106.

Tucker, A. C., & Edmondson, A. C. (2003). Why hospitals don't learn from failures: Organizational and psychological dynamics that inhibit system change. *California Management Review, 45*(2), 55–72.

Van Dyck, C., Frese, M., Baer, M., & Sonnentag, S. (2005). Organizational error management culture and its impact on performance: A two-study replication. *Journal of Applied Psychology, 90*, 1228–1240.

Weick, K., & Roberts, K. (1993). Collective mind in organizations: Heedful interrelating on flight decks. *Administrative Science Quarterly, 38*, 357–381.

Weick, K., & Sutcliffe, K. (2001). *Managing the unexpected: Assuring high performance in an age of complexity*. San Francisco: Jossey-Bass.

Weick, K., Sutcliffe, K., & Obstfeld, D. (1999). Organizing for high reliability: Processes of collective mindfulness. *Research in Organizational Behavior, 21*, 81–124.

Wiegmann, D. A., & Shappell, S. A. (2003). *A human error approach to aviation accident analysis. The human factors analysis and classification system*. Aldershot, UK: Ashgate.

8

Errors at the Top of the Hierarchy

Katsuhiko Shimizu and Michael A. Hitt

Strategic decisions made by the chief executive officer (CEO) and top management team (TMT) determine the direction of an organization over time. Such decisions are responsible for the success or failure of strategic initiatives and resulting organizational performance (Hambrick & Mason, 1984). Increasing globalization and technological development heighten competitive intensity; thus, organizations continuously need to plan and implement new strategic initiatives to revitalize their core businesses or develop new businesses (Brown & Eisenhardt, 1997; Nadkarni & Narayanan, 2007). However, new initiatives are not always successful. We observe various failures in such areas as acquisitions (Shimizu & Hitt, 2005; Weisbach, 1995); diversification (Johnson, 1996); international expansion (Chang, 1996); and even the core business (Perlow, Okhuysen, & Repenning, 2002).

In many cases, when such failures occur and performance declines, researchers simply assume that the errors are made at the TMT level, pay scarce attention to the errors, and move on to other antecedents and consequences of poor performance. For example, based on his review of restructuring activities, Johnson (1996) concluded that organizational performance, the amount of slack resources and change in top management are important antecedents of restructuring (also see Shimizu & Hitt, 2005; Weisbach, 1995). Boeker (1997) and Lant, Milliken, and Batra (1992) also found that negative performance is an important antecedent of major organizational change. Research that directly focuses on errors in the context of the TMT across different strategic failures is less common, although research on strategic decision making is abundant (e.g., Eisenhardt & Zbaracki, 1992). It is the aim of this chapter to shed a specific light on errors made at the TMT level.

Understanding why errors were made at the TMT level is increasingly important. However, future directions of technology, market needs, and competition are difficult to predict; thus, analysis of the past events does not guarantee future favorable performance (Adner & Levinthal, 2004; Brown & Eisenhardt, 1997). Even with rigorous planning, new decisions and initiatives often encounter unexpected challenges in their implementation (Denrell & March, 2001; Greve, 1998). Further, even the best CEOs and the most effective TMTs who understand the risks of erroneous decisions can fail to manage the errors effectively (McGrath, 1999). In an interview with *Wall Street Journal* reporters, Juergen Schrempp, former CEO and chair of Daimler Chrysler, stated, "My principle always was ... move as fast as you can and [if] you indeed make errors, you have to correct them. ... It's much better to move fast, and make errors occasionally, than move too slowly" (Simison & Miller, 1999). However, he later received significant criticism for the acquisition of Chrysler and was eventually forced to resign as CEO at the end of 2005. This and other examples suggest that we need more efforts to understand the unique characteristics of errors made by TMTs and how to manage those errors.

In this chapter, we examine research on errors made by the CEO and TMT members (top of the organizational hierarchy). Our primary focus here is on error management issues for top managers as opposed to error prevention. First, we elaborate on errors at the TMT level in contrast to other types of errors and discuss unique issues associated with TMT errors. After concisely reviewing sources of such errors, we examine research on error management from two different perspectives: how the TMT detects and responds (or fails to respond) to errors and how the TMT learns (or fails to learn) from the errors. Finally, we discuss future research directions.

Errors at the TMT Level

Research on Errors

Although errors and the resulting negative outcomes have been a major issue in managing organizations for some time, researchers have tended to emphasize successful firms (Denrell, 2003; McGrath, 1999). It is partly because success bias is almost immortalized in society; success is highly praised and errors and failures receive often severe treatment with irreparable harm to the reputations of those assumed to have made them (McGrath, 1999). However, a growing number of researchers have examined errors and failures in organizations. Such errors include the *Challenger* tragedy (Vaughan, 1996), train wrecks (Baum & Dahlin, 2007); airline accidents (Haunschild & Sullivan, 2002); and somewhat less-severe operational problems such as automobile recalls (Haunschild & Rhee, 2004), operational violations in nuclear plants (Carroll, 1998), and errors in hospitals (Cannon & Edmondson, 2005; Edmondson, 1999). Researchers have examined not only the causes of the errors but also how organizations respond to or learn from the errors (Cannon & Edmondson, 2005; Haunschild & Sullivan, 2002). Although errors are negative and should be avoided when possible, managers and organizations can learn from errors that are made and utilize this learning experience to avoid such errors in the future (Hofmann & Frese, 2011; Sitkin, 1992).

Errors at the Top of Hierarchy

Hofmann and Frese (2011) defined actions as erroneous "when they unintentionally fail to achieve their goal where this failure was potentially avoidable." In applying this definition at the TMT level, top managers' decisions and actions that fail to deliver intended outcomes when the failure was potentially avoidable can be regarded as erroneous. We do not include top managers' scandals, such as those in Enron and World Com. Those are arguably intentional activities, thus are not errors. We explain three important issues in errors at the TMT level.

First, in many cases, the TMT makes strategic decisions under uncertainty (Eisenhardt, 1989). Here, *uncertainty* refers to the perceived inability to predict something accurately because of a lack of pertinent information or the inability to distinguish relevant from irrelevant information (Milliken, 1987; Weick, 1995). Andy Grove, former CEO of Intel, said, "None of us has a real understanding of where we are heading. ... Investment decisions or personnel decisions and prioritization don't wait for the picture to be clarified. You have to make them when you have to make them. So you take your shots and clean up the bad ones later." Peter Drucker also said, "Every decision is risky: It's a commitment of present resources to an uncertain future" (Byrne, 2005, p. 100). Accordingly, some of the errors may reside in the gray area between "avoidable" and "unavoidable." Yet, the TMT (particularly the CEO) is responsible for the errors and resulting outcomes (Wiersema, 2002). It is partly because, in retrospect, there is almost always a way to achieve a better outcome (March & Shapira, 1987).

To be prepared for and be able to minimize the negative effects of uncertainty, TMTs often use strategic planning. Strategic planning refers to systematic, analytical, formalized approaches to strategy formulation (Grant, 2003; Miller & Cardinal, 1994). By analyzing environments rigorously, TMT members should be able to make better strategic decisions. Ironically, it is also argued that rapid environmental change and resulting higher uncertainty make strategic planning less effective (Brown & Eisenhardt, 1997; Miller & Cardinal, 1994). Given that analysis usually focuses on past data, not the future, the benefits of strategic planning approaches may be severely limited when environments change rapidly and continuously. Moreover, it is also possible that the established plan becomes a "sacred cow," thus hindering the TMT from changing the decisions and actions, disallowing flexibility. For these reasons, Mintzberg (1987) argued that, "Setting oneself on a predetermined course in unknown water is the perfect way to sail straight into an iceberg" (p. 26). Practitioners and consultants also report that strategic planning in an organization tends to become "a primitive tribal ritual" (e.g., Kaplan & Beinhocker, 2003). Accordingly, the value of strategic planning has been debated (Miller & Cardinal, 1994). The empirical results have also been largely mixed. While studies show positive effects of strategic planning on performance (e.g., Grant, 2003), others still report no relationships (Falshaw, Glaister, & Tatoglu, 2006). It may be more practical to think that the real value of strategic planning is not eliminating uncertainty by rigorous analyses but is in establishing a "prepared mind" among top managers by facilitating the understanding of the current conditions of an organization and its environments, thereby helping them be better prepared for the future changes (Kaplan & Beinhocker, 2003).

Second, errors at the TMT level are determined by outcomes. Whether a certain strategic decision is erroneous can only be determined by its outcomes, not by analyzing the decision alone. In contrast, errors in hospitals (Edmondson, 1999), airline operations (Haunschild & Sullivan, 2002), and nuclear plants (Carroll, 1998) have been studied as deviations from predetermined procedures. In this sense, many errors, particularly at an operational level, have clear criteria that signify when a certain action or decision represents a deviation from the norm; thus, such actions apparently become erroneous at the time they occur (Hofmann & Frese, 2011). However, in decisions at the TMT level, "appropriate procedure" or "error signal" cannot be easily identified at the time of decision or action, if they even exist. Moreover, the TMT is challenged to develop a creative alternative that is inimitable by competitors. In fact, many successful strategies and innovations are based on "outrageous" ideas. For example, it is well known that the Japanese government (Ministry of Trade and Industry) tried to stop Honda from entering the auto business in the early 1960s. (Obviously, the government action would have been a major error if it had been successful in stopping Honda.)

Governance and control researchers have long made similar arguments (e.g., Jensen & Meckling, 1976; Walsh & Seward, 1990). To the extent that there is no one right answer to executives' decisions, it is difficult to protect shareholders' interests by monitoring executives' decisions and actions (Walsh & Seward, 1990). Instead, outcome-based compensation systems have been suggested as a more effective governance mechanism to align executives' interests and the interests of shareholders (Jarrell, 1993). Dramatic increases in the use of stock and stock options during the last decade coincided with the greater complexity and uncertainty facing the TMT members; under such conditions, outcome-based evaluations are more appropriate.

Outcome-based assessment, however, also poses serious challenges (cf., Hofmann & Frese, 2011). Besides the ambiguity of the feedback information about the outcomes (Hofmann & Frese, 2011, their Figure 3), other issues include the following questions: What is an appropriate standard below which an outcome is inappropriate? What is an appropriate time frame to assess? How much of the outcome can be attributable to the TMT? For example, if the time frame to assess the outcome is too long, one might argue "a strategy is never an error until we admit" that it is (Adner & Levinthal, 2004). A large amount of resources may continue to be invested and wasted, which has been referred to as "throwing good money to bad investments" or an "escalation of commitment" (Staw, 1997). Meanwhile, a short time frame may prematurely terminate valuable ideas. It is well known that Xerox lost a significant opportunity for its future by abandoning the personal computer too quickly (Smith & Alexander, 1988).

Third, errors at the TMT level always result in significant negative consequences. This conclusion is not surprising because "errors" are determined by the outcomes. Accordingly, errors made by the TMT are salient and attract a large amount of attention from stakeholders. In this sense, assessment and influence of errors at the TMT level are not limited to the TMT level. Instead, errors by TMTs are examined by various stakeholders as well as mass media, resulting in high social pressure.

Attribution biases of observers (e.g., shareholders, mass media, and mass media customers) enhance the pressure on TMTs. Attribution biases result from systematic skewness in understanding causality by people (Fiske & Tayler, 1991). In many cases, people attribute a certain event to something that is (a) readily identifiable, (b) vivid, and (c) salient (Fiske & Tayler, 1991). Because a TMT (particularly the CEO) meets all these conditions nicely, top management turnover is often associated with errors and negative performance (Weisbach, 1995; Wiersema, 2002) to remove the causes of the errors. For example, at Maytag the board of directors fired a new CEO after only 15 months, even though he did not have enough time to make or implement substantial changes (Callahan, 2000). Because the CEO is visible and salient, "when bad things happen on your watch, you're responsible—even if all the factors were not in his control" (comment of an analyst cited in Callahan, 2000, p. 138). The tendency to overly attribute causality to dispositional human factors is called a "fundamental attribution error" (Ross, 1977).

Based on the discussion so far, errors at the TMT level can be categorized and contrasted with other errors as shown in the framework in Figure 8.1.

The y-axis represents whether the decision or action has a predetermined state (i.e., a procedure), from which deviation becomes an error. When there is such a clear state, we call the decision or action a *routine*. In contrast, when it is difficult to specify such a predetermined state, partly

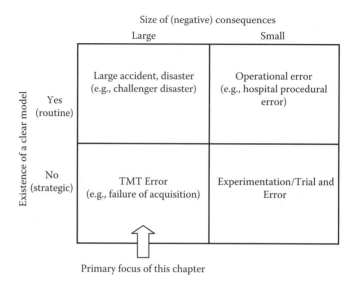

FIGURE 8.1
Categorization of errors.

due to uncertainty and unpredictability of their environmental context, we call the decisions or actions *strategic*. The *x*-axis represents the size of consequences of an error. It is difficult to determine whether a particular error per se is major or minor. Instead, we use the size of negative consequences of an error to determine the extent.

In this framework, errors categorized in the upper left box are those that deviate from routines and result in major problems. Typically, errors that lead to major accidents such as the *Challenger* disaster (Vaughan, 1996) and railroad wrecks (Baum & Dahlin, 2007) are categorized in this box. The upper right box includes errors resulting from a deviation from operational routines. These errors include operational errors (Carroll, 1998; Edmondson, 1999) that result in relatively minor problems.

Errors that are categorized in the lower left box are strategic ones that have no clear state and whose consequences are major for an organization. Examples of such errors include failure of acquisitions (Hayward, 2002; Shimizu & Hitt, 2005), exit from new businesses (Johnson, 1996), and retreat from international expansion (Chang, 1996). As discussed, errors in this category are attributed to the TMT and are the primary focus of this chapter. We use errors made by the TMT and strategic errors interchangeably hereafter.

Finally, the lower right-hand box includes strategic errors whose consequences are not large. Although strategic errors typically result in major damage to an organization, uncertain environments led some organizations to engage in experiments before initiating a full strategic move

(Adner & Levinthal, 2004; Brown & Eisenhardt, 1997). For example, Brown and Eisenhardt (1997) report that firms using a variety of low-cost probes to explore the possibilities of future products are more successful than firms that bet on a single view of the future. With small investments or experiments, an organization maintains the possibility to fully implement a strategy with an option to abandon the strategy when initial outcomes are unfavorable (Adner & Levinthal, 2004; McGrath, 1999). Although negative outcomes from those experiments are not intentional, the goal of those experiments is not necessarily success. Instead, organizations that utilize experiments try to accumulate knowledge from the experiments and use it to design and implement their future moves (Brown & Eisenhardt, 1997). We revisit this topic in the final section as a future research issue.

Sources of Errors at the TMT

There are many sources of errors at the TMT level. As Hofmann and Frese (2011) suggest, the sources include cognitive limitations and biases, collective group dynamics processes, and coordination problems through implementation processes within an organization. In this section, we briefly review each of the sources.

Cognitive Limitations and Biases

In *Administrative Behavior*, Simon (1976) pointed out three practical limitations to making rational decisions and achieving goals: (a) unconscious level skills, habits, and reflexes; (b) values; and (c) knowledge and information. Although the second values are also important, we here focus on the first and third as important sources of errors.

Regarding the unconscious level skills, habits, and reflexes, psychology and decision-making researchers have identified various cognitive biases that often make information processing more efficient but also entail the risk of distorting information interpretation (e.g., Fiske & Tayler, 1991; also Hofmann & Frese, 2011). Over time, people develop a certain way of perceiving their environment and interpreting information (Louis & Sutton, 1991); this phenomenon is referred to as the development of cognitive maps or schemas (Walsh, 1995). Although the notion of cognitive maps is focused on individuals, researchers have noted that groups and organizations also develop similar maps through interaction (Hofmann & Frese, 2011; Walsh, 1995). Cognitive maps facilitate information processing by categorizing new information into an existing framework (Dutton & Jackson, 1987; Walsh, 1995). Incomplete information about an issue is usually filled with category-consistent information and can be distorted (Dutton & Jackson, 1987). Thus, sometimes top managers are accustomed to viewing the environment with an old lens and fail to perceive important changes (Louis & Sutton, 1991). Other times, they select an alternative that was successful in the past but not appropriate in a new setting (also see Finkelstein & Haleblian, 2002). In addition, one well-known cognitive bias among others at the TMT level is "self-serving attributions," by which top managers attribute successes to themselves and failures to external factors (Staw, McKechnie, & Puffer, 1983).

The knowledge and information problem is referred to as bounded rationality, by which "rational actors are significantly constrained by limitations of information and calculation" (Cyert & March, 1992, p. 214). To achieve an appropriate decision, a TMT needs to collect information and analyze it (Eisenhardt & Zbaracki, 1992). However, it is almost impossible to collect all information needed or to digest and understand all the analyses (Simon, 1976). In fact, research suggests that top managers are often overloaded with information; in response, they take a simplified approach (e.g., not assessing all the possibilities, selecting an alternative from readily available options) to make decisions (Anderson, 1983; Cyert & March, 1992). The chance of information overload and thus selecting a suboptimal decision is high in an uncertain environment; as the degree of uncertainty increases, the need for information and difficulty in analyzing and interpreting information increase (Milliken, 1987; Weick, 1995).

Group Dynamic Processes

Groupthink (Janis, 1972) is one of the most famous problems in group dynamics in relation to errors at the TMT level. Observing such critical decision errors as the one related to the Bay of Pigs, Janis (1972) defines *groupthink* as "a mode of thinking that people engage in when they are deeply involved in a cohesive in-group, when the members' striving for unanimity override their motivation to realistically appraise alternative courses of action" (p. 8). Groupthink often dominates group-level decision processes and leads the group to suboptimal decisions. Janis (1972) and other researchers suggest that a group is more prone to groupthink when an illusion of invulnerability of the group is shared, and pressure on members who express questions is strong. In applying this concept to TMTs, Priem (1990) argues that a TMT with moderately high levels of consensus are likely to engage in effective implementation and obtain favorable performance, while very high consensus reduces their effectiveness and negatively influences performance because of groupthink.

In addition, politics can significantly distort TMT decisions. *Organizational politics* are defined as "activities taken within organizations to acquire, develop, and use power and other resources to obtain one's preferred outcomes in a situation in which there is uncertainty or dissensus about choices" (Pfeffer, 1981, p. 7). Because the TMT is a coalition of people who have different beliefs, interests, and specialties (Cyert & March, 1992; Hambrick, 1998), it is no surprise that decision making is often influenced by the power structure within a team (Eisenhardt & Zbaracki, 1992). To Hambrick (1998), managers in the TMT are "individuals who rarely come together (and then usually for perfunctory information exchange), who rarely collaborate, and who focus almost entirely on their own pieces of the enterprise" (p. 123). When organizational performance is favorable, potential conflicts of interest are not evident. In contrast, failures provide an opportunity for less-powerful parties to initiate political behaviors more overtly (Eisenhardt & Bourgeois, 1988). The TMT is more likely to be subject to internal power games by downplaying or hoarding the negative information (Edmondson, 2002) or by blaming other managers (Boeker, 1992; Wagner & Gooding, 1997). A particular interpretation (e.g., cause of a failure) may be shared not because it is valid but because it is advocated by a powerful manager (Pfeffer, 1981).

Coordination Problem in the Implementation Processes Within an Organization

New strategic initiatives often encounter various types of problems and challenges in their implementation (Denrell & March, 2001; Greve, 1998). Moreover, as ex ante predictions of the market changes and competitive moves become difficult, an organization should adjust its strategy throughout the implementation efforts (Ghoshal, Bartlett, & Moran, 1999). An excellent strategy can fail because of poor implementation. In contrast, a seemingly "erroneous" decision may result in a major success with significant efforts to implement and adjust the strategy. For example, analysts almost unanimously agreed that Sony's acquisition of Columbia Pictures in 1989 was a major error. When Sony announced a $2.7 billion write-off from a $3.4 billion purchase price in 1994, a headline on the first page of the *Wall Street Journal* stated that, "Last Action: Sony Finally Admits Billion-Dollar Error: Its Messed-up Studio—Columbia Was Mismanaged From the Very Start; Public Sale May Be Next—In a League of Their Own." However, as of 2005, the movie business at Sony provided more than one third of the operating profit of the company (with only 10% of the total sales). This success contributed to the selection of Howard Stringer as the new Sony CEO. Similarly, the decision at Renault to invest $5.4 billion in Nissan in 1999 was initially called a "waste" or "a marriage of desperation." In contrast, the *Wall Street Journal* also reported the "Daimler-Chrysler Merger to Produce $3 Billion in Savings, Revenue Gains" on May 8, 1998. We all know what happened to these auto companies: Nissan has done well, and Daimler finally sold off Chrysler after highly disappointing results.

Compared to such decisions and actions that are tightly coupled with the outcomes, such as those at a health care nursing station (Edmondson, 1999), strategic decisions typically require lengthy periods for outcomes to become obvious. Moreover, throughout the time period, various factors affect the outcomes, including the quality of implementation and changes in environments (Bossidy & Charan, 2002; Denrell & March, 2001). Accordingly, successful implementation requires effective coordination of various parts within an organization to achieve total organizational optimization, as opposed to local optimization (Hofmann & Frese, 2011). Coordination problems in the implementation process can result in disastrous outcomes, thereby labeling what appeared to be an initial good decision as a serious error.

In this section, we examined three major sources of TMT errors at three different levels: individual and group cognitive biases, group dynamics, and implementation coordination problems within an organization. It is important to note that these causes are more influential in highly uncertain environments (cf., Nadkarni & Narayanan, 2007; Shimizu & Hitt, 2004). The more uncertain the environment, the more information a TMT requires, which may lead to information overload. The more uncertain the environment, the more that interpretations and alternatives are possible. These conditions may increase political conflicts or enhance the potential for groupthink to occur. Finally, the more uncertain the environment, the more important is the feedback from frontline managers and staff to help in adjusting the strategy. To the extent that implementation is not heeded or not well coordinated by top management, the important information may not be communicated or utilized.

Error Management at the TMT Level

Given an increasingly uncertain environment, it is difficult for TMTs to completely avoid errors (Adner & Levinthal, 2004; Brown & Eisenhardt, 1997). In fact, Honda has been using "No error, no play" (i.e., no error means you do not challenge enough) as a company-wide slogan. To this end, our primary focus in this chapter is error management, as opposed to error prevention (cf., Hofmann & Frese, 2011). As there is a large amount of uncertainty and there is no clear ideal state, detection of an error and responding to the error play an important part in minimizing negative results (Hofmann & Frese, 2011). Moreover, how a TMT learns from the initial error and prevents secondary errors or produces future success is critical (Denrell, 2003; Haunschild & Sullivan, 2002; Sitkin, 1992). These are the two topics in this section.

Error Detection and Response

In many cases, new initiatives encounter various types of resistance and challenges in their implementation that must be overcome to be successful (Denrell & March, 2001; Lynn, Morone, & Paulson, 1996). Without strong commitment and patience, their potential may never be realized (Ghemawat, 1991). There are numerous examples of heroic leaders and innovators who achieved their final victory by maintaining a strong commitment to overcome multiple obstacles. Sony's acquisition of Columbia Pictures is one example. Another example is that Corning took more than 10 years and $100 million—dealing with high market skepticism and middle management resistance—to launch its optical fiber business with much eventual success (Lynn et al., 1996). In contrast, a firm that frequently changes its strategy and course of actions may vacillate, waste resources, and eventually fail (Hambrick & D'Aveni, 1988). Yet, being overly committed to an erroneous decision can also be disastrous (Ross & Staw, 1993; Staw, 1997). Correctly balancing commitment and timely detection of errors can maximize potential benefits and minimize losses, while achieving the correct balance is undoubtedly challenging (Gersick, 1994).

Extending the initial discussion of errors in the previous section, this section examines barriers that make error detection and response difficult at three different stages: (a) barriers to pay attention to negative feedback (attention stage), (b) barriers to assess negative data objectively (assessment stage), and (c) barriers to initiate a change in a timely fashion (action stage) (Hofmann & Frese, 2011; Perlow et al., 2002; Shimizu & Hitt, 2004).

Barriers to Attention: Insensitivity to Negative Feedback

Regarding strategic initiatives, top managers need to be sensitive (maintain attention) to feedback from the market, particularly negative feedback, so that they can respond to this feedback in a timely fashion (Hofmann & Frese, 2011; Nadkarni & Narayanan, 2007; Ocasio, 1997). Unfortunately, both research and anecdotal evidence suggest that managers often ignore early signs of strategic errors until much damage is done (e.g., Ross & Staw, 1993; Shimizu & Hitt, 2005).

It is discussed that top managers develop a particular cognitive map along with a set of decision rules and heuristics based on their experiences, and this cognitive map is a common source of strategic errors (Louis & Sutton, 1991; Prahalad & Bettis, 1986). Ironically, the cognitive map that leads to an error also makes error detection difficult. As successful managers are promoted within an organization and successful initiatives are repeated, prior successful experiences often control attention to the new issues (Cyert & March, 1992). The cognitive map is self-reinforcing, such that top managers select and interpret only information that is consistent with their cognitive map, and they tend to attribute success to their own credit (Dutton & Jackson, 1987; Staw et al., 1983). This situation nurtures managerial overconfidence and complacency (Roll, 1986). Moreover, successful experiences often attract media attention and praise, thus supporting managerial hubris (Hayward & Hambrick, 1997). As a result, overconfident managers assume that their decisions are unlikely to fail and unconsciously ignore signs pointing to potentially negative decision outcomes (Hayward & Hambrick, 1997; Ocasio, 1997).

Furthermore, top managers' cognitive maps and decision rules are often shared, routinized, and accepted within the organization (Walsh, 1995). This process ensures that the same type of information will be collected using the same methods, and the information collected will be analyzed using taken-for-granted assumptions with routinized approaches (Huff, Huff, & Thomas, 1992; Nelson & Winter, 1982). Such organizational inertia makes it less likely that the TMT will receive and be attentive to new information (e.g., negative feedback from the market) (Shimizu & Hitt, 2005). Instead, this type of information will be either ignored or assumed to be an exception and not analyzed further (Levitt & March, 1988).

Barriers to Assessment: Self-Serving Interpretation

Even if managers are vigilant and a negative signal is recognized at an early stage, they may not necessarily initiate a response. Managers are often reluctant to admit that they made an error (Ross & Staw, 1993). To justify their decision, they may attribute the poor outcomes to external factors (Staw et al., 1983). Alternatively, they may reemphasize their commitment to make the initiative a success (Ross & Staw, 1993; Staw, 1997). Research has shown that people in situations that are likely to produce a loss are more willing to take risky actions to create positive returns (Kahneman & Tversky, 1979; Shimizu, 2007). An early negative signal may be interpreted as the result of insufficient time or inadequate implementation efforts; thus, even more resources may be invested (Levitt & March, 1988). Therefore, when confronted with undesirable outcomes, managers may take a risk by making a further commitment to the initiative as opposed to abandoning it (Kahneman, 1992; Shimizu, 2007). These managers will continue to invest with the hope of a dramatic turnaround in the initiative; this is a well-examined phenomenon called *escalation of commitment* (Ross & Staw, 1993).

The political context within TMT also plays an important role in assessing the negative outcomes of the strategic decisions. Negative outcomes commonly change the power balance in an organization (Cyert & March, 1992; Eisenhardt & Bourgeois, 1988). The potential power struggles and the potential career-limiting effects of strategic errors exacerbate a manager's unwillingness to admit errors (Staw, 1997). When the supporters of the original decision have power, they prefer to retain the project and avoid admitting an error to maintain their power. Those who have less power may either support the idea or contest the idea to increase their power. Therefore, while a poor outcome signals an error and the need for repair, organizational politics often prevent assessing the outcome correctly or in a timely fashion.

Barriers to Action: Uncertainty and Resistance

In addition to the psychological and organizational biases, another type of barrier hinders or slows taking an action to change. It is the uncertainty and resulting resistance associated with the future of the strategic initiative.

To the extent that evaluation of a particular initiative involves forecasts of the future environment, it is difficult to predict the outcomes confidently, especially for a loss-generating initiative (Adner & Levinthal, 2004). While simple net present value calculations or strategic planning provide seemingly objective assessments, the results are dependent on assumptions, which can easily change in a dynamic environment (Miller & Cardinal, 1994; Perlow et al., 2002). Moreover, even if the current performance is poor, an initiative still may have potential (Adner & Levinthal, 2004). Adner and Levinthal (2004) argue that this "impossibility of proving failure" is "an inherent feature of firm initiative under uncertainty" (p. 77).

Further, change often creates resistance (Huff et al., 1992). In general, people resist change because of familiarity with the current conditions and a fear of the unknown. People prefer the status quo because change disrupts the established routines and produces uncertainty, thereby involving risks (Greve, 1998; Lant et al., 1992). It is likely that even some managers at TMT resist change. Combined with organizational politics, a decision to initiate a change may be excessively delayed (Huff et al., 1992).

Shimizu and Hitt (2004) point out that these barriers are not independent. Rather, the barriers often interact with each other and create a vicious cycle that makes it more difficult for TMTs to detect errors and respond to them, as shown in Figure 8.2. As explained, established cognitive maps coupled with overconfidence/complacency often hinder managers' attention to early signals of errors. Coupled with unwillingness to admit errors, managers' assessments are positively skewed. When negative outcomes are either ignored or interpreted optimistically, it is unlikely for managers to confidently initiate a new action. Instead, managers often prefer the status quo as the easiest path until the outcomes become extremely bad (Huff et al., 1992; Shimizu & Hitt, 2005). In this case, uncertainty is seemingly "controlled" by ignoring it (Bourgeois, 1985). When no action is taken, the existing cognitive maps and current routines are further reinforced

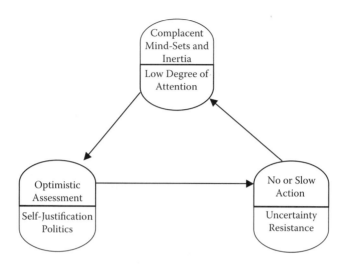

FIGURE 8.2
Vicious cycle of rigidity. Adapted from Shimizu and Hitt (2004). Strategic flexibility: Organizational preparedness to reverse ineffective strategic decisions. *Academy of Management Executives, 18*(4), 44–59.

(Louis & Sutton, 1991). This cycle becomes institutionalized and exacerbated as suggested in Figure 8.2 (Shimizu & Hitt, 2004).

Learning From Errors at the TMT Level

Even if strategic failures result in significant performance problems, top managers may be able to learn important lessons that can be used in the future (Haunschild & Sullivan, 2002; Schulz, 2002). As success becomes a source of complacency and errors, failure can be an important source of future success (Hofmann & Frese, 2011). The effectiveness of learning depends on how accurately or intelligently the causality in the decision-negative outcome relationship is inferred (Sitkin, 1992). By identifying the major causes of the error, an organization can correct the errors, avoid making the same errors, and improve future strategic initiatives (Sitkin, 1992).

Learning from errors is different from learning from successes. Whereas a TMT continues to repeat actions that resulted in prior success (Cyert & March, 1992), a TMT that experiences errors needs to initiate changes to correct the errors. However, recognizing an error provides little insight in terms of causality or an appropriate new direction (Levinthal, 1991). Causal relationships in strategic errors are often ambiguous (Levitt & March, 1988; Weick, 1995). In the following sections, the general issues of and barriers to learning from strategic errors and failures are elaborated.

Barriers to Learning

While admitting the importance of learning, many researchers have noted problems in learning from experiences (e.g., Baumaard & Starbuck, 2005; Levinthal & March, 1988, 1993). Researchers provided various types of barriers that can occur at multiple levels and interactively hinder valid causal inferences. These barriers are explained briefly next.

Event-Level Barriers

Ambiguous Causality

A failed initiative involves various factors, including individual and organizational factors, planning and implementation factors, or management and environmental factors. Because of the time lag between the inputs and the outputs (Levitt & March, 1988), the existence of "noisy" variables (Levinthal, 1991; Lounamaa & March, 1987), and dynamism associated with inputs and outputs (Lant & Montgomery, 1987), there are frequently a number of plausible explanations for a given outcome, and it may be difficult for TMT members to identify one or a few valid ones. As Weick (1995) points out, this is the problem of ambiguity, with people confused by too many explanations.

Individual-/Group-Level Barriers

Cognitive Biases

As discussed, individual cognitive biases are not only the sources of error but also a barrier to detecting an error. Moreover, the cognitive biases also act as a barrier to learning from the error. Even when errors are obvious, top managers are not completely free from their cognitive maps and thus may interpret the causality from a predetermined perspective. Attributing a failure to external factors is one example (self-attribution bias; Staw et al., 1983). Other biases include automatic scanning, hindsight biases, and availability biases, all of which can lead to illusionary conclusions about causation (Kiesler & Sproull, 1982). Thus, "believing is seeing" (Weick, 1995, p. 133).

Group-/Organizational-Level Barriers

Inertia

Once a particular cognitive map is established and shared, it develops structural rules and routines that often remain the same even when the perspective becomes obsolete and excludes new information and knowledge systematically (Cohen, 1991; Nelson & Winter, 1982). New causal understandings can be hindered by inertial forces resulting from extant perspectives, structures, and rules (Hannan & Freeman, 1984). Because failures are not consistent with the plan and the organizational processes that created the plan (Kiesler & Sproull, 1982), interpretation and knowledge inferred from failures are often adjusted to accommodate the existing rules and shared assumptions, not vice versa.

Different Views and Politics

While organizational inertia anchors causal inferences narrowly to established rules and assumptions, diversion of interests and resulting politics fragment or distort causal inferences. Because an organization is a coalition of people who have different beliefs, interests, and specialties (Cyert & March, 1992), it is no surprise that the same phenomenon is interpreted differently by organizational members (Weick, 1995). March, Sproull, and Tamuz (1991) state that "organizational experience leads to a variety of interpretations, and an organization's repertoire may come to include several different, possibly contradictory, story lines" (p. 3). A particular interpretation may be shared not because it is valid but because it is advocated by a powerful manager (Eisenhardt & Bourgeois, 1988).

Social-Level Barriers

Fundamental Attribution Error

As discussed, strategic errors are commonly attributed to top management (Wiersema, 2002). To the extent stakeholders and mass media expect top management to be responsible, a causal relationship embedded in the failure will be assumed to be the result of the incompetence of top management (Ross, 1977). In response, boards of directors often respond to the strategic error by replacing top management, regardless of the real reason for the error (Lerner & Tetlock, 1999).

Success Bias

Individuals and organizations are embedded in a social environment that has been obsessed with a success bias (Denrell, 2003; Levinthal & March, 1993). In the face of strategic errors, top management in an organization usually receives strong social pressure to account for and correct the error (McGrath, 1999) from the perspective of norms of rationality and progress (Abrahamson, 1996). To do so, relying on socially accepted and visible explanations and actions can be effective in reducing the pressure but may not resolve the problem (Lerner & Tetlock, 1999). Thus, some reactions not only may have symbolic value in accommodating social pressure but also can potentially exacerbate the situation (McGrath, 1999).

Learning From Errors: Reality

People learn, but an organization does not (Crossan, Lane, & White, 1999). Individual learning becomes organizational learning only through appropriately sharing the learned lessons among members (Crossan et al., 1999). It is important to recognize that, while individual-level learning is driven by the intuitions and insights of an individual, organizational learning is driven by simplification and institutionalization processes (Crossan et al., 1999).

The rich insights gained by an individual are hard to communicate because errors are often unexpected to individuals as well as to organizations; thus, there may be little common language. Kiesler and Sproull (1982) note:

> Memories of past crises will tend to be aschematic and poorly remembered in terms of any organized plan, theory, or norm of the organization. ... Because of the memory discordances and the absence of an appropriate, shared framework of the events that occurred, there will also tend to be arguments in the organization over what actually occurred, what was accomplished, and what applies to the situation at hand. (p. 563)

Without either a common language or a framework, sharing new insights will be slow and challenging (Crossan et al., 1999). Because of success biases, individuals are less inclined to disclose and elaborate on error experiences, which further disrupts integration of individual knowledge (Edmondson, 2002). Moreover, under strong shareholder pressure to resolve and learn from the failure, managers will not have the luxury to spend enough time and effort to explore the causal relationships by carefully integrating diverse and possibly incongruous knowledge across individuals (Crossan et al., 1999). As a result, only visible and easy-to-understand inferences are shared at the organizational level. Thus, rich insights learned at individual levels are lost in the attempted process of integrating knowledge at the organizational level.

Two types of organizational reactions typically emerge from the simplified learning processes based on strategic errors. First, failures are often attributed to the errors of the decision makers or those in charge of the initiative (Ross, 1977; Wiersema, 2002). Attributing errors and failures to the incompetence of top management is well accepted in society; thus, replacing top management becomes a conventional, safe, and legitimate response to a strategic error (Boeker, 1992; Wiersema, 2002). The ousting of the CEO from Maytag in 15 months is one example (Callahan, 2000).

Second, to comply with and account for social norms and pressures, the TMT may simply label the initiative as an error and abandon it, which is what Sitkin (1992) called "learning by avoidance" (p. 236). Although a failed initiative can be recoverable by closely examining the causes and resolving the problems, abandoning a failed project also provides a strong signal that the TMT seriously considered the failure, and that the TMT took a tough and decisive action not to make the same error again (Abrahamson, 1996; Boeker, 1992). Divestitures of acquired businesses can be a good example (Porter, 1987; Shimizu & Hitt, 2005).

The general conclusion from the discussion is that organizations often rely on socially accepted accounts, attributing the failure to incompetence of top management and abandoning the initiatives or people in charge (Boeker, 1992; Wiersema, 2002). Under uncertainty and various biases, conventional, existing general knowledge is tapped to develop a solution to the error. By doing so, the TMT shows decisive commitment in response to the error, seals off the painful experience, and directs the organization in a different direction for the future (e.g., Brunsson, 1982). Unfortunately, adopting a conventional solution represents a poor substitute for important learning from errors (cf., March, 1991).

Toward Better Management of Errors

While we discuss various barriers and difficulties in managing errors at TMT levels, there are some remedies to overcome the barriers and provide better management of errors. At this point, however, we believe that those remedies are primitive and need to be examined further in future research. Next, we list three major points that have been discussed.

First, as we emphasized repeatedly, one of the key sources of errors and barriers to managing errors effectively is the rigidity of cognitive maps (Dutton & Jackson, 1987). One way to break the rigidity of cognitive maps is by incorporating new and different ideas by consciously examining the issues in the process of decision making (Hofmann & Frese, 2011). There are at least two possibilities for a TMT to do so: utilizing a board of directors and incorporating a devil's advocate in the process of decision making. Governance has been important to monitor and control TMTs as a way to avoid major strategic errors (in addition to self-interest maximizing behaviors). In addition to control duties, a board of directors can provide advice and suggestions (Johnson, Daily, & Ellstrand, 1996), reduce information overload, and unfreeze rigid cognitive maps shared by TMT members. For example, Shimizu and Hitt (2005) found that the addition of new outside directors increased the possibility of divesting a poorly performing unit. Another way to change established cognitive maps is by using a devil's advocate to consciously examine the strategic issues (Amason, 1996; Anderson, 1983). After the failure of the Bay of Pigs, President Kennedy is reported to have intentionally questioned various assumptions in the discussion and to avoid repeating similar errors (Janis, 1972).

Another factor that makes TMT error management difficult is uncertainty associated with assessing the outcomes. Unfortunately, strategic planning has limited effects (e.g., Miller & Cardinal, 1994). To better deal with the uncertainty, Gersick (1994) and Brown and Eisenhardt (1997) suggest that top managers can use time, as opposed to event, as a trigger to examine the results. To the extent that it is hard to set a clear criteria for event-based examination (e.g., initiatives are examined when performance falls below acceptable norms), time-based assessment may be more appropriate under conditions of uncertainty (Gersick, 1994). In addition, Simonson and Staw (1992) recommend some techniques that can help actors deescalate: setting minimum target levels and evaluating decision makers on the basis of their decision processes rather than outcomes.

Finally, implementation problems can be much lower for TMT members who have a "prepared mind" (Kaplan & Beinhocker, 2003). To the extent that they are promoted by their past success, top managers tend to be overly confident (Hayward & Hambrick, 1997; Roll, 1986). However, no CEO or TMT is error free, and no strategy is guaranteed to be successful, particularly in the environment in 2010. Thus, top managers need to be more sensitive by admitting that errors are possible (Hofmann & Frese, 2011), and that most strategic decisions must be modified in the process of implementation (Denrell & March, 2001; Shimizu & Hitt, 2004). To do so, TMTs need to pay more attention to strategy implementation and feedback from the frontline employees who actually implement the strategy (Ghoshal et al., 1999; Mintzberg, 1990).

Future Research Issues

In this section, we propose future research directions regarding errors at the TMT level.

First, as we pointed out in the beginning of this chapter, studies that directly focus on errors at the TMT level are surprisingly limited. Some studies have examined failures in such specific events as divestiture of acquisitions (Shimizu & Hitt, 2005; Weisbach, 1995), restructuring after diversification (Johnson, 1996), and retreat from international markets (Chang, 1996). But, in many cases, research on such strategic failures implicitly assumes the errors of the TMTs and fails to elaborate on the fine-grained explanation for why such errors occur. Using an analogy of the personal computer, we have a number of excellent applications, but a more basic operating system is not well developed. Given that there have been a large number of studies on cognitive biases (e.g., Fiske & Tayler, 1991; Louis & Sutton, 1991; Staw et al., 1983); decision making (e.g., Eisenhardt, 1989; Eisenhardt & Zbaracki, 1992); and organizational politics (e.g., Cyert

& March, 1992; Eisenhardt & Bourgeois, 1988), scholars should be able to integrate these lines of research and develop a rich theoretical foundation to examine errors of TMTs. This chapter provides at least a preliminary base for future research in this area.

One promising area for future research is on the identification of "errors" at the TMT level. As we discussed here, while TMT errors can be defined as "top managers' decisions and actions that fail to deliver intended outcomes when the failure was potentially avoidable," the definition and actual standard to identify errors can be elaborated. Particularly, given the increasing uncertainty in and dynamism of the environment, it is important for an organization to set a performance standard below which an initiative is regarded as an error. Unless a performance standard or "intended outcome" is clearly defined, assessment of whether an initiative is an error can also be subjective, possibly biased, resulting in either premature termination or escalation of commitment. In addition, it is important to elaborate on what "potentially avoidable" means in the context of TMT decisions. Future research may be able to distinguish "avoidable errors" from "inevitable errors" or even "good errors" from "bad errors" (cf., Sitkin, 1992, who discusses small "intelligent" errors).

Second, due to the rapid change and uncertainty in an environment, strategic experimentation (Brown & Eisenhardt, 1992) or making intentional small errors (Sitkin, 1992) becomes increasingly important (see Figure 8.1). Rather than relying on analysis of past data and resulting strategic planning, top managers can benefit from actual and often-unexpected results from experimentation. Although a growing number of studies have started to examine this issue in relation to real options (e.g., Adner & Levinthal, 2004; McGrath, 1999), we need more theoretical development and empirical work in this area (Adner & Levinthal, 2004). Moreover, allowing or promoting experimentation needs to encourage generation of new ideas not only from top management or planning divisions but also from rank-and-file employees. Because those frontline employees directly deal with operational concerns, customers, and competitors, they are likely to identify subtle but important changes in customer needs or a change in the rules of the current competitive game (Burgelman, 1994). Some research on these issues can be found in the corporate entrepreneurship literature (e.g., Kuratko, Ireland, & Hornsby, 2004), but broader attention and further development are needed. Combined with the growing importance of implementation, the roles of rank-and-file employees as well as TMT members in the context of experimentation need to be examined further.

Another important issue involves the mechanisms for learning from errors at the TMT level. Although organizational learning has been a major concern in management, and studies on learning from errors and failures have been conducted (e.g., Hayward, 2002), the mechanisms for learning from errors, particularly at the TMT level, remain in a black box (e.g.,

Levitt & March, 1988). For example, how does a TMT learn new knowledge or new decision-making processes from the error? When errors are not corrected in a timely manner, is it because no learning took place or that the TMT did not learn the appropriate knowledge from the error? To the extent that barriers (e.g., causal ambiguity, cognitive biases, politics) to effective learning are influential (Levitt & March, 1988), we cannot simply assume that the individual learning mechanisms are in place at the TMT level, particularly to correct strategic errors. More fine-grained research on actual mechanisms of learning at the TMT level will not only enhance our understanding but also provide insights to improve the effectiveness of the learning.

Fourth, while we point out that hubris is a critical source of decision errors and barriers to error management, some researchers suggest that strong confidence can be helpful in executing difficult tasks (Russo & Schoemaker, 1992). If top managers do not show confidence in their strategic initiatives, organizational members may not trust the validity of the initiatives that they take (Brunsson, 1982). As new initiatives often have unexpected problems, these initiatives are unlikely to be completed successfully unless organizational members are willing to work hard to implement them successfully (Russo & Schoemaker, 1992). Yet, top managers must still perceive the organization and its environment accurately to avoid making erroneous decisions. Regarding this critical balance between realism and confidence, Pfeffer and Sutton (2006) suggest that "you can privately describe doubts and uncertainties and fully recognize the limitations of your knowledge and abilities, while still projecting the confidence required to get others to commit their energy and effort" (p. 183). More research is needed to understand better how TMTs achieve this balance.

In conclusion, how to manage strategic errors at the TMT level is a critical issue that requires further research. Errors at the top of a hierarchy are extremely important and result in significant consequences. Meanwhile, increasing uncertainty makes the errors more common; thus, we need more research on errors and how they can be effectively managed.

References

Abrahamson, E. (1996). Management fashion. *Academy of Management Review, 21*, 254–285.

Adner, R., & Levinthal, D. L. (2004). What is not a real option: Considering boundaries for the application of real options to business strategy. *Academy of Management Review, 29*, 74–85.

Amason, A. C. (1996). Distinguishing the effects of functional and dysfunctional conflict on strategic decision making: Resolving a paradox for top management teams. *Academy of Management Journal, 39*(1), 123–148.

Anderson, P. A. (1983). Decision making by objection and the Cuban Missile Crisis. *Administrative Science Quarterly, 28*, 201–222.

Baum, J. A. C., & Dahlin, K. B. (2007). Aspiration performance and railroads' patterns of learning from train wrecks and crashes. *Organization Science, 18*, 368–385.

Baumaard, P., & Starbuck, W. H. (2005). Learning from failures: Why it may not happen. *Long Range Planning, 38*, 281–298.

Boeker, W. (1992). Power and managerial dismissal: Scapegoating at the top. *Administrative Science Quarterly, 37*, 400–421.

Boeker, W. (1997). Strategic change: The influence of managerial characteristics and organizational growth. *Academy of Management Journal, 40*, 152–170.

Bossidy, L., & Charan, R. (2002). *Execution: The discipline of getting things done.* New York: Crown Business.

Bourgeois, L. J., III. (1985). Strategic goals, perceived uncertainty, and economic performance in volatile environments. *Academy of Management Journal, 28*, 548–573.

Brown, S. L., & Eisenhardt, K. M. (1997). The art of continuous change: Linking complexity theory and time-paced evolution in relentlessly shifting organizations. *Administrative Science Quarterly, 42*, 1–34.

Brunsson, N. (1982). The irrationality of action and action rationality: Decisions, ideologies and organizational actions. *Journal of Management Studies, 19*, 29–44.

Burgelman, R. A. (1994). Fading memories: A process theory of strategic business exit in dynamic environment. *Administrative Science Quarterly, 39*, 24–56.

Byrne, J. A. (2005). The man who invented management: Why Peter Drucker's ideas still matter. *Business Week*, November 28, 96–106.

Callahan, P. (2000, November 11). Maytag chairman Ward resigns after tenure lasting 15 months. *Wall Street Journal*, 138.

Cannon, M. D., & Edmondson, A. C. (2005). Failing to learn and learning to fail (intelligently): How great organizations put failure to work to innovate and improve. *Long Range Planning, 38*, 299–319.

Carroll, J. S. (1998). Organizational learning activities in high-hazard industries: The logics underlying self-analysis. *Journal of Management Studies, 35*, 699–717.

Chang, C. J. (1996). An evolutionary perspective on diversification and corporate restructuring: Entry, exit, and economic performance during 1981–89. *Strategic Management Journal, 17*, 587–611.

Cohen, M. D. (1991). Individual learning and organizational routine: Emerging connections. *Organization Science, 2*, 135–145.

Crossan, M. M., Lane, H. W., & White, R. E. (1999). An organizational learning framework: From intuition to institution. *Academy of Management Review, 24*, 522–537.

Cyert, R. M., & March, J. G. (1992). *A behavioral theory of the firm* (2nd ed.). Cambridge, MA: Blackwell Business.

Denrell, J. (2003). Vicarious learning, undersampling of failure, and the myths of management. *Organization Science, 14*, 227–243.

Denrell, J., & March, J. G. (2001). Adaptation as information restriction: The hot stove effect. *Organization Science, 12*, 523–538.

Dutton, J. E., & Jackson, S. E. (1987). Categorizing strategic issues: Links to organizational action. *Academy of Management Review, 12*, 46–90.

Edmondson, A. (1999). Psychological safety and learning behavior in work teams. *Administrative Science Quarterly, 44*, 350–383.

Edmondson, A. (2002). The local and variegated nature of learning in organizations: A group-level perspective. *Organization Science, 13*, 128–146.

Eisenhardt, K. M. (1989). Making fast strategic decisions in high-velocity environments. *Academy of Management Journal, 32*, 543–576.

Eisenhardt, K. M., & Bourgeois, L. J., III. (1988). Politics of strategic decision making in high-velocity environments: Toward a midrange theory. *Academy of Management Journal, 31*, 737–770.

Eisenhardt, K. M., & Zbaracki, M. J. (1992, Winter). Strategic decision making. *Strategic Management Journal, 13* (Special issue): 17–37.

Falshaw, J. R., Glaister, K. W., & Tatoglu, E. (2006). Evidence on formal strategic planning and company performance. *Management Decision, 44*, 9–30.

Finkelstein, S., & Haleblian, J. (2002). Understanding acquisition performance: The role of transfer effects. *Organization Science, 13*, 36–47.

Fiske, S. T., & Taylor, S. E. (1991). *Social cognition* (2nd ed.) New York: McGraw-Hill.

Gersick, C. J. (1994). Pacing strategic change: The case of a new venture. *Academy of Management Journal, 37*, 9–45.

Ghemawat, P. (1991). *Commitment, The dynamic of strategy*. New York: Free Press.

Ghoshal, S., Bartlett, C. A., & Moran, P. (1999). A new manifesto for management. *Sloan Management Review, 40* (Spring), 9–20.

Grant, R. M. (2003). Strategic planning in a turbulent environment: Evidence from the oil majors. *Strategic Management Journal, 24*, 491–517.

Greve, H. R. (1998). Performance, aspirations, and risky organizational change. *Administrative Science Quarterly, 43*, 58–86.

Hambrick, D. C. (1998). Corporate coherence and the top management team. In D. C. Hambrick, D. A. Nadler, & M. L. Tushman (Eds.), *Navigating change* (pp. 123–140). Cambridge, MA: Harvard Business Press.

Hambrick, D. C., & D'Aveni, R. A. (1988). Large corporate failures as downward spirals. *Administrative Science Quarterly, 33*, 1–23.

Hambrick, D. C., & Mason, P. A. (1984). Upper echelons: The organization as a reflection of its top managers. *Academy of Management Review, 16*, 193–206.

Hannan, M., & Freeman, J. (1984). Structural inertia and organizational change, *American Sociological Review, 49*, 149–164.

Haunschild, P. R., & Rhee, M. (2004). The role of volition in organizational learning: The case of automotive recalls. *Management Science, 50*, 1645–1560.

Haunschild, P. R., & Sullivan, B. N. (2002). Learning from complexity: Effects of prior accidents and incidents on airlines' learning. *Administrative Science Quarterly, 47*, 609–643.

Hayward, M. L. A. (2002). When do firms learn from their acquisition experience? Evidence from 1990–1995. *Strategic Management Journal, 23*, 21–39.

Hayward, M. L. A., & Hambrick, D. C. (1997). Explaining the premiums paid for large acquisitions: Evidence of CEO hubris. *Administrative Science Quarterly, 42*, 103–127.

Hofmann, D. A., & Frese, M. (2011). Errors, error taxonomies, error prevention, and error management: Laying the groundwork for discussing errors in organizations. In D. A. Hofmann & M. Frese, M. (Eds.). *Errors in organizations* (pp. 1–43). New York, NY: Routledge.

Huff, J. O., Huff, A. S., & Thomas, H. (1992). Strategic renewal and the interaction of cumulative stress and inertia. *Strategic Management Journal, 13*, 55–75.

Janis, I. L. (1972). *Victims of groupthink.* Boston: Houghton Mifflin.

Jarrell, G. A. (1993). An overview of the executive compensation debate. *Journal of Applied Corporate Finance, 5*, 76–82.

Jensen, M. C., & Meckling, W. H. (1976). Theory of the firm: Managerial behavior, agency costs and ownership structure. *Journal of Financial Economics, 3*, 305–360.

Johnson, J. L., Daily, C. M., & Ellstrand, A. E. (1996). Boards of directors: A review and research agenda. *Journal of Management, 22*, 409–438.

Johnson, R. A. (1996). Antecedents and outcomes of corporate refocusing. *Journal of Management, 22*, 439–483.

Kahneman, D. (1992). Reference points, anchors, norms, and mixed feelings. *Organizational Behavior and Human Decision Processes, 51*, 296–312.

Kahneman, D., & Tversky, A. (1979). Prospect theory: An analysis of decisions under risk. *Econometrica, 47*, 263–289.

Kaplan, S., & Beinhocker, E. D. (2003, Winter). The real value of strategic planning. *MIT Sloan Management Review*, pp. 71–76.

Kiesler, S., & Sproull, L. (1982). Managerial responses to changing environments: Perspectives on problem sensing from social cognition. *Administrative Science Quarterly, 27*, 548–570.

Kuratko, D. F., Ireland, R. D., & Hornsby, J. S. (2004). Corporate entrepreneurship behavior among managers: A review of theory, research, and practice. In J. A. Katz & D. A. Shepherd (Eds.), *Corporate entrepreneurship* (pp. 7–45). Oxford, UK: Elsevier.

Lant, T. K., Milliken, F. J., & Batra, B. (1992). The role of managerial learning and interpretation in strategic persistence and reorientation: An empirical exploration. *Strategic Management Journal, 13*, 585–608.

Lant, T. K., & Montgomery, D. B. (1987). Learning from strategic success and failure. *Journal of Business Research, 15*, 503–517.

Lerner, J. S., & Tetlock, P. E. (1999). Accounting for the effects of accountability. *Psychological Bulletin, 125*, 255–275.

Levinthal, D. A. (1991). Organizational adaptation and environmental selection—interrelated processes of change. *Organization Science, 2*, 140–145.

Levinthal, D. A., & March, J. G. (1993). The myopia of learning. *Strategic Management Journal, 14*, 95–112.

Levitt, B., & March, J. G. (1988). Organizational learning. *Annual Review of Sociology, 14*, 319–340.

Louis, M. R., & Sutton, R. I. (1991). Switching cognitive gears: From habits on mind to active thinking, *Human Relations, 44*, 55–76.

Lounamaa, P., & March, J. G. (1987). Adaptive coordination of a learning team. *Management Science, 33*, 107–123.

Lynn, S., Morone, J. G., & Paulson, A. S. (1996). Marketing and discontinuous innovation: The probe and learn process. *California Management Review, 38*, 8–37.
March, J. G. (1991). Exploration and exploitation in organizational learning. *Organization Science, 2*, 71–87.
March, J. G., and Shapira, Z. (1987). Managerial perspectives on risk and risk taking. *Management Science, 33*, 1404–1418.
March, J. G., Sproull, E. G., & Tamuz, M. (1991. Learning from samples of one or fewer. *Organization Science, 2*, 1–13.
McGrath, R. G. (1999). Falling forward: Real options reasoning and entrepreneurial failure. *Academy of Management Review, 24*, 13–30.
Miller, C. C., & Cardinal, L. B. (1994). Strategic planning and firm performance: A synthesis of more than two decades of research. *Academy of Management Journal, 37*, 1649–1665.
Milliken, F. J. (1987). Three types of perceived uncertainty about the environment: State, effect, and response uncertainty. *Academy of Management Review, 12*, 133–143.
Mintzberg, H. (1987). The strategy concept II: Another look at why organizations need strategies. *California Management Review, 30*(1): 25–32.
Mintzberg, H. (1990). The design school: Reconsideration the basic premises of strategic management. *Strategic Management Journal, 11*, 171–195.
Nadkarni, S., & Narayanan, V. K. (2007). Strategic schemas, strategic flexibility, and firm performance: The moderating role of industry clockspeed. *Strategic Management Journal, 28*, 243–270.
Nelson, R. R., & Winter, S. G. (1982). *An evolutionary theory of economic change.* Cambridge, MA: Belknap Press.
Ocasio, W. (1997). Towards an attention-based view of the firm. *Strategic Management Journal, 18S*, 187–206.
Perlow, L. A., Okhuysen, G. A., & Repenning, N. P. (2002). The speed trap: Exploring the relationship between decision making and temporal context. *Academy of Management Journal, 45*, 931–955.
Pfeffer, J. (1981). *Power in organizations.* Boston, MA: Pitman.
Pfeffer, J., & Sutton, R. I. (2006). *Hard facts, dangerous half-truths, and total nonsense: Profiting from evidence-based management.* Cambridge, MA: Harvard Business School Press.
Porter, M. E. (1987, May–June). From competitive advantage to corporate strategy. *Harvard Business Review*, pp. 43–59.
Prahalad, C. K., & Bettis, R. (1986). The dominant logic: A new linkage between diversity and performance. *Strategic Management Journal, 7*, 485–501.
Priem, R. L. (1990). Top management team group factors, consensus, and firm performance. *Strategic Management Journal, 11*, 469–478.
Roll, R. (1986). The hubris of corporate takeovers. *Journal of Business, 59*, 197–216.
Ross, J., & Staw, B. M. (1993). Organizational escalation and exit: Lessons from the Shoreham nuclear power plant. *Academy of Management Journal, 36*, 701–732.
Ross, L. (1977). The intuitive psychologist and his shortcomings: Distortions in the attribution process. In L. Kowitz (Ed.), *Advances in experimental social psychology* (pp. 173–220). New York, NY: Academy Press.

Russo, E. J., & Schoemaker, P. J. H. (1992). Managing overconfidence. *Sloan Management Review, 33* (Winter), 7–17.

Schulz, M. (2002). Organizational learning. In J. A. C. Baum (Ed.), *Companion to organizations* (pp. 415–441). Boston: Blackwell.

Shimizu, K. (2007). Prospect theory, behavioral theory, and threat-rigidity thesis: Combinative effects on organizational divestiture decisions of a formerly acquired unit. *Academy of Management Journal, 50,* 1495–1514.

Shimizu, K., & Hitt, M. A. (2004). Strategic flexibility: Organizational preparedness to reverse ineffective strategic decisions. *Academy of Management Executives, 18*(4), 44–59.

Shimizu, K., & Hitt, M. A. (2005). What constraints or facilitates divestitures of formerly acquired firms? The effects of organizational inertia. *Journal of Management, 31,* 50–72.

Simison, R. L., & Miller, S. (1999, September 24). Making "digital" decisions. *The Wall Street Journal,* 131.

Simon, H. A. (1976). *Administrative behavior* (3rd ed.). New York: Free Press.

Simonson, I., & Staw, B. M. (1992). Deescalation strategies: A comparison of techniques for reducing commitment to losing course of action. *Journal of Applied Psychology, 77,* 419–426.

Sitkin, S. B. (1992). Learning through failure: The strategy of small losses. In B. M. Staw and L. L. Cummings (Eds.), *Research in organizational behavior* (Vol. 14, 231–266). Greenwich, CT: JAI Press.

Smith, D. K., & Alexander, R. C. (1988). *Fumbling the future: How Xerox invented, then ignored, the first personal computer,* New York, NY: HarperCollins.

Staw, B. M. (1997). The escalation of commitment. In Z. Shapira (Ed.), *Organizational decision making* (pp. 191–215). New York: Cambridge University Press.

Staw, B. M., McKechnie, P. I., & Puffer, M. (1983). The justification of organizational performance. *Administrative Science Quarterly, 28,* 582–600.

Vaughan, D. (1996). *The Challenger launch decision.* Chicago: University of Chicago Press.

Wagner, J., & Gooding, R. (1997). Equivocal information and attribution: An investigation of patterns of managerial sense making. *Strategic Management Journal, 18,* 275–286.

Walsh, J. P. (1995). Managerial and organizational cognition: Notes from a trip down memory lane. *Organization Science, 6,* 280–321.

Walsh, J. P., & Seward, J. K. (1990). On the efficacy of internal and external corporate control mechanisms. *Academy of Management Review, 15,* 421–458.

Weick, K. E. (1995). *Sensemaking in organizations.* Thousand Oaks, CA: Sage.

Weisbach, M. S. (1995). CEO turnover and the firm's investment decisions. *Journal of Financial Economics, 37,* 159–188.

Wiersema, M. (2002). Hoes at the top: Why CEO firing backfire. *Harvard Business Review, 80*(12), 70–77.

9

When Things Go Wrong: Failures as the Flip Side of Successes

Erik Hollnagel

> Ignorance of remote causes disposeth men to attribute all events, to the causes immediate and instrumentall: for these are all the causes they perceive.
>
> **Thomas Hobbes (1588–1679),** *Leviathan***, Chapter 11**

The Need of Error as a Cause

Humans have always had a need to find explanations for what happens to them and around them, not least when things go wrong. It is, of course, entirely possible to be either a devout believer or a fatalist and meekly accept that things sometimes go right and sometimes go wrong. But, ever since humans started to work together and organize their work, there has been a need to find explanations in practical situations when the outcome was different from what was planned. Although this may have happened as far back as the Natufian culture in 12,500 to 9,500 BC, it is only since the mid-1980s that there has been an explicit need systematically to understand why organized activities can go wrong and as a consequence of that learn to design ways to prevent it.

The concepts of organization, planning, and outcomes are by their very nature strongly linked, and the relation among them almost amounts to a definition of what an organization is. Indeed, the meaning of *to organize* is "to plan" or to arrange systematically for something to be done together or collectively. An organization can consequently be defined as a group of people that pursues collective goals, that controls its own performance, and that has a functional or structural boundary separating it from its environment.

Humans collectively organize their activities to achieve goals or objectives that they cannot achieve alone. In doing so, they try to find the actions that can bring about the desired outcome and to consider the possible problems or obstacles. In other words, they think ahead, and they plan. If in the end it turns out that the objective was not achieved, if it is the case that something went wrong, then they usually try to explain what happened by finding—or inventing—a cause. The fundamental way of thinking in constructing explanations is that if the outcome was wrong, then the process that led to the outcome must also in some way have been wrong. Something must have gone wrong in the organized activity, or something in the organization must have failed. The unintended and unwanted outcome is an effect, and since we know that the law of causality reigns supreme, we know that there must be a cause, and we furthermore believe that we can find it.

Although this line of reasoning seems eminently plausible, it is nevertheless fundamentally flawed. It commits at least two of the four "great errors" described by Friederich Nietzsche in *The Twilight of the Idols* (1895), namely, *the error of false causality* and *the error of imaginary causes*. (It possibly also touches on the two other errors, namely, *the error of confusing cause and effect* and *the error of free will*.) Nietzsche conveniently provided a psychological explanation for the error of imaginary causes, by noting that

> [t]o extract something familiar from something unknown relieves, comforts, and satisfies us, besides giving us a feeling of power. With the unknown, one is confronted with danger, discomfort, and care; the first instinct is to abolish these painful states.

In this chapter, I argue that the efforts to find and classify "organizational errors" in many ways is an error of imaginary causes based on oversimplified assumptions about causality that are inappropriate for both individual and collective behavior. Instead of trying to understand organizations as complex machines that sometimes can fail, we should understand them as complex systems that continuously adjust what they do to match the current demands and resources. These adjustments are in the majority of cases sufficient to bring about the desired outcomes but may in some cases lead to things going wrong. When this happens it is, however, not because the adjustments were wrong or because they failed, but because they were approximate rather than perfect.

The Three Ages of Safety

The realization that things can go wrong is as old as civilization itself. Probably the first written evidence is found in the Code of Hammurabi, created circa 1760 BC, which even includes the notion of insurance against risk ("bottomry"). It was nevertheless not until the late 19th century that industrial risk and safety became a more common concern. Hale and Hovden (1998) have suggested that the scientific study of safety can be described in terms of three distinct ages, which they named the age of technology, the age of human factors, and the age of safety management (Figure 9.1). To understand the concern for errors in organizations, it is useful briefly to consider the developments in organizations and societies that led to the third age of safety, which is the one we currently experience.

The First Age

In the first age, the age of technology, the main concern was to find the technical means to safeguard machinery, to stop explosions, and to prevent structures from collapsing. This period was introduced by the Industrial Revolution (usually dated to 1769) and lasted throughout the 19th century until after World War II. The focus on technology was underlined by the observation made by Hale (1978) that investigators in the late 19th century were only interested in having accidents with technical causes reported since other accidents could not reasonably be prevented.

One of the earliest examples of a discussion of a systematic risk assessment was the Railroad Safety Appliance Act, from 1893, which argued for the need to combine safety technology and government policy control. A railway is also an excellent early example of an organized activity that

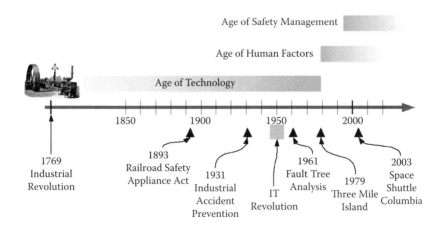

FIGURE 9.1
Three ages of safety. (After Hale, A. R. & Hovden, J., Management and culture: The third age of safety. A review of approaches to organizational aspects of safety, health and environment. In A. M. Feyer & A. Williamson (Eds.), *Occupational injury: Risk, prevention, and intervention*, CRC Press, Boca Raton, FL, 1988, pp. 129–165.)

includes all the facets that characterize today's organizational environment: the needs to plan activities and operations, to train staff, to procure materials and tools, to maintain equipment, to coordinate activities vertically and horizontally, to develop specialized functions and entities, to monitor and control daily operations, and so on. There were other prominent examples of a concern for safety, such as Heinrich's highly influential book, *Industrial Accident Prevention,* (1931). Yet despite the need for dependable equipment that exists in every industry, the need for reliability analysis only became widely recognized towards the end of World War II. One reason was that the problems of maintenance, repair, and field failures of the military equipment used during the World War II had become so severe that something had to be done. Another reason was that new scientific and technological developments made it possible to build larger and more complex technical systems that included extensive automation. Prime among these developments were digital computers, control theory, information theory, and the inventions of the transistor and the integrated circuit. The new systems were, however, often so complex that they challenged the human ability to describe and understand what went on inside them.

In the civilian domain, the fields of communication and transportation were the first to witness the rapid growth in complexity as equipment manufacturers adapted advances in electronics and control systems. In the military domain, the development of missile defense systems during the Cold War, as well as the beginning of the space program, relied on equally complex technological systems. This created a need for proven methods by which risk and safety issues could be addressed. Fault tree analysis was, for instance, originally developed in 1961 to evaluate the Minuteman Launch Control System for the possibility of an unauthorized missile launch. Other methods, such as failure mode and effects analysis (FMEA) and hazard and operability analysis (HAZOP), were developed not only to analyze possible causes of hazards (and later, causes of accidents) but also to identify hazards and risks, preferably before the systems were put into operation. (An FMEA is an analytical approach used to identify potential failure modes and their associated causes. It provides a way to determine the important risks that require actions to prevent that problems arise. A HAZOP is a systematic examination of a planned or existing process or operation to identify potentially harmful deviations.)

By the late 1940s and early 1950s, reliability engineering had become established as a new and separate engineering field that combined the powerful techniques of probability theory with reliability theory. This combination became known as probabilistic risk assessment (PRA), sometimes also called probabilistic safety assessment (PSA). The purpose of PRA is to evaluate risks associated with a complex engineered entity, such as an aircraft or a nuclear power plant. PRA was successfully applied to the field of nuclear power generation where the Atomic Energy Commission's

"Reactor Safety Study"—also known as the WASH-1400 study—became the defining benchmark (AEC, 1975). (The AEC is now called the Nuclear Regulatory Commission or NRC.) This study considered the course of events that might arise during a serious accident at a large, modern light water reactor; it used a fault tree/event tree approach. The WASH-1400 study established PRA as the standard approach in the safety assessment of modern nuclear power plants; the practice gradually spread to other industries with similar safety problems. The focus of PRA was, however, on the technology rather than on humans or the organization.

The Second Age

The second age was rather abruptly introduced by the accident at the Three Mile Island (TMI) nuclear power plant on March 28, 1979 (Kemeny, 1979). Before that accident, the safety of technical systems had been the main concern, and the established methods mentioned had been considered sufficient to ensure the safety of nuclear installations. After TMI, it became clear that something was missing in this approach, namely, the human factor. The human factor had already been considered in system design and operation through the discipline of human factors engineering, which had started in the United States as a specialty of industrial psychology in the mid-1940s. (In Europe, the history of human factors is a bit longer, as the journal *Le Travail humain* was started in 1937.) Human factors engineering had, however, focused mainly on the efficiency (productivity) side of system design and had paid little attention to safety issues. That changed completely after 1979.

Since PRA by that time had become established as the industry standard for how to deal with the safety and reliability of technical systems, it was also the natural starting point when the human factor needed to be addressed. The incorporation of human factors concerns in PRA led to the development of methods for human reliability assessment (HRA). This was first done by extending existing methods to consider "human errors" in the same way as technical failures and malfunctions but was soon replaced by the development of more specialized approaches. The details of this development have been described in several places (e.g., Hollnagel, 1998; Kirwan, 1994), but the essence was that human reliability became accepted as a necessary complement to system reliability—or rather that reliability engineering was extended to cover both the technological and human factors. The use of HRA quickly became established as the standard analysis for nuclear power plant safety, although there have never been any fully standardized methods (e.g., Dougherty, 1990), or even a reasonable agreement among the results produced by different methods (Poucet, 1989). The concern for human error soon started to appear in other industries, and the development of models and methods quickly took on a life of its own (e.g., Reason, 1990).

The Third Age

Whereas the first age of safety lasted more than 200 years, the second age lasted barely a decade. There were two main reasons for this. There was first an increasing dissatisfaction with the idea that health and safety could be ensured by a normative approach, simply by matching the individual to the technology (as in human–machine interaction design). Second, several accidents had made clear that the established approaches, including PRA-HRA and various human error methods, had their limitations. Although the change was less dramatic than in the aftermath of TMI, accidents such as the space shuttle *Challenger* disaster and the explosion of Reactor 4 at the Chernobyl nuclear power plant, which both happened in 1986, and in retrospect also the collision of two Boeing 747 airliners on the runway at Tenerife North Airport in Spain in 1977, demonstrated that the organization had to be considered over and above the human factor (Reason, 1997). One consequence was that safety management systems became a focus for development and research. This led to what is called the third age of safety, in which we still find ourselves.

The extension of the established basis for thinking about risk and safety (i.e., reliability engineering and PRA) to cover also organizational issues was, however, even less straightforward than in the case of human factors. Whereas the human in some sense could be seen as a machine, a tradition that goes back at least to la Mettrie (1748), the same was not the case for an organization. While it initially was hoped that the impact of organizational factors on, for example, the safety of nuclear power plants, could be determined by accounting for the dependence that these factors introduced among PSA parameters (Davoudian, Wu, & Apostolakis, 1994), it was soon realized that other ways of thinking were required. Pidgeon (1997) pointed out that organizational culture had a significant impact on the possibilities for organizational safety and learning, and that limits to safety might come from political processes as much as from technology and human factors. In a different context, the school of high-reliability organizations made clear that it was necessary to understand the organizational processes needed to operate complex, technological organizations (Roberts, 1990).

At present, the practice of risk assessment and safety management still finds itself in the transition from the second to the third age. On the one hand, it is realized by many, although not yet by all, that risk assessment and safety management must consider the organization as specific organizational factors (Van Schaardenburgh-Verhoeve, Corver, & Groeneweg, 2007), as safety culture, as "blunt-end" factors, and so on. Furthermore, if accidents are attributed to organizational factors, then any changes to these factors must obviously also be the subject of a risk assessment. On the other hand, it is still widely assumed that the established approaches inherited from engineering risk analysis can either be adopted directly or somehow be extended to include organizational factors and organizational issues. In other words, organizational "accidents" and organizational failures are seen as analogous to technical failures, just as human failures were in the aftermath of TMI. Since HRA had "proved" that the human factor could be addressed by a relatively simple extension of existing approaches, it seemed reasonable to assume that the same was the case for organizational factors. This optimism, however, was based on hopes rather than facts and has turned out to be completely unwarranted. It is becoming increasingly clear that neither human factors nor organizational factors can be adequately addressed using methods that follow the principles developed to deal with technical problems. There is therefore a need to revise or abandon the commonly held assumptions and instead take a fresh look at what risk and safety mean in relation to organizations.

From Sociotechnical Systems to Resilient Organizations

A study (Hollnagel & Speziali, 2008) looked at developments in accident investigation methods. It was found that although the sociotechnical systems that are the fabric of society continue to develop and to become more tightly coupled and complex, accident investigation methods—and a fortiori safety management methods—only change or develop occasionally. This means first that the methods we have and use today may be partly inappropriate. When the world around us changes and when the nature of the problems we face also changes, then we clearly cannot continue to use the same methods. But, it also means that new methods after some time will become underpowered, even though they may have been perfectly adequate for the problems for which they were developed in the first place. It is in this context a sobering thought that the predominant methods and models in both accident investigation and in risk assessment, and therefore also the mindset of people who use them, date from the 1970s, if not earlier. This goes for the individual (human) as well as the collective (organizational) domains alike.

The traditional safety models assume that events such as accidents and incidents can be represented as the end results of chains or sequences of causes and effects, either as simple linear progressions or as combinations of paths (e.g., Leplat, 1978). Accident investigations and risk assessments both typically proceed in a step-by-step fashion, gradually following links either backward or forward from the chosen starting point. In accident investigation, prominent examples are the domino model (Heinrich, 1931), which represents a simple linear model; the Swiss cheese model (Reason, 1990), which represents a complex linear model; and the man-technology-organization (MTO) model (Kecklund & Svenson, 1997), which also represents a complex linear model. In risk assessment, prominent examples are event trees (simple linear), fault trees (complex linear), and Petri nets (a tool for modeling discrete event systems, which also is complex linear).

Accidents and incidents can, however, sometimes occur in the absence of malfunctions and failures. In such cases, the accidents are explained as an outcome of coincidences and not as a result of causal relations. (Such outcomes are commonly called *emergent*.) The reason why this happens is that the increasing intractability of sociotechnical systems makes them uncontrollable. For a system to be controllable, it is necessary to know what goes on "inside" it, that is, to have a reasonably clear description or specification of the structure and functions of the system. The same requirements hold for a system to be analyzed and for its risks to be assessed. That this must be so is obvious if we consider the opposite. If we do not have a clear description or specification of a system, or if we do not know what goes on inside it, then it is clearly impossible both to control it effectively and to perform a risk assessment. We can capture these qualities by making a distinction between systems that can be completely described and systems that cannot be completely described or between *tractable* and *intractable* systems (see Table 9.1).

The established safety approaches all require that it is possible to describe the system in detail, for instance, by referring to a set of scenarios and a corresponding required functionality. (This is sometimes called the logical safety model.) Yet many, if not most, of today's sociotechnical systems are partly intractable. This means that the established methods and ways of thinking are unsuitable since they all presuppose that the system being analyzed is tractable. It also is not realistically possible to simplify the system descriptions so much that they become tractable in practice. It

TABLE 9.1

Tractable and Intractable Systems

	Tractable System	**Intractable System**
Number of details	Descriptions are simple with few details.	Descriptions are elaborate with many details.
Comprehensibility	Principles of functioning are known.	Principles of functioning are partly unknown.
Stability	System does not change while being described.	System changes before description is completed.
Relation to other systems	Independence.	Interdependence.
Metaphor	Clockwork.	Teamwork.

is therefore necessary to change approaches and to develop models and methods that recognize the following facts:

- *Performance conditions are always underspecified.* Since sociotechnical systems to some extent always are intractable, it is impossible to specify work in every detail; individuals and organizations must always adjust their performance to match the current conditions. Since resources and time are finite, such adjustments will inevitably be approximate. Performance variability is unavoidable but is the reason why things go right (source of successes) as well as why things go wrong (source of failures).
- *Many adverse events can be attributed to a breakdown or malfunctioning of components and normal system functions, but many cannot.* Such events are best understood as the result of unexpected combinations of the variability of normal performance. Adverse events thus represent the converse of the adaptations necessary to cope with real-world complexity.
- *Effective safety management cannot be based on hindsight or rely on error classification and the calculation of failure probabilities.* It is a general thesis of control theory that effective control cannot rely exclusively on feedback, except for very simple systems. Effective control requires that responses are prepared and sometimes executed ahead of time (i.e., feedforward). It also is insufficient to base safety management on a count of adverse outcomes since these are discrete events that do not represent the dynamics of the system.
- *Safety cannot be isolated from the core (business) process or vice versa.* Safety is the prerequisite for productivity, and productivity is the prerequisite for safety. Safety must therefore be achieved by improvements rather than by constraints.

Safety is often defined as, for example, the "freedom from unacceptable risk." This definition leads to a striving to eliminate failures and malfunctions as much as possible and therefore creates a need to be able to classify and describe failures and malfunctions. In agreement with the four principles listed in this section, safety should, however, be defined as "the ability to succeed under varying conditions." As a consequence, efforts are needed that improve the resilience of the organization.

The term *resilience engineering* (Hollnagel, Paries, Woods, & Wreathall, 2011; Hollnagel, Woods, & Leveson, 2006) represents an alternative way of thinking about safety. Whereas established risk management approaches are based on hindsight and emphasize error tabulation and calculation of failure probabilities, resilience engineering adheres to the four principles listed. Resilience engineering looks for ways to enhance the ability of organizations to create processes that are robust yet flexible, to monitor and revise risk models, and to use resources proactively in the face of disruptions or ongoing production and economic pressures. In resilience engineering, performance is based on the ability of groups, individuals, and organizations to adjust what they do to match present and near-future conditions, including threats and opportunities. This is the reason why things go right, as well as the reason why things sometimes go wrong. A consequence of this perspective is that there no longer is any need for a classification of failures or errors, whether on the individual or the collective level. What is needed instead is a way of describing or characterizing normal performance, in particular the variability of normal performance. The case of the "naked" machinery further in this chapter shows how this may be done.

Variability of Normal Performance and the Efficiency–Thoroughness Trade-Off Principle

In the study of organizational behavior, it has long been recognized that the theories of rational decision making are ill suited as descriptions of what happens in practice. People in organizations, from clerks to managers, do not make rational decisions but rather get by via a number of heuristics, strategies, or tricks that variously have been called muddling through (Lindblom, 1959), satisficing (March & Simon, 1993), prospecting (Kahneman & Tversky, 1979), and recognition-primed decision making (Klein, 1993). The use of these heuristics means that decisions most of the time are right (since otherwise the heuristics would have been abandoned), although they sometimes may be wrong. When they are wrong, the reason is not that the heuristic failed or malfunctioned, but rather that it was a heuristic rather than an algorithm and hence a trade-off between being "rational" on the one side and not spending more effort than necessary to make the decision on the other.

If decision making can be described in terms of a trade-off, then it seems reasonable to assume that other actions can be described in much the same way. Indeed, it is possible to propose that normal performance at work—and at leisure—can be described as if it followed an efficiency–thoroughness trade-off (ETTO) principle (Hollnagel, 2009a). The ETTO principle simply describes the fact that people in their daily activities routinely make a choice between being efficient and being thorough.

They do this because it rarely is possible to be both efficient and thorough at the same time. If demands for productivity or performance are high, thoroughness is reduced until the productivity goals are met. Conversely, if demands for safety are high, efficiency is reduced until the safety goals are met. Efficiency here means that the level of investment or amount of resources used or needed to achieve a stated goal or objective is kept as low as possible. Thoroughness means that an activity is carried out only if the individual or organization is confident that the necessary and sufficient conditions for it exist, so that the activity will achieve its objective and not create any unwanted side effects. For individuals, the decision about how much effort to spend is usually not made consciously but is rather a result of habit, social norms, and established practice. For organizations, it is more likely to be the result of a direct consideration.

If we look at individual work from this perspective, it does not take long to recognize a number of characteristic ETTO rules that people, from the factory floor to the boardroom, apply to cope with the complexity of their daily activities (for further details, see Hollnagel, 2009a). Some examples are as follows:

- "It looks fine"—so there is no need to do anything.
- "It is normally okay; there is no need to check"—it may look suspicious but do not worry.
- "(Doing it) this way is much quicker"—or uses fewer resources.
- "It normally works"—so it will probably also work now. This eliminates the effort needed to consider the situation in detail to find out what to do.
- "It must be ready in time"—so let us get on with it.
- "If you don't say anything, I won't either"—in this situation, one person has typically "bent the rules" to make life easier for another person or to offer some kind of service.
- "I am not an expert on this, so I will trust you. "—This is another kind of social ETTO rule; time and effort are saved by deferring to the knowledge and experience of another person.

If we look to the organization, it is possible to find collective counterparts to the individual ETTO rules. Some examples are as follows:

- Negative reporting, which means that only deviations or things that go wrong should be reported.
- The prioritizing dilemma or the visibility-effectiveness problem. Managers are often required by their bosses to be both efficient

in accomplishing their administrative duties and thorough in the sense that they are good managers (i.e., highly visible).
- Report and be good. Subcontractors and suppliers often feel under pressure to meet the standards of the organization for openness and reporting. At the same time, they may fear that they will be punished if they have too many things to report.

The ETTO principle provides a convenient way of characterizing typical behaviors from a work, an individual, or a collective perspective. In particular, it makes it possible to account for the variability of normal performance without invoking the notion of failure or error. In the following section, I use the ETTO principle to provide an alternative analysis of a work situation for which things went wrong.

The "Naked" Machinery

The example of naked machinery is taken from Kubota et al. (2001). It describes a case in which a large piece of heavy machinery was transported to the customer wrapped in a plastic sheet rather than the normal rustproof sheet. During the transportation, the plastic sheet came off, and the machinery arrived in a naked state that was unacceptable to the customer. This outcome was not an accident in the sense that someone was hurt or that material was destroyed. But, it was a serious event nevertheless since it concerned a large piece of machinery at considerable cost. Described in more detail, the event unfolded as described next. (Notice that the seven steps have been defined post hoc as places at which a "human error" occurred. A standard work analysis might therefore have resulted in a different breakdown.)

1. Design Department B (DD-B; design of hardware and equipment) placed a customer order for the manufacturing of the specific machinery to Manufacturing Department B (MD-B; manufacturing of products) on October 6, 1998. When the order was issued, it did not include specific packing instructions (i.e., that the machinery should be protected by a rustproof sheet during transportation).
2. In January 1999, MD-B noticed that DD-B had not provided any packing instructions. DD-B therefore submitted a request to Design Department A (DD-A; design of plan) to prepare a packing specification. DD-A did so and produced the packing specification during February, but there was a delay in issuing the specification.
3. After obtaining the packing specification from DD-A, DD-B issued the detailed instructions for packing on February 3, 1999.

This was done just before the machinery was due to be transported to the customer.
4. Manufacturing Department A (MD-A; management of schedule) received the specification on February 4, 1999. It gave instructions to MD-B, to the Inspection Department (ID), and to the Transportation Department (TD) but used the normal procedure for communication rather than one marking it urgent. Because of that, the specification did not reach the TD in time.
5. MD-B received the specification on February 4, 1999, but the department had too many specifications to process as part of its normal work. The department therefore missed the specification in question. Since the information did not reach the TD and MD-B in time, the packing of the machinery did not comply with the instructions from DD-B (see Step 3).
6. The ID received a request for inspection on February 9, 1999. It conducted the inspection to clear the machinery to be transported the following day. During the evening of the inspection, the ID found that parts of the machinery were not dimensioned according to the original design. The MD-B quickly made the necessary repairs and repainted the machinery. A second inspection of the machinery took place around 9 o'clock on February 11, 1999. Although the paint on the machinery was still not dry, it was inspected and cleared for transportation around 10 o'clock. The scheduled time for transportation to the customer was imminent.
7. The TD had been waiting since February 10, 1999. It packed the machinery around 10 o'clock on February 11. The Design Department had instructed that the machinery should be packed with a rustproof sheet, but the paint on the machinery was not yet dry. As there was little time left, the TD wrapped the machinery in a plastic sheet. The plastic sheet came off during transportation, and the machinery arrived at the destination in a naked state the next day, February 12.

Even this abbreviated description makes clear that this is a typical "organizational accident." It involved no less than six different departments, requiring the coordination of activities among these departments. The events also took place over an extended period of time, from October 6, 1998, to February 12, 1999. This means that the various steps had to fit into the ongoing activities and the busy work schedules of the six departments.

In the original analysis of this event, Kubota et al. (2001) identified seven human errors, which then were analyzed using an extended version of the cognitive reliability and error analysis method (CREAM; Hollnagel, 1998).

1. DD-B did not issue the instruction for rustproof packing when arranging an order to manufacture the machinery.
2. DD-A was late in issuing the packing specifications.
3. DD-B was late in issuing the instructions for rustproof packing for transportation of the machinery.
4. MD-A used the normal communication procedure for communication and did not mark it as urgent.
5. MD-B did not look at the specification.
6. The ID inspected and accepted the machinery for release, even though it was still wet.
7. The TD did not check the machinery in its final packed condition.

Each of these sharp-end human errors can be seen as an instance of a sharp-end failure caused by a blunt-end organizational error. As an example, Kubota et al. (2001) found that the first human error could be explained as the outcome of the unclear specification of organizational roles combined with inadequate quality control procedures, and that someone in DD-B assumed that the packing instructions would be done by DD-A. It is, however, also possible to explain each of the seven cases as the unexpected, and unfortunate, outcome of the variability of normal performance as shown in Table 9.2.

Looked at in this way, each and every step can be seen as a normal way of responding in a situation. If you think that a certain task is the responsibility of someone else, then there is no reason to do it yourself. When you are busy working, you deal with requests in the order they arrive unless they are clearly marked as special. If you are short of time, then you only deal with what you think is critical and assume that the rest is okay. And, if you do not have time to wait for something to be ready, you find a shortcut to the problem. Indeed, this way of responding not only is normal and tacitly accepted but also is sometimes even rewarded.

Work in an organization never happens in a vacuum but always takes place in a social context, in a complex fabric of interactions and dependencies. What one person does follows what another has done and is itself followed by what a third person will do. These couplings may either be more or less immediate or happen with long delays. In the current example, both relations can be found. Every person in the organization was a part of this social fabric. In relation to the work that each person does, thoroughness means that the input he or she receives from somewhere or from someone else is not simply accepted or taken on trust, but that efforts are made to confirm that it was correct. Thoroughness also means that people consider the possible side effects and secondary outcomes of what they produce

When Things Go Wrong: Failures as the Flip Side of Successes 239

TABLE 9.2

"Human Error" Versus ETTO Rules

Assumed Human Error	Applicable ETTO Rule
DD-B did not issue the instruction for rustproof packing when arranging an order to manufacture the machinery.	*It will be done (or checked) by someone else later.* At the time the machinery was ordered, the packing instructions were not considered since it was assumed that someone else would do that. Anyway, it was something that would happen much later.
DD-A was late in issuing the packing specifications.	*There is no time to do it now.* The request from DD-B came unexpectedly and was treated as just another task, which also had to compete with other work in DD-A.
DD-B was late in issuing the instructions for rustproof packing for transportation of the machinery.	*It is not really important.* The response to the request that had been issued earlier (end of January) was received, but there were too many other things to do, and it was treated as any other task (see next step.)
MD-A used the normal communication procedure for communication and did not mark it as urgent.	*It is not really important.* This was the first and only time MD-A was involved in this event. It handled this as any other assignment and followed normal routines.
MD-B did not look at the specification.	*There is no time or resources to do it now, so we will do it later.* The packing specifications were considered as just another part of the "normal" flow of work. For a time, the specification was "lost" in convergent tasks.
The ID inspected and accepted the machinery for release, even though it was still wet.	*It must be ready in time.* While the first inspection clearly worked, the second inspection was hurried due to a shortage of time. It may have been made more difficult because the paint on the machinery was still wet.
The TD did not check the machinery in its final packed condition.	*We must get this done (before time runs out). It is good enough for now. Doing it this way is much quicker.* There was no time to wait for the paint to dry, and it was important that the customer received the machinery on time. A temporary solution was therefore found.

as output, in a sense adopting the mindset of whoever is going to use the results. Similarly, efficiency means that people trust that the input they receive is correct, hence assuming that the previous person was thorough. Efficiency also means that the person assumes that the next person, whoever is going to use the results, will make the necessary checks and verifications (i.e., that the next person is thorough). This situation is illustrated in Figure 9.2.

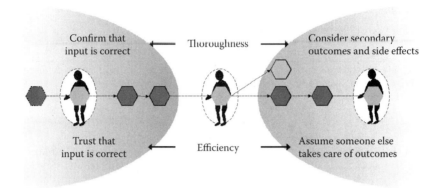

FIGURE 9.2
The ETTO principle in an organizational setting.

Most working conditions allow insufficient time to check everything. (This obviously goes for many situations outside work as well.) It is therefore necessary to trust what others do; indeed, organized work would be impossible without that. Trading off thoroughness against efficiency is in practice the only means available to save time and resources. In the best case, people may think ahead—or back—one step but only spend a small amount of effort to do more than that. It is as if everyone, including ourselves, reasons in the following way: "I can allow myself to be efficient because the others will be thorough." If some but not all do that, the system may be able to correct itself and to find a balance of functioning that is both reasonably effective and reasonably thorough. But, if everyone begins to work in this way, the net result may be that something goes wrong. Yet, the reason for that is not that anyone did anything that was manifestly wrong or incorrect (i.e., the explanation is not to be found in individual or organizational errors). The reason is more likely that everyone made an ETTO just as he or she normally does. The ETTO fallacy is that people are required to be both efficient and thorough at the same time—or rather to be thorough when with hindsight it was wrong to be efficient.

Safety by Constraints, by Design, and by Management

Safety is often pursued by imposing constraints on work. Examples of frequently used constraints are strict training, barriers of various kinds, procedures, standardization, rules, and regulations. This solution is based on the view that humans are a liability, and that their inability to perform in a reliable (or machine-like) manner is a threat. It follows from this premise that efficiency can be maintained, and that malfunctions/failures can be avoided by constraining all kinds of performance variability. Although this practice is widely used, it is both theoretically wrong and practically inefficient.

Safety can also be pursued by design, which means that all possible, or practicable, precautions needed to ensure an acceptable level of safety are taken ahead of time. This can be when the system is conceived of or designed, when detailed plans for operation are made, or when the system is made ready for operation. The technical parts of the system are designed in detail and must perforce be configured, operated, and maintained according to meticulously prepared instructions. In such cases, safety by design is not only an option but also a requirement. Even so, there are limitations to what this can accomplish:

> Clearly not all accidents can be prevented by design. There are some consequences of technology that cannot reasonably be predicted at the design stage, particularly in new technologies, using new materials and scientific principles. However, once these have led to accidents, there is a clear responsibility for designers to prevent them in future designs. (Hale, Kirwan, & Kjellen, 2007, p. 310)

The nontechnical parts of the system, namely, the people and the organization, are not susceptible to design. Indeed, an organization can never be explicitly designed in the same way that a piece of machinery can, one fundamental reason being that the "components" by their nature are variable and flexible, regardless of whether they are considered individually or collectively. Humans and social systems simply do not function like machines, despite courageous attempts to make them do so through training, interaction design, and automation. In these cases, the alternative is to focus on the operation of the system and on how it is possible to keep the variability within acceptable limits (i.e., by managing safety).

The principle of safety by design presupposes that something fails or has failed, so that it can be corrected. But, in the resilience engineering perspective things that go wrong are usually not due to failures. The design issue therefore becomes how to manage performance under varying conditions. This can be done by identifying situations in which the variability of normal performance, described, for example, by the ETTO rules, may combine to create unwanted effects and continuously monitor how the system functions to intervene and "dampen" performance variability when it threatens to become out of control. (Similarly, performance variability should be accentuated or amplified when it can lead to successful outcomes.) This leads to the principle of safety by management, which is at the heart of resilience engineering (cf. Hollnagel, 2009b).

From Organizational Errors to Inappropriate Responses

Human's performance—individually and collectively—must always be adjusted to the working conditions. Because these adjustments normally work well, human performance is as a rule appropriate or adequate. However, there will be some situations for which human performance—individually or collectively—will be inappropriate. This is more often because the conditions have changed than because the performance changes. In other words, people do what they normally do, but this may under certain conditions turn out to be wrong. The conditions may change in the sense that they are different from what they normally are, hence different from what the person reasonably expected. Since this happens rarely, it is ecologically reasonable, in the Brunswikian sense, for people to do what they normally do.

People and organizations can only spend a limited amount of time or effort to verify that the conditions are as they are assumed to be. They only do it if something forces them to do so, whether it is the inability to succeed with something that normally works, the recognition that something has changed, or a feeling of uneasiness—that something is not quite right (Klein, Pliske, Crandall, & Woods, 2005)—and even then they usually look for confirming evidence. It makes considerable sense for people to assume that their expectations are correct. If they did not do that, they would constantly have to assess the situation; consequently, they would spend too much time ensuring that everything was right and too little time in actually doing something. The question, of course, is when and how often it pays to check whether the assumptions hold. When a person is on his or her own (i.e., outside an organization) and if the environment changes most or all of the time, it makes sense to do so often. (A simple example of that is driving a car.) When a person is part of a well-functioning organization, it makes sense not to do it too often since one purpose of the organization precisely is to ensure that the working environment is orderly and predictable. Indeed, organizations go to great length to make sure that is the case in order to improve productivity and efficiency. This means that an organization, by the logic of its very nature, indirectly encourages people to be efficient rather than to be thorough. If and when it turns out wrong, the reason is therefore not that an "organizational error" was made but rather that the responses were inappropriate for the situation.

References

Atomic Energy Commission. (1975). *Reactor safety study: An assessment of accident risks in U.S. Commercial power plants* (WASH-1400). Washington, DC: Author.

Davoudian, K., Wu, J.-S., & Apostolakis, G. (1994). Incorporating organizational factors into risk assessment through the analysis of work processes. *Reliability Engineering and Systems Safety, 45*, 85–105.

Dougherty, E. M., Jr. (1990). Human reliability analysis—Where shouldst thou turn? *Reliability Engineering and System Safety, 29*, 283–299.

Hale, A. R. (1978). *The role of HM inspectors of factories with particular reference to their training*. Doctoral thesis, University of Aston, Birmingham, UK.

Hale, A. R., & Hovden, J. (1998). Management and culture: The third age of safety. A review of approaches to organizational aspects of safety, health and environment. In A. M. Feyer & A. Williamson (Eds.), *Occupational injury: Risk, prevention, and intervention* (pp. 129–165). Boca Raton, FL: CRC Press.

Hale, A., Kirwan, B., & Kjellen, U. (2007). Safe by design: Where are we now? *Safety Science, 45*, 305–327.

Heinrich, H. W. (1931). *Industrial accident prevention*. New York: McGraw-Hill.

Hollnagel, E. (1998). *Cognitive reliability and error analysis method*. Oxford, UK: Elsevier Science.

Hollnagel, E. (2009a). *The ETTO principle: Why things that go right sometimes go wrong*. Aldershot, UK: Ashgate.

Hollnagel, E. (2009b). Extending the scope of the human factor. In E. Hollnagel (Ed.), *Safer complex industrial environments: A human factors approach*. Boca Raton, FL: CRC Press.

Hollnagel, E., Paries, J., Woods, D. D., & Wreathall, J. (Eds.). (2010). *Resilience engineering in practice: A guidebook*. Aldershot, UK: Ashgate.

Hollnagel, E., & Speziali, J. (2008). *Study on developments in accident investigation methods: A survey of the "state-of-the-art"* (SKI 2008:50). Stockholm, Sweden: Swedish Nuclear Inspectorate.

Hollnagel, E., Woods D. D., & Leveson, N. (2006). *Resilience engineering: Concepts and precepts*. Farnham, UK: Ashgate.

Kahneman, D., & Tversky, A. (1979). Prospect theory: An analysis of decision under risk. *Econometrica, 47*, 263–291.

Kecklund, L. J., & Svenson, O. (1997). Human errors and work performance in a nuclear power plant control room: Associations with work-related factors and behavioural coping. *Reliability Engineering and System Safety, 56*, 5–15.

Kemeny, J. G. (1979). *Report of the president's Commission on the Accident at Three Mile Island: The need for change: The legacy of TMI*. Washington, DC: Commission on the Accident at Three Mile Island. U.S. Government Printing Office.

Kirwan, B. (1994). *A guide to practical human reliability assessment*. London: Taylor & Francis.

Klein, G. A. (1993). *A recognition-primed decision (RPD) model of rapid decision making*. Norwood, NJ: Ablex.

Klein, G. A., Pliske, R., Crandall, B., & Woods, D. D. (2005). Problem detection. *Cognition, Technology and Work, 7*(1), 14–28.

Kubota, R., Kiyokawa, K., Arazoe, M., Ito, H., Iijima, Y., Matsushima, H., et al. (2001). Analysis of organisation-committed human error by extended CREAM. *Cognition, Technology and Work, 3*(2), 67–81.

La Mettrie, J. O. (1748). *L'Homme machine*. Retrieved April 27, 2010 from http://www.cscs.umich.edu/~crshalizi/LaMettrie/Machine/

Leplat, J. (1978). Accident analysis and work analysis. *Journal of Occupational Accidents, 1,* 311–340.

Lindblom, C. E. (1959). The science of "muddling through." *Public Administration Review, 19,* 79–88.

March, J. G., & Simon, H. A. (1993). *Organizations*. Cambridge, MA: Blackwell.

Nietzsche, F. (1895). *Die Götzen-Dämmerung—Twilight of the idols* [Text prepared from the original German; W. Kaufmann and R. J. Hollingdale, Trans.]. Retrieved August 15, 2009, from http://www.handprint.com/SC/NIE/GotDamer.html

Pidgeon, N. (1997). The limits to safety? Culture, politics, learning and man-made disasters. *Journal of Contingencies and Crisis Management, 5*(1), 1–14.

Poucet, A. (1989). *Human factors reliability benchmark exercise—Synthesis report* (EUR 12222 EN). Ispra, Italy: CEC Joint Research Centre.

Reason, J. T. (1990). *Human error*. Cambridge, UK: Cambridge University Press.

Reason, J. T. (1997). *Managing the risks of organizational accidents.* Aldershot, UK: Ashgate.

Roberts, K. H. (1990). Some characteristics of one type of high reliability organization. *Organization Science, 2,* 160–176.

Van Schaardenburgh-Verhoeve, K. N. R., Corver, S., & Groeneweg, J. (2007, April 25–26). *Ongevalonderzoek buiten de grenzen van de organisatie* [Accident investigation beyond the boundaries of organizational control]. Paper presented at NVVK Jubileumscongres, Sessie C, p. 1–11. The Netherlands.

10

The Link Between Organizational Errors and Adverse Consequences: The Role of Error-Correcting and Error-Amplifying Feedback Processes

Rangaraj Ramanujam and Paul S. Goodman

This chapter examines the link between latent errors and adverse organizational consequences. By latent errors, we refer to unintended deviations from prespecified expectations (e.g., rules, standard operating procedures, SOPs) that can *potentially* generate adverse outcomes of organizational significance (Ramanujam & Goodman, 2003). One example of a latent error in the securities trading operations of a bank would be the failure to review the trading transactions periodically as required by internal rules and external regulations. This error may allow a buildup of unauthorized trades that can potentially generate huge losses (Basel Committee on Banking Supervision, 2008). Another example of a latent error, in routine medication administration processes at a hospital, would be a failure to verify the correctness of the drug and its dose before administering it to a patient (Institute of Medicine, 1999). This error makes it more likely that a wrong drug or a wrong dose will be given to the patient, which can seriously harm or even kill the patient. Latent errors occur in a wide range of organizational settings, including nuclear power plants, aviation, coal mines, chemical plants, and space shuttle launch operations (cf. Reason, 1998). Given their prevalence and potential for undermining organizational effectiveness, there is growing interest in understanding how organizational structures and processes contribute to latent errors and adverse outcomes (Hofmann & Frese, Chapter 1, this volume; Reason, 2008).

In this chapter, we examine when and how latent errors actually generate organizationally significant adverse outcomes. Our core premise is that latent errors and adverse outcomes represent two separate concepts that are only loosely linked. For instance, although equivalent latent errors have been observed across the securities trading operations of several financial institutions, these errors contributed to multimillion dollar losses in only a few of these organizations, such as Barings (Ramanujam & Goodman, 2003). Similarly, although identical latent errors have been reported across the medication administration processes of several hospitals, these errors caused serious harm only in a few of these hospitals and to only a few patients (Institute of Medicine, 1999). Therefore, given that latent errors occur in the operations of all organizations, this invites a basic question: Which organizational processes cause latent errors to generate major adverse consequences in some organizations rather than in others?

By addressing this question, which is rarely discussed in the organizational literature, we intend to make several contributions. First, we clarify the frequently blurred distinction between latent errors and adverse outcomes. Currently, it is hard to distinguish the organizational explanations of errors from the organizational explanations of outcomes such as accidents. By formally separating these explanations, we identify several unexplored research questions that may help to understand better the organizational origins of the link between errors and adverse outcomes. Second, we offer a new conceptualization of the role of organizational antecedents that complements, but is different from, current approaches in a couple of important respects. We primarily focus on the antecedents of the feedback processes that link errors and adverse outcomes. Moreover, whereas prior discussions tended to focus exclusively on either the negative-feedback processes that reduce errors (e.g., Weick & Roberts, 1993) or the positive-feedback processes that amplify errors (e.g., Vaughan, 1996), we examine the interaction between these two sets of feedback processes. Our conceptualization can potentially help explain the puzzling phenomenon of even highly reliable organizations (i.e., organizations that have strong safety goals, safety climate, and error management processes) occasionally experiencing major adverse outcomes (Blatt, Christianson, Sutcliffe, & Rosenthal, 2006).

Third, in developing our arguments, we draw attention to the important need for conceptualizing latent errors at the organizational level of analysis. Given that errors entail the actions of individuals, most organizational studies understandably draw from prior research on individual-level errors. However, few studies have explored whether and how studying errors at the organizational level of analysis differs from studying the organizational antecedents and consequences of individual-level errors that occur in organizational settings. By proposing a couple of different ways to conceptualize errors at the organizational level of analysis, we wish to advance research about errors as an organizational-level phenomenon.

Finally, we present a conceptual framework that can potentially serve as the basis for bringing together the insights from the fragmented organizational studies of accidents (e.g., Perrow, 1984); high reliability (e.g., Bigley

& Roberts, 2001; Roe & Shulman, 2008); mindful organizing (e.g., Weick & Sutcliffe, 2006); safety climate (e.g., Hofmann & Mark, 2006); and error management (e.g., Keith & Frese, 2008; Heimbeck, Frese, Sonnentag, & Keith, 2005). In developing this framework, we draw on the findings from these disparate studies and point out the interconnections among these different literatures.

This chapter is organized as follows: First, we delineate latent errors from related concepts such as violations, risk, and adverse outcomes such as accidents, safety, and reliability and discuss the distinctive features of latent errors that are organizational. Next, we introduce a conceptual framework about the role of organizational feedback processes in linking latent errors and adverse consequences. We then present two contrasting cases of latent errors that produced adverse outcomes—one in an investment bank and the other in a hospital. In the investment bank, several precursors of errors and adverse outcomes that are frequently discussed in the literature (e.g., strong production goals, underdeveloped error management processes) were present. It was an accident waiting to happen. By contrast, the hospital was doing all the "right things" in terms of the recommendations for effective error management and high reliability that are discussed in the literature (e.g., strong safety goals, strong safety training). Yet, in each case not only were latent errors present but also they proliferated and eventually generated major adverse outcomes—$1.3 billion in losses and bankruptcy for the investment bank and preventable deaths of three prematurely born babies in the hospital's neonatal intensive care unit. We use these cases both to illustrate the framework and to generate research questions. We conclude with a discussion of the challenges and opportunities for future organizational research about the role of feedback processes in the linkage between latent errors and adverse consequences.

Delineating Latent Errors and Adverse Outcomes

Latent errors refer to unintended deviations from prespecified expectations that can potentially lead to adverse outcomes of organizational significance (Ramanujam & Goodman, 2003). Let us consider the various components of this definition, starting with expectations. It is meaningful to talk about errors only in reference to a prespecified standard or expectation. In an organizational context, such expectations are conveyed through rules, regulations, SOPs, and normative expectations that contain specific prescriptions and proscriptions about how work must or must not be carried out (Scott, 2002). Such expectations govern actions in the setting most relevant to studying errors—the daily operations of the technical core of an organization, where most actions tend to be rule based (March, 1997). However, even in these settings, there may be situations for which the expectations are unavailable, unknown, not well understood, or even incorrectly specified (i.e., conforming to the expectations can be detrimental to the organization; Reason, 1998). Although such situations also are a reality in organizational life, to simplify our initial analysis, we focus on situations for which organizational members have shared knowledge and understanding about expectations. For instance, without exception, employees in the back office operations of a financial institution understand that they are required to check every trading transaction, and nurses in a hospital understand that they are expected to verify medication details prior to administering medication to patients. Latent errors in such situations have contributed to major adverse outcomes in diverse organizational settings (Reason, 2008).

The second component of latent errors is *deviations*, which refer to the unintended actions of organizational members that do not conform to the prespecified expectations. We focus here on the actions of individuals who are acting in their formal organizational roles (e.g., a manager responsible for checking trading transactions, a nurse responsible for administering medications to patients) and oriented toward organizational goals (e.g., maintaining effective internal control over securities trading; ensuring the safety of patients). Deviations that are deliberately intended to subvert the organization (e.g., sabotage) or solely benefit the individual employee (e.g., employee theft) are beyond the scope of our discussion.

The third component underscores what is latent in these deviations: their potential, as yet unrealized, for causing organizationally significant adverse outcomes. These often are foreseeable outcomes that the prespecified expectations were specifically designed to help avoid. The rules requiring the verification of trading transactions in a bank or medications in a hospital are designed to avert the foreseeable adverse outcomes that may result if verifications are not carried out. The word *potential* signifies that adverse outcomes may or may not occur. A nurse's failure to verify the medication does not always produce an adverse consequence. If, as a result, however, the nurse administers an overdose of a high-risk drug to a high-risk patient, this error can seriously harm, even kill, the patient and generate additional adverse consequences for the hospital (e.g., litigation, reputation loss). This error can also generate even more serious outcomes if it combines or interacts with other latent errors. The point is that latent errors can potentially contribute to a wide range of organizationally significant adverse consequences, such as loss of life, injury, damage to physical equipment, disruptions to production schedule, costly product recalls and litigation, negative publicity, steep decline in sales, regulatory sanctions, financial losses, and bankruptcy (Reason, 2008).

Latent errors must be distinguished from related concepts, such as violations, risk, safety, reliability, and accidents. Whereas *errors* refer to deviations that are unintended, *violations* refer to deviations that are intentional (Hofmann & Frese, Chapter 1, this volume). Although errors and violations are hard to tell apart from the viewpoint of observable behaviors, their underlying intrapersonal psychological mechanisms are different. Errors result from problems in cognitive processes such as attention, memory, and understanding that cause individuals to forget rules, select the wrong rule, or incorrectly execute the correct rule (Rasmussen, 1987). In contrast, violations result from choice-based reasoning; individuals deliberately choose to deviate from known rules. Latent errors differ from risk in that errors refer primarily to the actions that deviate from a standard, while risk refers to the assessment, actual or perceived, of the likelihood and the magnitude of the adverse outcomes that could result from such actions (Hofmann & Frese, Chapter 1, this volume).

Finally, latent errors differ from but are related to organizational outcomes such as accidents (Perrow, 1984), high reliability (Roberts, 1990), and safety (Sagan, 1993). Latent errors sometimes precede and contribute to accidents, which are rare system-level events with adverse outcomes. However, accidents can also occur in the absence of errors because of violations, unexpected events, and so on. In other words, latent errors are neither necessary nor sufficient for the occurrence of accidents. Similarly, with their inherent potential for adverse consequences, latent errors pose a major threat to reliability (i.e., the extended absence of adverse outcomes in organizations that operate complex hazardous technologies) and safety (i.e., the sustained avoidance of physical harm to organizational stakeholders). However, it is conceivable that organizations may continue to deliver reliable and safe outcomes despite the widespread presence of latent errors or may produce unreliable or unsafe outcomes despite a low incidence of errors.

In our discussion so far, we have not made any explicit reference to the level of analysis. However, we are primarily interested in studying errors as an organizational-level phenomenon. Therefore, one important question is what it means to study errors—which are primarily the actions of individuals—at the organizational level of analysis. Surprisingly, this question is seldom discussed in the literature, in which studying errors as an organizational-level phenomenon is often implicitly equated to studying the organizational-level antecedents or the organizational-level consequences of individual-level errors in organizational settings. However, from the viewpoint of advancing the study of errors as an organizational-level phenomenon, it is necessary not only to identify their organizational-level causes and consequences but also to develop a meaningful representation of errors at the organizational level of analysis.

We propose two complementary ways to think about errors at the unit level of the organizational level of analysis. First, some errors possess features that render them inherently organizational. Second, the composition of errors may be helpful in systematically differentiating and comparing the incidence of latent errors between different work units or organizations. As previously discussed, latent errors in an organization entail the actions of *individuals* that deviate from prespecified expectations. In this sense, every latent error begins as an individual-level error. However, some latent errors acquire organizational characteristics. For instance, rather than a single individual deviating from expectations, multiple participants deviate from expectations. Moreover, individuals share a collective understanding (implicit or explicit) that others in the organization are deviating from these expectations. Further, the organizational conditions that give rise to the deviations and to the shared understanding persist over time.

Consider, for example, two different scenarios in a hospital unit with 10 nurses. In the first scenario, a single nurse fails to verify the medication as required by the operating procedures of the unit. In the second scenario, 7 of the 10 nurses fail to verify the medication as required. We would argue that the first scenario illustrates an individual-level error that could be potentially explained in terms of characteristics of the particular nurse committing the error (e.g., a distracting personal family situation). By contrast, the second scenario illustrates an organizational error in that the errors were committed by multiple nurses, and it is highly unlikely that these errors can be explained in terms of the idiosyncratic characteristics of the seven nurses. That is, organizational errors cannot be adequately explained without taking into account some organizational-level antecedents. Moreover, these antecedents operate through social processes that link the individual actions of the nurses to their shared understanding that others in the unit are also deviating from the expectation about medication verification. It could be that the unit is understaffed, and there is considerable time pressure to deal with patients quickly. These beliefs are shared by the unit members.

A second way to represent organizational errors is as a dynamic mix (e.g., frequency, severity, variety) of errors feeding or interacting with each other at the unit or organizational level. To start, the *frequency* of latent errors can differ between organizations. For instance, the number of instances that trading transactions should have been verified but were not can be higher in some financial institutions than in others. Next, latent errors can also differ in their *severity* or risk. The failure to verify trading transactions represents a more severe latent error in a financial institution in which the size and volume of trading transactions in relation to the capital base of the firm are large rather than small. In addition, the mix of latent errors can also be characterized in terms of their *variety*. For instance, latent errors can occur in the *execution* of primary work activities (e.g., a securities trader exceeding the limits while executing a trade); in the *monitoring* of the execution of work-related activities to detect and correct the errors in execution; or in the *infrastructure* (i.e., the stable set of prespecified arrangements for carrying out work). For instance, one well-known rule about structuring securities trading operations is that the responsibilities for trading and for booking the trades should be assigned to two independent sets of people. Assigning these responsibilities to the same person would be an example of an infrastructure error. Low variety means that there are fewer different types of latent errors in an organization, while high variety means that there are many more different types of latent errors.

Why introduce the three metrics of latent errors? The reason is that at the unit or organizational level these features may interact with each other and increase the magnitude, frequency, or variety of errors. Let us return to the financial example and the trading room. Some of the traders notice a number of trades are exceeding the limits. This continues over time. A number of the traders believe the limits may not be so fixed and begin to trade over the limits. This "normalization of deviance" (Vaughan, 1996) has at least two consequences. More traders begin to trade over limits, and the magnitude increases. In this case, initial frequency of deviations is contributing to greater frequency and the magnitudes of the deviations. In a related but different example, let us assume that the variety of errors increases. This means that not only are execution errors (i.e., trading limit deviations) increasing but also the same is true for monitoring and infrastructure errors. As monitoring errors increase, as an example, it would be more likely to observe more execution errors. Monitoring, to some extent, provides controls for detecting execution latent errors. Therefore, when monitoring errors result in reduced tracking of adherence to SOPs, an increase in the frequency and magnitude of trading deviations is more likely.

The theoretical picture we are drawing includes a system of latent errors that vary in frequency, magnitude, and variety. Each feature is interactive with the others at the unit or organizational level. Changes in frequency can lead to changes in magnitude or variety. The obverse also is true. Also, one should note that it is at least a two-directional system. Decreases (increases) in frequency of latent errors could lead to a decrease (increase) in magnitude or variety. We view this system of interacting types of errors as a unit- or organizational-level phenomenon.

Following are some of the implications of our conceptualization of latent errors, adverse outcomes, and organizational errors. First, focusing on latent errors moves us away from studying extremely rare error-related events such as accidents, which have been typically the focus of the literature. It is much harder to build and test theories about infrequent events. Second, a related point, it makes it possible to study errors using ex ante research design. By focusing on extremely rare events, prior studies tend to sample on the dependent variable (i.e., study only organizations that experienced adverse outcomes) and are prone to hindsight bias. The basic assumption underlying latent errors is that they occur frequently in all types of organizations. In other words, they are not rare events limited to organizations operating hazardous technologies such as nuclear power plants or air traffic control. By asserting that they occur in all types of organizations, one can design ex ante versus ex post studies. The former seems a more productive way to build and test a new theory. Third, we treat the concept of adverse consequences broadly. Initially in this literature, it was tied to physical loss (e.g., *Challenger*, Bhopal incidents). More recently, latent errors have been tied to other indicators of adverse consequences, such as financial loss and reputation loss (Ramanujam & Goodman, 2003). We adopt a broad view of organizational effectiveness, and substantial declines in any of these indicators would be measures of adverse consequences.

Last, this conceptual distinction between individual and organizational errors is important because these levels of analysis are often confused in the literature or not made explicit. In this chapter, we focus primarily on errors at the organizational level of analysis, which have received little attention.

Linking Latent Errors and Adverse Consequences: A Conceptual Framework

In the light of our foregoing discussion about latent errors and adverse consequences, we turn to the central questions in this chapter: When and how do latent errors generate organizationally significant adverse outcomes? We introduce a conceptual framework (see Figure 10.1) to capture the various elements and their interrelationships that are essential for addressing these questions. We briefly define the basic concepts in this model and then proceed to some examples, from which we develop a more complex representation of the model.

At the center of this framework are latent errors represented in terms of both their mix (i.e., frequency, variety, and severity) and their organizational features (i.e., involvement of multiple individuals). To its right, the framework contains organizationally significant adverse outcomes linked to these errors. The central part of our framework focuses on two mechanisms or processes: error-amplifying feedback or error-correcting feedback. These two processes act on the frequency, severity, and variety of errors. Error-correcting processes enable the organization to detect,

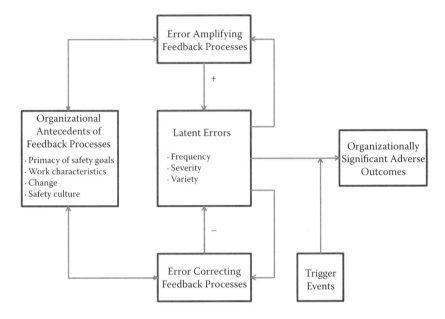

FIGURE 10.1
An organizational framework for the link between latent errors and adverse consequences.

correct, and contain latent errors. By contrast, error-amplifying processes promote an accelerated buildup of errors. As this mix of errors increases, so do the probabilities of adverse consequences, which would be activated by internal (to the organization) or external triggers. Most researchers have looked at the linkage between antecedent factors and adverse outcomes. Our framework links antecedent factors to the two feedback mechanisms, then to latent errors and eventually adverse outcomes. Being more specific about the mediating mechanisms can enhance our ability to predict latent errors and adverse outcomes.

Error-Correcting Feedback Systems

An important component of our model is error-correcting feedback processes. The key features of any error-correcting feedback system include (a) a prespecified standard, rule, or procedure; (b) a measurement system that detects deviations from that standard; and (c) an organizational mechanism that would eliminate or correct the deviation. All three features are critical. One must have an operational standard that is an explicit part of the input process or outcome systems of the organization. There needs to be a measurement system that detects deviations and makes that information available to the appropriate organizational unit. Then, there needs to be a review process that both diagnoses the reasons for the deviation and initiates organizational processes to rectify the deviation. Organizations have multiple error-correcting feedback systems, and their basic role is to keep the organization in equilibrium by reducing the number of latent errors (Reason, 1998). Consider the body shop in a computerized automobile assembly plant. Robots pick up sheets of metal and form the basic body of the automobile. What is critical in this process is achieving dimensional quality. That is, the different pieces forming the body must meet specific dimensional standards. If there are deviations, the automobile will be difficult to assemble. In this setting, there is an independent, real-time system measuring dimensional quality of specific components (e.g., door, hood). These data are sent to the operator of the robots and quality control and are reviewed. If deviations are identified, corrective actions are initiated at the end of this review meeting.

Given our premise that latent errors are found in all organizations, the error-correcting feedback systems are in place to identify and correct the causes of the deviations (see negative sign between error-correcting feedback systems and latent errors; Figure 10.1). Of course, the effectiveness of the error-correcting feedback systems depends on whether the standard is clear and shared, whether the measurement system captures the relevant deviations, and last, whether there is some system in place that will review the data and act to achieve the expected equilibrium. These are likely to be cases for which error-correcting feedback systems work effectively and reduce the number of latent errors. It also is likely that one or more of the three features of error-correcting feedback systems are not met, and the latent errors are not affected.

Early studies of high-reliability organizations identified three organizational characteristics that were especially seen as contributing to the organizational capabilities for error correction: redundancy, flexible structures, and culture. Roberts (1990), for example, refers to the "many pairs of eyes" on an aircraft carrier flight deck and Rochlin, La Porte, and Roberts (1987) to the large number of people "just watching" others perform their jobs. In high-tempo operations, this redundancy substitutes for time: Laborious checking by one or two individuals is precluded by the rapidly developing situations that must be managed, but redundancy allows quick checking by many individuals simultaneously. Moreover, there can be redundancy in ideas as well as resources. Bigley and Roberts (2001) describe how incident command systems

> appear able to structure and restructure themselves on a moment-to-moment basis, and ... oscillate effectively between various preplanned organizational solutions to the more predictable aspects ... and improvisation for the unforeseen and novel complications. (p. 1282)

Schulman (1993) argues that a high-reliability culture is especially critical for actions that are unconstrained by formal structures because it allows the possibility of interpretation, improvisation, and unique action.

In recent studies, *mindfulness* has emerged as a process that is central to high-reliability organizing (Weick & Sutcliffe, 2006). Mindfulness entails an

> enriched awareness ... [through] active differentiation and refinement of existing categories and distinctions ... creation of new discontinuous categories out of the continuous stream of events ... and a more nuanced appreciation of context and alternative ways to deal with it. (Weick, Sutcliffe, & Obstfeld, 1999, p. 90)

It enables organizations to more readily detect weak signals from interactively complex environments earlier and respond to them more effectively. It also loosens tight coupling by creating alternative paths of action. Studies have identified various specific processes that contribute to mindful organizing in work groups and organizations (e.g., reluctance to simplify interpretations, sensitivity to operations, commitment to resilience, underspecification of structures, and preoccupation with failure) (cf. Vogus & Sutcliffe, 2007; Weick & Sutcliffe, 2006). Together, the findings from studies of high-reliability organizations and mindful organizing provide a rich description of the feedback processes that enable organizations to detect, correct, contain, and effectively respond to errors.

Error-Amplifying Feedback Processes

The error-amplifying feedback component acts in the opposite way from error-correcting feedback systems. In this component, changes in one variable lead to changes in a second variable, which in turn lead to changes in the first variable. In our discussion of the features of latent errors, we proposed that these features might mirror an error-amplifying feedback system. Changes in frequency of latent errors might lead to changes in the magnitude of these errors, which in turn could increase the frequency. Also, the variety of errors may change as a function of changes in frequency or magnitude of latent errors.

The role of error-amplifying feedback systems and its contrast with that of error-correcting feedback systems is well illustrated in a study by Rudolph and Repenning (2002). They used a systems-dynamic model to explore the relationships among interruptions, stress, and performance. Initially, their model showed that interruptions that caused deviations from SOPs were recognized and resolved (error-correcting feedback system). Over time, if interruptions increased, the organization remained resilient and found new ways to resolve the interruptions. However, there was a tipping point at which the error-correcting feedback systems no longer worked, and the organization began to collapse. There were accumulative interruptions, which increased the level of stress, which in turn reduced the ability to resolve the deviations created by the interruptions, which led to a greater accumulation of interruptions, which led to greater stress. This vicious cycle accelerated with increased interruptions and stress and corollary declines in performance. In this accelerating downward cycle, there were no mechanisms to stop the downward cycle and move the organization to its original equilibrium. The final consequence was the collapse of the organization.

In discussing the role of positive-feedback loops, we focused on error-amplifying processes. That is, we discussed how increases in interruptions could increase stress, which in turn could decrease the ability to manage deviations from interruptions, which in turn could increase stress. However, positive-feedback loops can also operate in the opposite direction. That is, decreases in interruptions could decrease stress, which in turn could increase the ability to manage deviations, which in turn should reduce stress. Our primary focus will be on how these feedback systems increase latent errors and how they contribute to the link between increasing latent errors and adverse organizational consequences. The positive sign between error-amplifying feedback systems and latent errors signals this idea (see Figure 10.1).

Another feature of error-amplifying feedback systems is the rate of acceleration among variables. *Rate* refers to the amount of change and the timing of changes. In several studies (Rudolph & Repenning, 2002; Sterman, 1994), the initial reciprocal changes between variables are small, but over time, the frequency and magnitude of changes accelerate. The label *vicious downward cycle* means the changes (e.g., interruptions, stress, performance) are increasing in rate, and in magnitude, over time until the demise of the organization.

Organizational studies of accidents elaborate on several error-amplifying feedback processes. First, Vaughan's (1996) account of normalization of deviance details the processes that cause organizational members to deviate regularly from rules and procedures while viewing such deviations as normal. Checking for gas in a coal mine is a standard procedure to avoid explosions. Normalization of deviance would mean that miners in a particular mine do not regularly check for concentrations of gas. It is the normal thing to do. Snook (2000) proposes a similar explanation for how the challenges of communicating across highly specialized and differentiated work can make operational drift—gradual straying from standard procedures—not only more likely but also more difficult to detect. Second, Perrow (1984) suggests that multiple independent errors could produce accidents when they interact in unexpected ways. Such interactions are more likely when the technology is interactively complex as well as tightly coupled. Third, Rudolph and Repenning (2002) draw attention to the role of feedback loops in accidents. They suggest that dynamic relationship between errors and stress—when stress leads to errors, which in turn lead to more stress, and so on—can lead to a "quantity" effect by which the stress from a buildup of errors makes accidents increasingly likely over time.

The Interaction Between Error-Correcting and Error-Amplifying Processes

Error-correcting and error-amplifying processes interact. Although they have important independent effects, their interaction accounts for the development of organizational errors and the acceleration of latent errors. Consider the disaster in a West Virginia coal mine in which 29 people lost their lives. Mining is a dangerous work environment ("No Survivors Found," 2010). One of the risks is the level of methane gas, which can lead to explosions. One of the standardized procedures is to check for gas levels in multiple places at multiple times. Starting with the error-correcting mechanism, there are standards, measurement instruments, and corrective procedures if gas levels are too high. From an individual-level error perspective, one could see an individual miner not doing a check at a particular work area or routinely not doing gas checks. This represents an individual error. Throughout a mine, there are multiple people doing these gas checks, and there are federal and state inspectors doing these checks. In this case, there is redundancy in monitoring.

A different scenario occurs when other "monitors" see this miner not doing gas-level checks, and there are no consequences. Others then begin this practice. A related scenario is that the monitoring occurs, but there is no feedback or corrective action. In either scenario, latent errors are increasing. Over time, at least two things can happen. First, a general understanding develops that monitoring or corrective actions are not necessary. Second, deviations from other SOPs (e.g., checking the roof) begin to occur.

As feedback and corrective processes begin to diminish, the amplification processes begin to increase. As more monitors deviate from this standard practice, we are dealing with organizational errors, not individual-level errors. As the amplification processes become more predominant, the organizational latent errors begin to increase at an even faster rate. The probabilities for adverse consequences increase.

Let us clarify this picture of latent errors and adverse consequences. The failure of the miner to check for methane could lead to an explosion. But, we have argued that latent errors lead to potential adverse consequences, not actual consequences. On the other hand, as deviations from checking gas levels become more normal, the increase in these deviations puts the whole mine at risk, not a work section, and the chance of adverse consequences increases. The failure in the error-correcting mechanism can lead to negative outcomes. However, the combination of the failure of error-correcting mechanisms and error-amplifying processes spells danger for the organization.

Organizational Antecedents of Feedback Processes

Our analysis occurs in an organizational context. There are many possible organizational antecedents of errors (cf. Vaughan, 1999). Our goal is to illustrate the link between antecedents to the error-correcting and -amplifying mechanisms, which in turn affect the frequency of latent errors and eventual adverse outcomes. We want to explore the central linkages rather than do a comprehensive review of the literature.

Organizational goals are important signaling devices. Since organizations have multiple goals, one question is which goals are more salient or emphasized. If production goals have primacy over safety goals, this should indicate which work activities are dominant (Vaughan, 2005). In a trading company for which revenue or profitability is dominant, there may be fewer monitoring or corrective activities on trades over some prespecified limits. The more other traders see deviations from limits occurring, we would expect to see more deviations concerning this activity.

The structure of work is another class of antecedents (Ramanujam & Goodman, 2003). Take one dimension: work done face to face or distributed. The former case is an excellent setting for learning. One nurse sees other nurse not following the standard procedures for dispensing narcotic medications. Seeing relevant others not doing this can legitimate this nurse from not following the procedures. The visibility of the work setting facilitates normalization of deviance and stimulates error-amplifying processes. In the distributed setting, visibility is restricted. One can exchange verbal communications, but the visibility of others' behavior is limited, as is the opportunity to learn about deviations. In this case, there are at least two lessons: It is harder to observe others' behaviors, and normalization of deviation is more difficult. However, being in a distributed work setting also restricts monitoring behavior. So, deviations could be occurring without any opportunity for a monitoring corrective mechanism to work.

A different antecedent is change. All organizations are experiencing different forms and levels of change. Change is relevant in this context because it uses attentional resources (Ramanujam, 2003). One of our key mechanisms is error-correcting processes. If change demands many attentional resources and these are in limited supply, we would expect to see a decrease in monitoring behaviors, which in turn should increase latent errors. As argued, as error-correcting mechanisms decline, error-amplifying mechanisms can increase.

Another class of antecedents deals with perceived safety culture (Vaughan, 1996). This can include beliefs about the openness and supportiveness of discussing errors and finding new solutions (Edmondson, 1996). In the context of our framework, a strong culture to be open to errors and to find creative solutions should facilitate the error-correcting feedback process, reduce errors and not stimulate error-amplifying processes.

There are at least three lessons from this discussion of antecedents. First, we illustrated some linkages rather than generating an exhaustive list of antecedents. Second, we tied the antecedents to the two error-correcting and error-amplifying processes rather than to latent errors or adverse outcomes. We did this because we think these two processes directly affect the frequency, magnitude, and type of latent error. It is clear that the antecedents facilitate or hamper these processes. But, the first questions we would ask about the mine disaster, given that there were SOPs about monitoring, include the following: Were there deviations in monitoring? Were multiple miners deviating? Were corrective actions initiated when there were deviations from monitoring? Answers to these questions would indicate whether we were dealing with individual or unit or organizational errors and whether the errors were related to monitoring, feedback, or corrective action or all three. This information would direct us to possible relevant antecedents. Third, most research focuses on a specific antecedent, such as safety culture or change. But, the reality is that multiple antecedents are affecting two critical processes. The challenge is to trace whether the antecedents have a synergistic effect on the two main processes or are in conflict with each other.

Linking Latent Errors to Adverse Consequences: Two Cases

In this section we explore these and other issues in the context of two cases. They capture the relationship between organizational latent errors and adverse outcomes in different ways.

Barings Bank

Barings was the oldest investment bank in the United Kingdom, with total assets of $9.37 billion and 4,000 employees worldwide. After the deregulation of the financial markets in London in 1986, Barings set up a subsidiary to trade in securities and derivatives. In early 1992, Barings sent Nick Leeson to Singapore to set up a settlement process in their security trading subsidiary. Shortly after coming to Singapore, Leeson also had the responsibility for trading. By 1994, the Singapore operation was generating substantial profits for the firm (i.e., $30 million in the first 2 months of 1994 vs. $16 million for all of 1992). Much of this success was attributed to Leeson, and he was considered a star performer in the securities subsidiary. During this period, Leeson's trading volume increased. By the end of 1994, Leeson had accumulated over 28,000 contracts valued close to $29 billion. Most of these were unhedged positions betting on the upward movement of the Japanese stock prices and interest rates. On January 16, 1995, an earthquake in Japan led to a steep drop in Japanese stock prices and interest rates. Within a month, Barings was exposed to losses of $1.3 billion and was forced into bankruptcy.

This initial description sets the stage for some basic research questions. First, what were the latent errors in this case? When Leeson began trading, he also was in control of the settlement process. In Barings and in other financial institutions, settlement and trading are separate operations done by different people. This division of labor is really a control mechanism to ensure that the trader is following the standard rules. In this case, an internal audit team from Barings identified this deviation (infrastructure latent error), but in subsequent negotiations, Barings security subsidiary prevailed, and Leeson continued to trade and settle. At this time, Barings Securities and Leeson were major contributors to the profitability of Barings, and they wanted to maintain control over their operation.

There were other examples of latent errors. All traders had to respect certain trading limits. Also, traders are required to hedge positions. This means that at the end of the day if a trader had oversold a position, he had to buy additional securities so that there were no open positions. Leeson deviated frequently from these two rules and therefore generated execution errors. All during this time, there were deviations from standard monetary procedures, but there was no rectification of these deviations. There were indications of infrastructure, execution, and monitoring errors. These continued to accelerate over time. One reason is the interactive nature of the errors among themselves. If Leeson could settle and trade, it was easier to violate the trading limit and hedge SOPs. Infrastructure errors then facilitated the frequency and magnitude of execution errors.

Another question is whether this example is about individual or organizational errors. Some have construed the Barings case as an example of the "rogue trader"—an individual-level error. Our distinction between organizational and individual errors was based on the idea that there are multiple participants involved in errors, and these may occur over time. This was clearly the case at Barings. Leeson was clearly involved in trading errors. However, the management of Barings also was involved in deviations. They permitted Leeson to trade and settle. Also, when he had generated significant unhedged positions, Barings assured the Singapore Mercantile Exchange about the availability of funds to cover his position. The amount of funds provided to cover margins deviated from the level of funds permitted by Bank of England. Also, there were internal audit teams and account reconciliation procedures. None of these monitoring mechanisms was successful when Leeson was making profits or the hedge losses that brought down the bank. The point is that there were many deviations enacted by many players over time.

A third question concerns why the error-correcting feedback systems were not more effective. Error-correcting feedback systems require a clear standard, effective measurement systems, and organizational units and processes responsible for reviewing the deviations and moving the organization back to its equilibrium position. While the standards were clear, the effectiveness of the measurement systems and response mechanisms were less clear. In a power struggle, Barings Securities subsidiary was able to retain its control over settlement and trading. Why they won this battle is probably tied to the salience of profitability (for which the subsidiary was the major contributor), the star performer status of Leeson, and the desire of the subsidiary to maintain its independence from the parent company. While settling and trading gave Leeson the capacity to mask some of the deviations, there were still monitoring mechanisms in place to track the frequency and magnitude of trades. But, an argument about error-correcting feedback systems is that monitoring is not enough. One still needs an organizational unit and processes to rectify the deviations. Yet, these were absent. The primacy of non-"safety goals" (i.e., less incentive for vigilance and change), the distributed nature of work (i.e., Singapore is distant from London in both space and time), and the lack of a collaborative culture that has open discussions about why errors occur and how to eliminate them and improve the operation all contributed to an ineffective error-correcting feedback system. Without these systems, the organization becomes vulnerable to accelerating latent errors.

What is the role of error-amplifying feedback systems? As the effectiveness of the error-correcting feedback systems declined, the role of error-amplifying feedback systems became more dominant. We can see this at the individual and organizational levels. For Leeson, initial success in trading accelerated the frequency and magnitude of trades, many in violation of trading limits and hedging positions. As his luck changed and he began to experience losses, the same positive cycle continued. Losses in trading accelerated trading behavior to recoup the losses. This cycle of escalation of commitment led to greater frequency and magnitude of trades, most violating trading limits and hedged positions.

At the organizational level, senior management provided the subsidiary $790 million to cover margin requirements, thus reinforcing the position of the subsidiary and Leeson's trading behaviors. Other regulating institutions in both Europe and Singapore inquired about the large unhedged positions of the subsidiary. In these cases, senior management assured these regulatory institutions that there was no risk. All these behaviors by senior management at Barings accelerated the autonomy of the subsidiary and Leeson's trading behaviors.

What were the drivers of the positively accelerating spiral of deviations? The primacy of nonsafety goals contributed to the accelerated trading behaviors. Both more profits and more losses stimulated trading behavior. In the latter case, the motivation was to recoup losses and achieve profitability. This affected the behaviors of both Leeson and senior management. The organization of work created two independent entities: Barings and the subsidiary. There was no common work or need to coordinate. Only the bottom line involved both organizations. This independence for the Singapore subsidiary created the conditions for error-amplifying feedback systems to flourish. Also, the two entities did not embrace a common culture of cooperation and problem solving. The absence of this type of culture facilitated the error-amplifying feedback cycles in the subsidiary.

Over time we see (a) the frequency, magnitude, and variety of latent errors accelerating on their own; (b) an error-correcting feedback system, which should decrease latent errors, becoming less effective; (c) an error-amplifying feedback system accelerating the frequency, magnitude, and variety of errors; and (d) all these changes supported by an organizational context that supports nonsafety or noncompliance goals, an organization of work that makes monitoring and redesign more difficult, an inadequate control system, and a culture that does not support focusing on and solving deviations from standard operating values and procedures.

It clearly was a trigger event—the 1995 earthquake in Japan—that affected the economic system and led to a drop in the stock market and interest rates. This exposed Barings to over $1 billion in losses. On the one hand, this exogenous trigger event led to the collapse of Barings. On the other hand, the four themes in the preceding paragraph indicate that Barings was headed for disaster. In this particular case, it was the earthquake that finalized it. But, one could postulate that other external regulatory agencies would have created serious adverse consequences for Barings. Or, there could have been internal forces such as a change in senior management that acknowledged the problem and accepted large losses as a way to make the bank viable.

The basic argument is that the conditions in the bank were out of control, and internal or external triggers might have caused large negative consequences.

The following case is loosely based on actual events at a hospital, which we refer to as Mid-Western Hospital (MWH).

Mid-Western Hospital

Mid-Western Hospital (MWH) is one of three hospital facilities operated by a health care group in a midwestern state. MWH is a tertiary care facility that offers specialty treatment for its patients. Many of its units appear routinely in the *U.S. News and World Report* rankings of top hospital departments. Over a weekend in 2006, five different nurses administered a thousand-fold overdose of the blood thinner heparin to six infants in the neonatal unit at the hospital. As a result, three infants died. Three other infants recovered subsequently, but the effects of the overdose on their long-term health are unclear.

Before we analyze the specific situation in the hospital and its neonatal unit, let us look at the broader context. For several years before this incident, many of the government or standard-setting officials had taken public positions on the risks of heparin. The Institute of Medicine, the joint commission that accredits hospitals, U.S. Pharmacopoeia (a standard-setting agency for drugs manufactured in the United States) all had issued advisories about heparin and, in general, how to reduce medication errors. An Institute of Medicine panel had identified heparin as one of the five drugs contributing to 28% of medication errors. From an institutional perspective, hospitals and their employees were receiving many warnings about heparin.

Heparin is a high-risk medication for a variety of reasons. First, it is a colorless liquid, and different levels of the drug are indistinguishable. At the time of the incident, the packaging of the drug also did not differentiate dosage levels. So, reading the label was critical to determine the dosage on hand. Second, different levels of the dosage were used to treat different types of patients with different medical conditions. In the neonatal unit, 10 units/ml heparin solution were used to flush intravenous catheters to prevent closing. A dose of 10,000 units/ml would be appropriate for treating adults, but it would be dangerous or fatal to an infant in a neonatal unit.

MWH was one of the first hospitals in its region in 2000 to set up a safety program on medical errors. There was a major review and modification of the processes for medication distribution. At that time, nurses on the floor were responsible for selecting the correct dosage from a set of vials that contained different levels of heparin. Part of the new safety program was a campaign to sensitize nurses to the need to verify medication.

In 2001, a nurse administered an overdose of heparin to two infants. Fortunately, following some intensive medical procedures, the infants recovered. An investigation followed, and a set of new recommendations to prevent these errors was developed. These included the following: (a) Patient floors would no longer stock multiple doses; (b) medication carts of the neonatal unit only would be stocked with dosages of 10 units/ml; (c) the pharmacy would stock different dosages in different bins; and (d) the pharmacist would carry out multiple verifications in stocking and dispensing.

In addition to these specific actions, which were well publicized within the hospital, MWH initiated a series of other safety initiatives, including 2-day safety training programs for nurses. All of these efforts were to make safety more salient. One aspect of safety was medical verification. The picture we want to draw is a hospital with clear goals on safety and a strong climate of safety culture. When the heparin incidents occurred, there were instant diagnosis and new interventions to improve safety. The hospital and the neonatal unit were proactive and reactive about safety.

The big question is why multiple nurses administered the wrong dosage, and three children died. One important factor can be tied to the error-correcting feedback systems. The verification process was not measured. While this is a formal expectation, measuring verification would require some monitoring system to identify if the nurse matched the vial to the medication requirements for the patient. This happens in the patient's room, and the nurse verification process is not a highly visible activity and is harder to measure. The absence of measuring this verification process leaves the system in a vulnerable position. We do not know if verification was happening intermittently or not at all. But, we do know that multiple nurses administered incorrect doses to different infants. It is unlikely that they all failed to verify on one particular day. A more likely scenario is that the multiple nurses were deviating over time, and in this particular instance, the pharmacist sent the fatal dosage.

The role of the error-amplifying feedback systems is less clear in this case. In Barings, there were clear examples of accelerating latent errors. We have no information about this for the hospital, although it is apparent that at some point in time multiple nurses in the neonatal unit stopped routinely checking the heparin dose prior to administration. Similarly, there was no external trigger event in this case comparable to the earthquake in the Barings case. Instead, the trigger event that linked the errors to adverse outcomes was itself an error that occurred in a different part of the organization (i.e., the pharmacy technician stocking the neonatal unit medication cart with vials containing an incorrect dose of heparin).

The organizational antecedents play an important role in understanding this situation. First, there were clear goals for safety and a climate that was proactive and reactive about safety. The organizational arrangements provided an interesting clue. There is strong interdependence between the pharmacy and the neonatal unit. The new procedure of storing different levels of heparin in different bins and requiring the pharmacist to do a different verification was a workable solution. Over time, the pharmacy delivered the vials with 10 units/ml only to the neonatal unit all the time. It was perceived to be a highly reliable innovation. In MWH, all of these antecedent features predicted no adverse consequences. Yet, three infants died, and the other three were injured.

Discussion

The central question in this chapter is: What are the mechanisms that predict and explain the relationship between latent errors and adverse organizational consequences? We begin with latent errors because they provide a newer approach to the research on organizational errors. Latent errors are found in all types of organizations and vary across units in some systematic ways. Actual errors, which lead to immediate adverse consequences, are rarer events. Also, when they occur, researchers are forced into ex post explanations of why the adverse consequences occurred. The basic assumptions of latent errors are that (a) they can cause adverse effects and (b) they occur with sufficient frequency that we can study them ex ante versus sampling on the dependent variable.

Given this position, the central question of this chapter becomes important. Our framework (Figure 10.1) provides the mechanisms to explain when latent errors will have no adverse effects. There are at least two key mechanisms or feedback processes. First, the error-correcting mechanism is most important. Basically, this says (a) there are some prespecified standard rules or procedures, (b) a measurement system exists that detects deviations from that standard, and (c) there is an organizational mechanism that would eliminate and correct the deviation. All three features are critical. If an error-correcting system measures the deviation but there is no mechanism to correct the deviation, the error-correcting system is not operable. Or, if there is no measurement of the desired behaviors but strong mechanisms to return the organization to its equilibrium again, the system will not be operable.

The other mechanism is error-amplifying systems. In this scenario, an increase in latent errors stimulates the increase of other latent errors, which in turn leads to an increase in other latent errors. Changes in the frequency of errors could lead to an increase in the magnitude of errors, which could stimulate greater frequencies. The key idea, well illustrated in the Barings case, is the amplification of errors, and over time the amplification increases.

There is a common condition underlying both of these mechanisms: They both lead to more latent errors if the error-correcting mechanism is not working and the error-amplifying system is working. Consider nurses in a unit who are not verifying medication levels to the patient's requirements. If there is no error-correcting mechanism, these behaviors will continue. If other nurses who rotate across units observe this type of deviation, latent errors can spread throughout the hospital. The normalization of deviance would be a form of error amplification, and increasing numbers of latent errors would appear in the second example.

What are the consequences of this increase in latent errors? One consequence is that the probability of adverse effects increases. In the case of the nurse who is not verifying medication level to patient requirements, the chances of a mismatch over time is more likely, which can have adverse implications for the patient and the hospital. In MWH, when the pharmacist sent the wrong dosage and the nurses were not verifying, the adverse consequences were devastating. In the Barings case, in which there was a huge accumulation of errors, any event external or internal to the bank could have led to its demise. In this case, it was an earthquake and a drop in the Japanese stock index, but there could have been many triggers. Barings was becoming more vulnerable over time.

Table 10.1 frames this analysis in terms of our two cases. Barings represents one extreme. The error-correcting system was not operative, while the error-amplifying system was. All of the antecedents supported the increase in latent errors. The goals were focused on financial performance versus reliability. Work was distributed, which made monitoring difficult. Also, there was no culture supporting vigilance and high reliability.

At MWH, the opposite was true. They had strong safety goals and culture. Work was face to face, and in general, they had a strong control

TABLE 10.1

Contrasting Barings and Mid-Western Hospital

Mechanism	Barings	Mid-Western Hospital
Error correction	No	No
Error amplification	Yes	No
Antecedent safety goals	No	Yes
Nature of work	Distributed	Face to face
Control	Weak	Strong
Safety culture	No	Yes

system. What was missing was an error-correcting system tied to the medication verification process. Specifically, there was no measurement of the verification process. There was a standard and an organizational mechanism to correct deviations. The lack of a measurement system is not surprising. Organizations cannot measure all relevant behaviors, and this particular behavior, whether nurses verify, is hard to assess. You really have to be in the patient's room.

What can we learn from Table 10.1 relevant to our central question? First, MWH had all the features of a highly reliable and safe system. There were strong goals and culture about safety. There had been a series of structural and process interventions to minimize errors. Yet, three infants died, and others required extra treatments. Second, strong safety culture can have dysfunctional effects. If you know structural changes to minimize errors (e.g., sorting different levels of heparin in bins) have been put in place *and* they work well, personal vigilance may decline, and latent errors or verification may increase. In one sense, a prevention system can be too successful. Remember that a nurse's job is full of activities, interruptions, and time demands. The nurse needs to make choices for time allocations. If delivery from the pharmacy is accurate over time, then verification of medication might not be a priority. Third, the absence of an error measurement process meant that there was no knowledge whether medicine verification was ongoing and, therefore, no ways to rectify the situation.

A minimal condition for latent errors to lead to adverse consequences is when the error-correcting system is not working. An important qualification is that the error-correcting system needs to be focused on a high-risk, central behavior. In the case of MWH, it was administering a high-risk drug to a vulnerable patient. We can think of other latent errors in a hospital (e.g., unlocked medication drawer), which are relevant but not high-risk challenges to the patient.

The Barings case demonstrates optimal conditions for major adverse consequences. The error-correcting system did not work, and the error-amplifying system was operating. Also, all the antecedent factors facilitated the accumulation of errors. This situation was ripe for adverse consequences. What was difficult to predict was the time frame or triggering agent.

Our analysis has focused on the error-correcting and error-amplifying mechanisms. The literature has placed more emphasis on the direct antecedents of errors (cf. Vaughan, 1999). A review of Table 10.1 shows that many of the factors in the literature, such as emphasis on safety goals and safety climate, should lead to little or no adverse consequences. But, despite these conditions, serious adverse consequences occurred. Our argument is that one needs to understand in detail the error-correcting and error-amplifying mechanisms with regard to high-risk or vulnerability behaviors. At MWH, there were no working correction mechanisms for the heparin administration.

Research Opportunities

One contribution of this chapter is identifying some areas for future research. Instead of providing a list of questions, we explain a few opportunities. There are many others than the ones enumerated. The framework in Figure 10.1 is our starting point. We think it explains the processes between latent errors and adverse outcomes. However, we do not think testing the model itself is productive. Rather, our preference is to select relationships within the framework for more empirical investigation.

One possible issue concerns the content of the latent error. For example, in MWH, the general behavior was verification of medicines. But, the specific behavior was verification of heparin, a drug related to fatalities in hospitals. This was a high-risk transaction, particularly in a neonatal unit. We pointed out that there were other SOPs, such as locking the medication drawer so others (nurses or patients) could not access the medication or not leaving the room during administration of medications to ensure that patients take their correct medications. Both of these examples can be consequential to the hospital, but not at the same risk level as giving the wrong dose of heparin to a patient because of lack of verification. The challenge here is to classify latent errors on their propensity to create adverse consequences and then collect data on this classification scheme to verify the content effects on adverse consequences. More generally, there is an opportunity to think about focusing on latent errors in terms of frequency, severity, and variety. For instance, what are the antecedents of frequency rather than variety?

On a related note, the link between errors at the individual and at the organizational levels of analysis provides interesting opportunities for future research. For instance, through what processes do individual-level errors give rise to organizational errors? What are the organizational conditions that facilitate or impede the link between individual-level and unit- or organizational-level errors? Such questions remain largely unaddressed in the organizational research on errors and point to a second set of research opportunities. A related issue is whether organizational errors are more likely to lead to more adverse consequences than individual-level errors. There are at least two considerations. First, there are more people deviating. Second, the implicit understanding that others are deviating from expectations with respect to some activities might lead to latent errors in other activities. These factors can increase the frequency, severity, and variety of errors that can facilitate adverse consequences.

The MWH case provides an interesting illustration of dysfunctional learning. After the 2001 heparin-overdose incident, there was a flurry of safety activities. One initial lesson was to be sure to verify heparin administrations. Another later lesson was that the changes were successful, the units always received the correct vials, and over time, the nurses might have assumed that they need not be so vigilant. That is, if the system is working 100% (which it was), one lesson is that administering heparin is straightforward. There is no need to be as vigilant about what the pharmacy sends. We do not know exactly how the learning took place. One scenario is that frequent conversations among the nurses about the high quality of the new system might have begun the creation of a collective understanding about not verifying. Another scenario is that one nurse observed another skipping verification, and that person eventually adopted this practice. The underlying concept is the normalization of deviance (Vaughan, 1996) via learning.

Another interesting issue, again from the MWH case, is the interplay between the antecedents and nurse behaviors. There were strong goals and culture supporting safe behavior. Yet, we know that nurses were not doing the verification process. Our theories about strong climates and culture are that they do influence employee behavior. In the literature we cited, the implications are that building an organization with safety goals and safety culture should lead to fewer errors, safety mishaps, or other adverse consequences. But, that did not occur at MWH. Indeed, one possible inference from our analysis is that the system was too reliable. The doses of heparin always were correct. This might have led to less vigilance in the verification process. But, there obviously are other possible explanations. High levels of stress could divert the nurses' attention from the verification process. Given the organization of a hospital, we could visualize a strong headquarters culture but variations in unit culture with respect to safety. The research challenge is acknowledging the strong safety goals and cultures as desirable, but we have mentioned alternative factors that could offset the effects of culture. Exploring the effects between organizational antecedents with conflicting impacts would be a different research option.

There are dynamic and temporal dimensions underlying our analysis. We know little how they work. The picture of Barings was an accelerating accumulation of latent errors. Our assumption was that this acceleration would continue until a trigger led to adverse consequences. But, another scenario might have been that the acceleration of latent errors slowed and then moved in the opposite direction. One could imagine that changes in managerial personnel or reorganization of the trading operation could create such changes. A related issue is predicting when adverse consequences would occur. Is there some tipping point at which triggers are more likely to create adverse consequences? If the earthquake had not occurred and the accumulation of errors had continued, when would have Barings collapsed? Or, if the pharmacy at MWH had not delivered the high dose of heparin and nurses did not verify, when would adverse consequences have occurred? Here, we are not talking about specific times (e.g., dates) but ranges of time. A related question is whether we can predict the emergence of adverse consequences.

Another potentially interesting research opportunity is examining the link between organizational antecedents and feedback processes. In Figure 10.1, the lines connecting these variables are indicated with arrows that point in both directions. In our discussion, we focused on how antecedents such as emphasis on production goals can weaken error-correcting feedback processes and promote error-amplifying feedback processes. It is also conceivable that the feedback processes may alter the antecedents. For instance, given a strong production goal, if negative-feedback processes detect several severe errors, the organization might respond by reducing the emphasis on production. Or, if the positive-feedback processes are dominant and if the buildup of errors is accompanied by increased productivity or profits but not adverse outcomes, then the organization may further increase its emphasis on production goals. Therefore, one interesting research opportunity is about identifying the conditions under which feedback processes affect antecedents.

A last and very different question is: When do latent errors lead to positive outcomes? Let us think of a situation for which there are strong SOPs in an organization. Latent errors are prevalent for some of these procedures. An error-correcting system is in place. It identifies the deviations and implements a corrective action. However, during the analysis of the latent errors, evidence is presented that the SOP is no longer functional. It was functional in the past, when the context and technology were different. This discovery leads to abandoning the SOP, giving people more time for other critical activities. The research question then is, Can we predict where latent errors no longer lead to adverse outcomes?

Conclusion

Our goal in this chapter was to draw attention to complex links between latent organizational errors and adverse outcomes in organizations. From an organizational research perspective, it is important to focus on when and how frequently occurring latent errors result in rare but organizationally significant adverse outcomes. We presented a framework to explain how organizational mechanisms interact dynamically to shape the link between latent errors and adverse outcomes. An important feature of this framework is that it provides a basis for bringing together the findings and insights from currently fragmented organizational research on errors, accidents, safety climate, and reliability. Our discussion also identified several implications for future research. Understanding the role of organizational antecedents and mechanisms in the link between latent errors and adverse outcomes is critical to understanding what is "organizational" about errors.

References

Basel Committee on Banking Supervision. (2008). *Consultative document: Guidelines for computing capital for incremental risk in the trading book*. Basel, Switzerland: Bank for International Settlements.

Bigley, G. A., & Roberts, K. H. (2001). The incident command system: High-reliability organizing for complex and volatile task environments. *Academy of Management Journal, 44*, 1281–1300.

Blatt, R., Christianson, M. K., Sutcliffe, K. M., & Rosenthal, M. M. (2006). A sensemaking lens on reliability. *Journal of Organizational Behavior, 27*, 897–917.

Edmondson, A. (1996). Learning from mistakes is easier said than done: Group and organizational influences on the detection and correction of human error, *Journal of Applied Behavioral Science, 32*, 5–32.

Heimbeck, D., Frese, M., Sonnentag, S., & Keith, N. (2003). Integrating errors into the training process: The function of error management instructions and the role of goal orientation. *Personnel Psychology, 56*, 333–361.

Hofmann, D. A., & Mark, B. A. (2006). An investigation of the relationship between safety climate and medication errors as well as other nurse and patient outcomes. *Personnel Psychology, 59*, 847–869.

Institute of Medicine. (1999). *To err is human: Building a safer health system*. L. Cohn, J. Corrigan, & M. Donaldson (Eds.). Washington, DC: National Academy Press.

Keith, N., & Frese, M. (2008). Effectiveness of error management training: A meta analysis. *Journal of Applied Psychology, 93*(1),: 59–69.

March, J. (1997). How decisions happen in organizations. In Z. Shapira (Ed.), *Organizational decision making* (pp. 9–34). New York: Cambridge University Press.

No survivors found after West Virginia mine disaster. (2010, April 10). *New York Times*, p. A1.

Perrow, C. (1984). *Normal accidents: Living with high-risk systems*. New York: Basic Books.

Ramanujam, R. (2003). The effects of discontinuous change on latent errors: The moderating role of risk. *Academy of Management Journal, 46*, 608–617.

Ramanujam, R., & Goodman, P. S. (2003). Latent errors and adverse organizational consequences: A conceptualization. *Journal of Organizational Behavior, 24,* 815–836.

Rasmussen, J. (1987). Cognitive control and human error mechanisms. In K. Rasmussen, J. Duncan, & J. Leplat (Eds.), *New technology and human error* (pp. 53–61). London: Wiley.

Reason, J. (1998). *Managing the risks of organizational accidents.* Aldershot, UK: Ashgate.

Reason, J. (2008). *The human contribution: Unsafe acts, accidents, and heroic recoveries.* Aldershot, UK: Ashgate

Roberts, K. H. (1990). Some characteristics of one type of high reliability organization. *Organization Science, 1,* 160–176.

Rochlin, G. I., La Porte, T. R., & Roberts, K. H. (1987). The self-designing high reliability organization: Aircraft carrier operations at sea. *Naval War College Review, 40,* 76–90.

Roe, E., & Schulman, P. (2008). *High reliability management.* Stanford, CA: Stanford University Press.

Rudolph, J. W., & Repenning, N. P. (2002). Disaster dynamics: Understanding the role of quantity in organizational collapse. *Administrative Science Quarterly, 47,* 1.

Sagan, S. D. (1993). *The limits of safety: Organizations, accidents, and nuclear weapons.* Princeton, NJ: Princeton University Press.

Schulman, P. R. (1993). Negotiated order of organizational reliability, *Administration and Society, 25,* 356–372.

Scott, R. W. (2002). *Organizations: Rational, natural, and open systems* (5th ed.). Upper Saddle River, NJ: Prentice-Hall.

Snook, S. A. (2000). *Friendly fire: The accidental shootdown of U.S. Black Hawks over northern Iraq.* Princeton, NJ: Princeton University Press.

Sterman, J. D. (1994). Learning in and about complex systems. *Journal of the System Dynamics Society, 10,* 291–330.

Vaughan, D. (1996). *The Challenger launch decision: Risky technology, culture, and deviance at NASA.* Chicago: University of Chicago Press.

Vaughan, D. (1999). The dark side of organizations: Mistake, misconduct, and disaster. In J. Hagan & K. S. Cook (Eds.), *Annual review of sociology* (Vol. 25, pp. 271–305). Palo Alto, CA: Annual Reviews.

Vaughan, D. (2005). Organizational rituals of risk and error. In B. Hutter & M Power (Eds.), *Organizational encounters with risk* (pp. 33–66). Cambridge, UK: Cambridge University Press.

Vogus, T. J., & Sutcliffe, K. M. (2007). The impact of safety organizing, trusted leadership, and care pathways on reported medication errors in hospital nursing units. *Medical Care, 45,* 997–1002.

Weick, K., Sutcliffe, K., & Obstfeld, D. (1999). Organizing for high reliability: Processes of collective mindfulness. *Research in Organizational Behavior, 21,* 81–124.

Weick, K. E., & Roberts, K. H. (1993). Collective mind in organizations: Heedful interrelating on flight decks. *Administrative Science Quarterly, 38,* 357–381.

Weick, K. E., & Sutcliffe, K. M. (2006). Mindfulness and the quality of attention. *Organization Science, 17,* 514–525.

11

Cultural Influences on Errors: Prevention, Detection, and Management

Michele J. Gelfand, Michael Frese, and Elizabeth Salmon

On April 26, 1986, a chain of problems caused by design flaws and exacerbated by human error culminated in the worst nuclear power plant disaster in history. When Reactor 4 at the Chernobyl plant exploded, it released 400 times the radioactive fallout as the atomic bombing of Hiroshima (Stone, 2006). While an accurate death count is impossible because of Soviet efforts to cover up the effects of the fallout, 31 people died instantly after the reactor explosion ("The Chernobyl Accident," 2000), at least 28 workers who were diagnosed with acute radiation sickness died in the 4 months following the accident (United Nations Scientific Committee on the Effects of Atomic Radiation, 2000), and the Chernobyl Forum (2006) estimated that the accident could cause another 4,000 cancer deaths among those who experienced the highest levels of exposure.

In 2002, nearly 300 people died and hundreds were injured after a Tanzanian passenger train lost power on a hill and rolled back into a freight train. A government report found that the driver of the train failed to apply the manual brakes, a mistake attributed to human error and inexperience ("Human Error Blamed," 2002).

In 2003, after the crash of the *Andrew J. Barberi* ferry, in which 10 people died, a review of the U.S. Coast Guard safety records showed that more than 30 of the 50 accidents that have occurred on Staten Island ferries "have been blamed on what investigators deemed to be mistakes or acts of negligence by captains, mates, deckhands or other ferry employees" (McIntire, 2003).

These stories illustrate an intuitive conclusion: Errors are universal. Errors, whether they cause thousands of deaths or minor inconveniences, are a global phenomenon. Yet, despite the fact that errors are a human universal, a careful look at this volume illustrates that scholarship on the topic is generally a Western enterprise, with theories and research generated largely in the United States and Western Europe (for notable exceptions, see Helmreich, 2000; Helmreich & Merritt, 1998;

Helmreich, Wilhelm, Klinect, & Merritt, 2001; Jing, Lu, & Peng, 2001; Li, Harris, Li, & Wang, 2009). Examining cultural influences on errors is critical for theory and practice. Theoretically, cross-cultural research on errors will help to elucidate further what is universal (i.e., *etic*) and culture specific (i.e., *emic*) about error processes while expanding error theories, constructs, and measures to be globally relevant. Practically, a cultural perspective on errors is critical to help identify how to prepare best for and manage errors in ways that are targeted to specific cultural contexts. For example, there are large differences in aircraft accidents across different nations even though most airplanes are similar in make and age and are often serviced by the same specialized service firms (Civil Aviation Authority, 1998). Although such differences are likely multiply determined, cultural characteristics of the pilots and the crews in error prevention, detection, or management may at least partially explain such cultural variation. Cross-cultural research is also needed to help identify ways in which multicultural teams can better manage cultural differences in responses to errors. Finally, the study of errors also enhances our understanding of culture itself. As Freud (1901/1954) has noted, errors often point out critical *system characteristics* that may go unnoticed. In other words, errors may tell us something about fundamental characteristics of cultural systems themselves. For example, as we discuss in this chapter, errors in high-power-distance cultures can occur when low-power members fail to communicate openly with their superiors—a phenomenon that is a key defining feature of such cultures. More generally, a cultural perspective on errors has much to offer the science and practice of errors in organizations.

In this chapter, we integrate research on culture with research on errors to identify key propositions for future research. We first discuss critical distinctions regarding the construct of errors, and we advance a process model of errors that includes error prevention, error detection, and error management. We then discuss how key cultural dimensions, including uncertainty avoidance, humane orientation, tightness-looseness, fatalism, power distance, and individualism-collectivism differentially affect each stage of the error process. We identify numerous "cultural paradoxes" regarding the error process for each of these cultural dimensions that we expect could have important short- and long-term consequences for organizations. We also discuss error management in culturally diverse groups, identifying group compositions that are the most ideal in managing the three stages of the error management process. Finally, we conclude with implications of this perspective for theory, research, and training to manage errors within a global context.

Key Conceptual Distinctions

Defining Errors

Based on the presentation by Hofmann and Frese in Chapter 1 of this volume, we define actions as *erroneous* when they unintentionally fail to achieve their goal when this failure was potentially avoidable (i.e., did not arise from some unforeseeable chance agency; Reason, 1990; Zapf, Brodbeck, Frese, Peters, & Prumper, 1992). Inherent to this definition are some critical distinctions. The first regards the *intentionality of errors*. As discussed in the introduction by Hofmann and Frese, we think of errors as unintended deviations from achieving goals and standards. After an error has been made, people have the sense that they should have known better. In contrast, violations are intentional deviations from goals and standards.

The differentiation of errors from risk also needs our attention. A risk is part of an objective situation, and it can be analyzed in terms of probability (Hofmann & Frese, Chapter 1, this volume). If an individual acts as a result of such a risk analysis, he or she will probably say that even if the goal was not achieved, the same action would have been called for (thus, there is no feeling that one should have known better, as in the case of an error). However, there may be miscalculations of risks, and that would be an error within a risk situation (Reason, 1995). For example, the economic depression that resulted from the mortgage crisis in the United States in 2008 and that led to the default of Lehman Brothers was based on miscalculations of risks that often occur in risky situations. In this case, it constituted the issue of common mode error, which means that the risk analysis was done on single probabilities that were perceived to be unconnected, but those probabilities were conditional on each other and turned out to be connected: People lost trust, banks lost trust in each other, and negative expectations fed into lower buy rates, which in turn led to lower sales, and so on. All of this then led to the first dramatic economic downturn of the 21st century. Thus, the miscalculation of risk was particularly important in this situation.

In sum, what follows from the definition of these concepts is that cross-cultural research needs to differentiate between errors (nonintended), violations (intended deviations), and risks (depth and veracity of risk analysis).

The Error Process

Errors will always appear because the human mind works in such a way that easy principles of action are chosen that are fast, frugal, and most often right (Gigerenzer & Todd, 1999; Kahneman & Klein, 2009; Reason, 1990). But, that means that there are also a number of decisions and actions that are wrong. As discussed in Chapter 1 of this volume, a comprehensive approach to error processes includes error prevention, detection, and management. First, errors can be avoided through *error prevention* mechanisms. Second, once errors have been made, they need to be *detected* (Allwood, 1984; Sellen, 1994). Finally, all processes that appear after an error has been detected fall under a rubric of *error management*. As we discuss throughout this chapter, different aspects of culture (e.g., uncertainty avoidance, humane orientation, tightness, fatalism, power distance, individualism-collectivism) have implications for error prevention, detection, and management.

Error Prevention

The error process starts with an intention that will lead to a certain action that may result in (most often negative) error consequences. One strategy to deal with errors is to prevent an error from occurring, what we refer to as an *error prevention* strategy. Error prevention implies that one attempts to analyze the risks adequately, develops the right intentions (from the standpoint of long-range intentions), and prevents doing the wrong action. As discussed in Chapter 1 (this volume), the factors that facilitate error prevention include individual anticipatory strategies of prevention, technology to detect errors, and social/contextual factors that afford communication and planning for errors.

Error Detection and Error Management

Even if there are elaborate prevention mechanisms in place, errors still occur and need to be detected, through human and machine mechanisms alike (see Chapter 1 of this volume for error detection modes). For error detection, the actor needs to have a clear representation of the goals of actions, needs to expect that an error could occur, and needs to get feedback to be able to detect an error. After errors have been detected, they can be managed in ways to interrupt the occurrence of the negative error consequences. After an error is detected, it is necessary to act quickly and to have adequate actions in one's repertoire to manage the error. A number of issues are important here, such as general orientations toward errors (positive or negative attitudes toward errors); error competence (e.g., error self-efficacy); emotional reactions toward errors (e.g., defensive versus open); and communication processes regarding the management of errors (e.g., the psychological safety for communicating about errors once they have occurred). The error detection process has to be quick because undetected errors tend to have increased negative consequences and can turn into more catastrophic errors if they are unaddressed (Reason, 1995).

An important component of error management is maximizing learning from errors. Learning from errors is an approach that can ultimately reduce future error consequences. This can be done by learning how to develop strategies of error prevention to avoid the error in the future or by learning how to develop strategies of error management, such as processes designed to identify the error quickly and stop further consequences from accruing. A good example is crew resource training (Helmreich, Merritt, & Wilhelm, 1999). This training starts with the assumption that the pilot may make an error; once this error is made, it has to be detected. Since it is less likely that the pilot detects the error, the error detection of the copilot becomes important. Since the copilot is usually not able alone to do the actions to prevent the occurrence of the negative error consequences, he or she needs to communicate the error to the pilot. The pilot then has to accept the error report (i.e., has to accept that the copilot has the authority to talk about an error by the pilot) and has to go into additional secondary preventive action (check the veracity of the copilot's report and do the appropriate action so that the airplane is not endangered). This may then lead to a wider learning process in the whole company, or it may not (again, it depends on how the colleagues and the company react to the error, whether the error is communicated, and whether there is an acceptance of learning from these errors by the whole company). All of these processes can be influenced by the culture; therefore, we discuss important cultural variation in relevant issues such as blaming, accepting errors, compensatory strategies, the acceptability to criticize a hierarchically superior pilot, and so on.

Latent Errors

A final distinction is that of latent errors: errors that exist within an organization but are not acknowledged and not handled (e.g., safety procedures that produce errors or an organizational culture that allows certain errors to go unnoticed or without correction). Latent errors can arise because of faulty error prevention or detection and error management. One factor is *error explanation*: An error needs to be explained to be corrected. Error explanation is far from easy in an organization. People may start to blame others for making an error. After the error explanation, actions need to be put into effect to deal with the error consequences and to manage the error. Here again, the question is whether the organization has routines available to deal with an error, whether there is enough competency to deal with the situation, whether people communicate effectively about errors, and whether the action can be put into effect quickly. All of these factors also contribute to whether there are latent errors in an organization and, if so, whether they are conditioned by culture as discussed further in this chapter.

Time and Errors

The longer error detection takes, the stronger are the negative consequences of errors. Many errors need to be detected quickly to eliminate the negative consequences; other errors allow more time. For example, a management mistake, such as employing the wrong person, does not need to be corrected within a few seconds. But, over the period of a few weeks, employing the wrong person may be damaging to the business. On the other hand, a mistake while driving a car (e.g., not seeing a red light) needs to be corrected within seconds to reduce the negative consequences; otherwise, an accident may happen. What is true of most errors (both the management error and the driving error), however, is that the longer the error is not detected, the more negative can be the consequences of the error. Thus, error detection is crucial.

In sum, as we build theory and a research agenda for the study of culture and errors, it is important to note that error prevention, detection, and management are all critical processes for organizations in all cultures. These processes share the same goal: to reduce the short- and long-term negative consequences of errors. While error prevention is aimed at trying to avoid having errors occur in the first place, error detection and management processes accept that errors will occur and are designed to catch and deal with errors after they occur. Latent errors can transpire due to faulty processing in either stage. As we discuss further in the chapter, although we expect that error prevention, detection, and management are universal processes, culture influences the extent to which there is a focus on each of these processes and can affect the ways these processes can have culture-specific manifestations. Put simply, cultural differences, we believe, arise from differences in attention to prevention, detection, or management and the ways these practices are implemented. We discuss not only cross-cultural variations in the error process but also the implications for these variations for cultural interfaces, particularly in multicultural teams.

Cultural Influences on Error Processes

There are many definitions of the term *culture*, each with its own implications for the study of the subject. Kroeber and Kluckhohn (1952) claimed that "the essential core of culture consists of traditional (i.e., historically derived and selected) ideas and especially their attached values" (p. 357), while Campbell (1965) suggested that elements of culture are made up of useful ideas that were adopted by increasing numbers of people. Hofstede

(1980) defined culture as "the collective programming of the mind," which distinguishes human groups from one another and influences how they respond to their respective environments (p. 25). Triandis (1972) suggested that culture consists of both the objective and the subjective elements of the human-made environment and defined subjective culture as "a cultural group's characteristic way of perceiving that man-made part of its environment" (p. 4). Triandis and Suh (2002) later defined elements of culture as "shared standard operating procedures, unstated assumptions, tools, norms, [and] values" (p. 136). Similarly, the Global Leadership and Organizational Behavior Effectiveness (GLOBE) study defined culture as common practices and common values (House & Javidan, 2004). While there is variation in the definitions of culture, many point to the shared nature of culture, its ability to impart adaptive (or once adaptive) knowledge, and its transmission across time and generation (Triandis, Kurowski, & Gelfand, 1994).

Culture has received attention in numerous areas in organizational behavior, including job attitudes, motivation, leadership, conflict and negotiation, teams, human resource management practices, and organizational culture (see Gelfand, Erez, & Aycan, 2007, for a review). To date, however, there have been few efforts to tie specific cultural dimensions to different stages of the error process (for exceptions, see Helmreich, 2000; Helmreich & Merritt, 1998; Helmreich et al., 2001; Jing et al., 2001; Li et al., 2009). To begin filling this void, we discuss numerous cultural dimensions that we see as particularly relevant to the error processes articulated, namely, error prevention, detection, and management. We theorize that organizations within and across cultures can differ in their "trajectories" of each stage of the error process, and there is considerable variation in the importance placed on different stages across cultures. Generally, one culture might be highly attentive to error prevention and not at all attentive to detection and error management, whereas another cultural might be just the reverse.

Although there are numerous cultural dimensions, we discuss six that have specific implications for the error management process. First, how people react to uncertainty in the environment, as measured by the cultural value of uncertainty avoidance, plays an important role in how people in different cultures address the unpredictability associated with the occurrence of errors. Second, humane orientation, or how people value affiliation with and support of others over the fulfillment of personal needs, affects the interpersonal aspects of error management, such as error communication. Third, tightness-looseness, a measure of the strength of norms and consistency of punishment for their violation, has an impact on how people plan for and address the inherent norm violation that occurs when an error, even if unintentional, is committed. Fourth, fatalism, or the belief that outcomes are dictated by external forces rather than personal decisions or behaviors, plays a role in how people try to control and explain errors. Fifth, power distance, which relates to the extent to which people expect and accept power inequalities between levels in the social hierarchy, has far-reaching implications for communication between people of different organizational levels about error management. Finally, the extent to which people feel autonomous versus interdependent with their respective groups, or individualism-collectivism, has an impact on how individuals and groups differentially address issues surrounding error prevention, detection, and management. In sum, different parts of the "cultural elephant" are theorized to influence different aspects of the error process.

Table 11.1 provides a summary of our main points, which are elaborated in the following sections, as well as several cultural paradoxes that we have identified regarding each cultural dimension and the error process. We note several important caveats of our theoretical analysis. First, while we focus on these particular dimensions of culture, we recognize that there are other dimensions that could also be relevant for errors that require analysis in future work. Second, although we discuss each dimension separately for exposition purposes, it is critical to note that cultures are complex wholes that vary in multiple dimensions; thus, an analysis of any particular cultural system should take this into account. Third, it is critical to note that there is much within-culture variation on all of the dimensions discussed, and that organizational cultural differences within national cultures can vary dramatically depending on industry, region, strategy, and the like, which also have critical influences on error detection and error management. While we do not discuss these distinctions vis-à-vis societal culture and their potential interactions in this chapter, they are fertile ground for future research, as we point out in the general discussion. Finally, we should hasten to add that due to the general lack of studies, many of our comments are theoretical and speculative—future research will have to tell whether our suggestions are right. With these caveats in mind, we now define the cultural dimensions of interest in this chapter; discuss their hypothesized relationship with error prevention, detection, and management; and highlight the implications of our analysis for error processes in multicultural teams.

Uncertainty Avoidance

The cultural dimension of uncertainty avoidance (UAI) refers to how well people tolerate ambiguous, unstructured situations (Hofstede, 1980). People in cultures with high UAI, such as Sweden and Germany (as reported by House, Hanges, Javidan, Dorfman, & Gupta, 2004), view uncertainty with anxiety and attempt to minimize their exposure to it by establishing policies, rules, and laws to impose certainty across many

TABLE 11.1
Summary of Theory on Culture and Errors

Cultural Dimension		Error Prevention	Error Detection	Error Management
Uncertainty avoidance	High	High on individual planning and strategizing High on technology High on communication and information search Negative attitude toward errors engenders creation of formalized rules, regulations, and procedures	Less monitoring Less communication and feedback	Negative emotional reaction Stress interferes with deployment of skills and resource Defensive attribution Lower reporting of individual error Formalized and restricted communication and information exchange delays error handling
	Low	Fewer individual attempts to plan or strategize Technology not developed to same degree Less communication and information search Fewer bureaucratic measures to prevent errors	Greater monitoring Less negativity surrounds errors, facilitating open communication and feedback	Fewer and less-intense negative emotional reactions Lower stress levels allow individuals to use skills to address error Fewer defensive attributions Greater reporting by individuals Open communication fosters speedy error handling
Humane orientation	High	Greater tolerance for errors results in fewer individual and collective planning activities and preventive strategies	Lower monitoring Slower detection	Fewer and less-intense negative emotional reactions Low expectations of serious consequences of errors reduces defensiveness Greater consideration of situational context of errors, attenuating negative consequences and leading to less blaming Supportive interpersonal relationships help individuals deal with errors

(continued)

TABLE 11.1 (Continued)

Summary of Theory on Culture and Errors

Cultural Dimension		Error Prevention	Error Detection	Error Management
	Low	Greater emphasis on individual planning and strategizing Increased defensiveness and conflict surrounding errors decreases collective error planning	Greater monitoring Stress surrounding errors inhibits communication and feedback	Negative emotional reactions Greater defensive reactions because of high potential for negative consequences Errors attributed to the individual rather than to the situation and more blaming
Tightness-looseness	High (tight)	More attention to error prevention Greater felt accountability increases accessibility of norms and prevention-focused guides Greater expectations for punishment for deviations amplify individual error prevention	High individual self-regulation enhances individual ability to detect errors Greater sensitivity to deviations from norms increases attention paid to feedback that error might occur High degree of social monitoring facilitates detection of errors	Negative reactions and attitudes toward errors because of implication of deviant behavior Expectation of monitoring and punishment increases desire to suppress errors and limit their visibility to others Increased negative social consequences for errors, including bullying of deviant
	Low (loose)	Less attention to error prevention Lower felt accountability and expectation for punishment reduce individual error prevention	Lower self-regulation inhibits individual error detection Lower acceptability of social monitoring decreases error detection Lowered sensitivity to norm deviation decreases attention to feedback about errors Lower degree of social monitoring inhibits detection of errors	Higher error tolerance results in less-intense negative reactions to errors Lower expectations for monitoring and punishment decrease attempts to suppress errors Fewer social and interpersonal consequences for deviation

		Prevention	Detection	Management
Fatalism	High	Fewer prevention strategies because of low feelings of control over outcomes	Belief that error occurrence is inevitable and due to external factors decreases detection	External attribution for errors and fewer negative or defensive reactions Less blaming may facilitate communication about errors, but there is little motivation to communicate because little can be done to mitigate damage or prevent future errors
	Low	Sense of control over process and outcome facilitate creation of prevention strategies	Belief that error occurrence can be affected and potentially halted by human efforts increases error detection	Internal attributions for errors because under personal or organizational control Negative and defensive reactions, which inhibits communication
Power distance	High	Planning and preventive measures more "top-down," with little input from subordinates	Managers may work to detect errors, but subordinates may be hesitant to step outside their roles to detect errors and alert their superiors	Potential for face loss over error makes high-status members reluctant to deal with errors Communication about errors likely to be stilted, with subordinates uncomfortable expressing opinions and emotions about error Subordinates unlikely to request supervisor feedback about actions that may have contributed to error
	Low	Error prevention a participatory endeavor, with both management and subordinates engaged in the process	Participative nature of workplace induces all organizational members to monitor for errors	Lower concern for face makes high-status members more comfortable confronting error Participative workplace makes subordinates more comfortable expressing opinions and emotions about errors Subordinates more likely to seek supervisor feedback about behavior
Individualism-collectivism	High (individualistic)	Promotion focus leads to less focus on potential losses, failure, or errors and thus fewer error prevention measures	Tendency to self-enhance reduces likelihood of detection and communication of individual errors	Accountability and blame fall on individual, increasing defensiveness Lack of face concerns facilitates communication about error

(Continued)

TABLE 11.1 (Continued)
Summary of Theory on Culture and Errors

Cultural Dimension	Error Prevention	Error Detection	Error Management
Low (collectivistic)	Prevention focus enhances desire to avoid failures or mistakes, thus facilitating the creation of error prevention strategies	Lower motivation to self-enhance increases self-criticism, thus facilitating the detection and communication of individual errors Interdependence and high network density may also facilitate detection of others' errors	Proclivity to make external attributions decreases defensiveness Dialectical thinking decreases surprise at error occurrence and hindsight bias Less communication about errors because of potential for individual or collective face loss Accountability falls on collective and, by proxy, the representative or leader of the collective Conformity and group think processes may lead to collective denial of error

domains of life. In contrast, people in cultures with low UAI like Greece, Guatemala, and Russia are more comfortable with ambiguity and less likely to establish structures to control or predict uncertain situations. The most important approaches to dealing with uncertainty are to plan well and to anticipate potential negative events in the future to prevent them from occurring.

Uncertainty Avoidance and Error Processes

A number of features of cultures with high UAI lead us to predict that these societal contexts will be both simultaneously high on error prevention yet, ironically, slower on error detection and worse on error management. As discussed, people in cultures with high UAI manage uncertainty through planning and the establishment of rules that make situations and behavior more predictable. Research has shown, for example, that managers from these cultures spend more time planning and scheduling activities than those from cultures with low uncertainty avoidance (Hofstede, 1984). As the basis of planning activities, information search may also be more important in cultures with high UAI, such that workers in the same occupation spend a greater proportion of their working hours on information searching depending on levels of cultural uncertainty avoidance. Beckmann, Menkhoff, and Suto (2008) found that asset managers from Japan, a country with a high UAI culture, spent the highest percentage of their working hours on research, while asset managers from the United States, which has a culture of comparatively lower UAI, spent the smallest proportion of their working hours on research. People in high UAI cultures may also be more likely to rely on "tried-and-true" methods of planning and prevention. For example, travelers from high UAI cultures purchase planned, prepackaged tours more frequently than travelers from moderate UAI cultures (Money & Crotts, 2003), highlighting the connection between UAI and the desire to minimize the possibility of unanticipated events, even on vacation. Similarly, Ryan, McFarland, Baron, and Page (1999) suggested that organizations in high UAI cultures may use fewer types of selection methods because of the need to rely only on proven methods. By planning for different possible situations and depending on reliable methods, people in high UAI cultures establish a sense of control over their environments. The sense of control that results from planning is particularly critical to attenuate the higher anxiety and stress levels experienced by people in high UAI cultures (Hofstede, 1980; Lynn & Hampson, 1975; Millendorfer, 1976).

Based on this discussion, we suggest that, given the higher stress level and greater tendency to worry and plan for uncertain situations, there will be a greater focus on error prevention in cultures with high versus low uncertainty avoidance. In particular, given their concern for avoiding uncertain events, people in high UAI cultures will have enhanced individual anticipatory strategies of prevention, and organizations will seek to provide structured training to improve individuals' preparatory strategies. However, the effectiveness of the individual and organizational attempts to plan may be diminished in high UAI cultures. Burke, Chan-Serafin, Salvador, Smith, and Sarpy (2008) suggested that managers and trainers in high UAI cultures may prefer standardized and structured approaches (e.g., lectures) to safety training over more engaging strategies like role-playing. While the reliance on structured approaches reduces the variability in the training process, these methods do not provide workers with as great an opportunity to critically assess the knowledge or engage in counterfactual, "what if" thinking. Thus, workers in high UAI cultures may be more reliant on standard safety protocols and less able to anticipate and respond to unusual events. Burke et al. supported these propositions with a meta-analysis, finding that UAI was negatively related to worker engagement in safety training, and that UAI moderated the relationship between training and negative safety outcomes, such that as UAI increases, the effectiveness of safety training in reducing negative outcomes decreases. Thus, although planning and prevention may receive greater attention in high UAI cultures, these planning activities may not impart individuals with the flexibility needed actually to deal with errors as they occur.

In addition to the focus on planning in high UAI cultures, people will readily talk about potential errors and attempt to avoid them by planning for the "correct action." This implies that there is a high degree of communication about potential errors, and all of this communication is under the rubric of *avoiding* the occurrence of these errors. This has several important implications for this analysis. First, errors are seen as negative and must be avoided at all costs as they indicate a dangerous deviation from norms (Hofstede, 1991). Second, if there is inadequate planning, there will be more errors. Third, because the appearance of errors implies that people have not really dealt with future problems adequately, it is the fault of the people making an error. Thus, high UAI leads not only to high communication about potential errors but also to a negative attitude toward errors and a higher negative attitude toward people who make errors.

Therefore, in high-UAI cultures, the concern about managing uncertainty in the future will result in greater attention to error prevention than in low-UAI cultures. Thus, at least in the stage of error prevention, there is "worry work"—individuals in high-UAI cultures are expected to reduce ambiguity through a high degree of communication, which can provide important clarifying information about the situation and relevant behaviors (Sully de Luque & Sommer, 2000). Organizations in high UAI cultures should likewise develop and use error prevention technology to the extent that it provides a reliable method of reducing ambiguity in the environment (Hofstede, 1980). For example, pilots in high UAI cultures show more acceptance of and reliance on automation (Helmreich & Merritt, 1998). However, high UAI may actually impede the adoption of innovative technology (Erumban & de Jong, 2006) since these unproven methods may actually increase uncertainty in the organization.

However, ironically, we expect, particularly when work has become routinized, that the high emphasis on error prevention in high UAI cultures will lead to a consequent slower ability to recognize errors and to manage errors. These individuals are expected to be focused on designing strategies and routines to prevent errors. Once a prevention strategy is seen as effective, anxiety is reduced, and perceptions of control over the environment are enhanced. Accordingly, there is less reason to monitor and a decreased need for communication and feedback; thus, high-UAI cultures should be slower to recognize if errors do occur and more surprised when they do occur. One anecdotal example (in which one of the authors was indirectly involved) is that both *Business Week* and its German equivalent *Wirtschaftswoche* had a cover story about dealing with management and errors within a few weeks of each other in August 2006. Both magazines asked chief executive officers (CEOs) of important companies to talk about their "biggest error"; by and large, the German CEOs described errors by others that they had not detected early enough (thus, they saw their fault in low error detection but not for making errors themselves), while the American CEOs reported about real errors they themselves made.

Moreover, high-UAI cultures are expected to be poorer at error management for a variety of reasons. First, individuals are expected to have *highly negative attitudes* when errors are discovered due to the fact that they were unexpected and were seen as highly dangerous and stressful deviations (Hofstede, 1991). Because errors are seen as a sign of insufficient preparation and planning, there will be more *defensive attributions*: more blaming and less reporting of individual errors (Hofstede, 2001; Merkin, 2006). High UAI may also be associated with less-flexible problem-solving coping to deal with managing errors. As discussed in the error prevention section, workers in high UAI cultures may be less able to respond flexibly to unanticipated events due to the standardized safety training they receive (Burke et al., 2008). Furthermore, workers in high UAI cultures may adhere to standard operating procedures or organizationally provided standards regardless of whether flexibility is allowed (Beckmann et al., 2008; Helmreich & Merritt, 1998). The lack of ability or desire to deviate from standard procedures may make the occurrence of errors even more stressful and ultimately have a negative impact on workers' ability to deal flexibly with the uncertainties and ambiguities associated with errors after they occur.

Social contextual factors in high-UAI cultures also likely inhibit the collective detection and management of errors. For example, the high degree of formalized communication associated with high-UAI cultures may delay the handling of errors because workers have to rely on a more time-consuming process to disseminate formal communication about the error to those involved. Moreover, high UAI cultures tend to encourage restrictions on the exchange of information between people not directly involved in managing the specific situation (Gray, 1988; Salter & Niswander, 1995). The restriction of information exchange protects the security of the organization by minimizing conflict and competition (Gray, 1988) but may ultimately reduce the information on error occurrence disseminated throughout the work group or organization. In addition, when communication does occur, it may be hindered by the ritualistic behaviors and high degree of stress in uncertainty avoidance cultures. Merkin (2006) found that in embarrassing situations, people from high UAI cultures utilized more ritualistic face-saving strategies, harmonized less, and employed more aggressive communication than people from low UAI cultures. Because of their inability to deviate from ritualistic face-saving scripts and extend conciliatory expressions, error communication in high UAI cultures may be overly stiff, negative, and adversarial and thus may not foster the information exchange necessary to manage errors. In all, these processes should lead to a lower degree of collective error detection and hindered error management in high UAI cultures.

Slowness in error detection may lead to the increased likelihood that catastrophes (very negative error consequences as a result of latent errors) occur given that a certain risk exists. However, once the error is accepted by high-UAI organizations and cultures, there is a higher degree of error prevention again—with conscious attention to improving systems, products, safety regulations, and so on. Ryan et al. (1999) found that organizations in high UAI cultures were more likely to audit their selection procedures to ensure their effectiveness; it may also be the case that, after adequate time has elapsed and the anxiety surrounding an error subsides, these organizations may likewise examine the effectiveness of their error prevention process. After this examination, the organization may go the route of ever-increasing bureaucracy (ever more rules, regulations, procedures, standards, etc). In all, this suggests that high-UAI societies may be slower to detect errors and manage them and thus may actually see short-term consequences that are more disastrous. Once errors have been recognized, however, they might prompt a deeper analysis of the error situation and activate secondary error prevention, which may help in long-term error prevention but not necessarily reduction of catastrophes.

Proposition 1: Uncertainty avoidance will be positively related to error prevention and negatively related to error detection and error management. Specifically, error prevention will be higher in high- versus low-UAI cultures, yet error detection and error management will be lower in high- versus low-UAI cultures. As a result, errors will have short-term consequences that are more disastrous in high- than low-UAI cultures.

This proposition leads to an interesting *cultural paradox:* High-UAI cultures plan for errors more but have potentially more negative and disastrous consequences due to poor error detection and error management processes.

Humane Orientation

Humane orientation is related to how societies reward expressions of altruism, fairness, and caring (House et al., 2004, p. 13). Cultures that value a humane orientation tend to focus on the fulfillment of needs related to affiliation rather than self. Therefore, people in cultures that endorse a humane orientation, such as those in Egypt, Malaysia, and Ireland, express more prosocial values and emphasize the importance of providing support and help to others, spending time with others, and sharing information. In contrast, people in cultures that are low in humane orientation, like those in Germany, Singapore, and South Africa, tend to be more self-centered and spend their time and resources in the pursuit of personal enjoyment, comfort, wealth, and power.

Broadly, levels of humane orientation are associated with the quality of the human condition; as survival becomes more difficult and requires greater resources, humane orientation increases. For example, societies with low levels of modernization and economic development tend to have high levels of humane orientation, whereas greater modernization and economic development are related with lower levels of humane orientation (Schlösser et al., 2007). This relationship exists because as the hardships of survival increase, there is more need for solidarity and support. High humane orientation is also associated with helping behaviors, such as providing financial and material help to others. As an example of this point, humane orientation is negatively related to the role of the state in caring for its citizens (as in welfare societies). In high humane orientation cultures, citizens care for people they know. In contrast, governments in low humane orientation cultures develop welfare institutions as a substitute for personal support and protection (Schlösser et al., 2007). Overall, high humane orientation is associated with societal-level variables related to ensuring human survival through personal relationships rather than external intervention.

Humane orientation accepts human nature as it is and therefore also accepts and is more tolerant of errors. Indeed, the item of the GLOBE study that inquired about the practice of error tolerance and intolerance in the society turned out to be part of the humane orientation scale in the factor analyses done in the GLOBE study (Hanges & Dickson, 2004). However, humane orientation should not be confused with a high degree of protection for anyone in society; humane orientation is positively related to racism and authoritarianism (Schlösser et al., 2007).

Humane Orientation and Error Processes

Humane orientation has important implications for the error process. First, high humane orientation is related to increased compassion and acceptance and thus acceptance of mistakes, which may affect the attention given to prevention and error detection. If mistakes are expected and are tolerated, there is a reduced need for prevention strategies. Likewise, if errors are not particularly stressful or important, errors should be slower to be detected once they have occurred in cultures with high humane orientation. However, once errors have occurred, features of humane-oriented societies make them more prepared to deal with them in a variety of ways. First, there should be less-negative attitudes toward errors once they have occurred in cultures with high humane orientation. The tolerance of mistakes should be related to less defensiveness when errors are discovered since tolerance for errors would lead workers to expect fewer serious consequences for them. Humane orientation is also characterized by greater consideration for individual circumstances. Managers in cultures with high humane orientation are more likely to take individual workers' situations into account when making decisions (Schlösser et al., 2007); thus, when errors are handled, there may be a greater consideration of the individual's situation when the error occurred (e.g., fewer dispositional attributions and blaming of individuals), which may attenuate negative consequences. Finally, once errors have been identified, the emphasis on supportive relationships may help individuals to deal collectively with errors. However, it is important to note that people in cultures with high humane orientation may be reluctant to point out the errors of others because of the potential for such interactions to cause rifts in relationships between organization members. Accordingly, we predict the following:

> *Proposition 2*: Humane orientation will be negatively related to error prevention and error detection yet positively related to error management. Error prevention and detection will be lower in cultures with high as compared to low humane orientation, yet error management will be higher in cultures with high as compared to low humane orientation.

This proposition leads to another interesting cultural paradox: Cultures high on humane orientation are less attentive to errors and therefore have less error prevention and detection. However, once errors have occurred, they may have more positive attitudes and strategies that are less defensive as compared to cultures with low humane orientation.

Tightness-Looseness

The cultural dimension of tightness-looseness reflects the degree to which a society provides clear norms and consistent sanctions for deviation from norms (Chan, Gelfand, Triandis, & Tzeng, 1996; Gelfand, Nishii, & Raver, 2006; Pelto, 1968; Triandis, 1989). Tight societies, like those of Japan, Singapore, and Pakistan, tend to provide strong norms and sensitive monitoring systems to detect deviations, which are severely punished. As such, these societies tend to value order, formality, discipline, and conformity. In contrast, norms in loose societies, like those of Brazil, Israel, or the United States, tend to be more ambiguous, and the monitoring of behavior is less stringent. Deviations from norms are more likely to be tolerated, and punishments for deviations are less severe. Thus, loose societies tend to value innovation, openness to change, tolerance, risk taking, and variety (Gelfand et al., 2006).

Tightness-looseness is associated with a variety of ecological and sociopolitical correlates. Pelto (1968), who first introduced the concept after anthropological investigations into traditional societies, suggested that tightness is associated with dense populations, unilineal kinship systems (i.e., kinship is traced through either the male or female parent), and economic systems based on agriculture. Looseness, in contrast, is associated with sparse populations, bilateral kinship systems (i.e., kinship is traced through both parents), and economic systems based on hunting, fishing, or gathering. Triandis (1977) expanded the proposed antecedents of the construct, arguing that social differentiation, or the extent to which specialized roles have developed in a society, may predict tightness-looseness in preliterate societies. Societies with differentiated social organization tend to be tighter since the greater complexity of the social structure requires stronger norms to ensure the survival of the group. Chan et al. (1996) also pointed to homogeneity, isolation, interdependence, and cultural stability as important antecedents to the development of tight societies. In a 33-nation study of modern societies, Gelfand and colleagues (2011) found that tightness is indeed a separate cultural dimension than others studied previously (e.g., uncertainty avoidance, collectivism) and is related to such factors as population density, percentage of arable land, environmental vulnerability, order (low crime), conservative attitudes, and less tolerance for deviant groups.

Tightness-Looseness and Error Processes

Tightness-looseness has an impact on a wide variety of social and organizational factors associated with the error process. Tight societies tend to have higher levels of felt accountability (Frink & Klimoski, 1998; Tetlock, 1985), or the belief that one's actions will be evaluated and that evaluation may lead to punishment (Gelfand et al., 2006). Two outcomes of higher levels of felt accountability in tight societies are increased cognitive accessibility of norms and ought-focused self-guides (Gelfand et al., 2006). That is, people in tight societies are expected to have higher accessibility of the public self, stronger associations between situations and the appropriate norms in the eyes of *generalized others* (as compared to *close others*, as in the case of collectivism, discussed in a separate section), and should be more focused on avoiding failure or mistakes than achievement of goals. In addition, people in tight cultures (e.g., Japan) are concerned about a broader array of consequences that their actions might have on others as compared to people in loose cultures (Maddux & Yuki, 2006). Accordingly, people in tight societies are more likely to create and enforce strategies to prevent potential errors at the individual, team, and organizational levels. On the other hand, people in loose societies are not expected to engage in extensive error planning since expectations are less well defined, felt accountability is lower, and tolerance for mistakes is expected.

Error detection is also expected to be enhanced in tight versus loose societies. From a cognitive perspective, error detection is facilitated through *internal goal comparisons*, by which individuals complete an internal comparison between feedback signals and the goals of the action (Hofmann & Frese, Chapter 1, this volume). Tight cultures are theorized to have higher degrees of self-regulation, by which individuals monitor their own goals vis-à-vis important standards to avoid making mistakes, as compared to loose cultures, which have less self-regulation (Gelfand et al., 2006). As such, they should be better able to detect errors that they make. Tight cultures also experience greater monitoring of social behavior than loose cultures, in part due to a high population density, which makes monitoring by others more possible than in loose cultures. Monitoring is not only more possible but it is also more expected and accepted in tight versus loose cultures. That is, given the importance of abiding by social norms in tight cultures, it is much more typical, as compared to loose cultures, to communicate to others what they have done wrong, thus making error detection through external communication more prevalent in tight versus loose cultures. For example, Syrian children inform their parents of anyone who may be violating government mandates (Hopwood, 1988), and Chinese children monitor their peers' behavior to ensure it is appropriate (Chen, 2000). Organizations in tight societies also monitor their employees (e.g., Aoki, 1988; Jennings, Cyr, & Moore, 1995; Morishima, 1995).

In contrast, such monitoring of employees leads to more negative reactions in loose societies (Frink & Klimoski, 1998). For example, research in the United States suggests that employee monitoring causes negative emotional reactions, such as feelings of loss of control and perceptions of being oppressed (Aiello & Kolb, 1995; Martin & Freeman, 2003), and disrupts performance (Sutton & Galunic, 1996). Because of this increased negative reaction to monitoring, loose societies are expected to place less emphasis on error detection than tight societies. Particularly when errors are ambiguous (Hofmann & Frese, Chapter 1, this volume), tight cultures, which are highly sensitive to deviations, will take more seriously any feedback that signals an error might be occurring. By contrast, loose cultures have much more tolerance for deviance and will have a much more lenient response criterion, such that it must be that much clearer and definite that an error has occurred before corrective actions occur.

At the same time, once errors have been detected, error management processes are likely to be hindered in tight versus loose cultures. Because violations are punished more severely in tight cultures, people will likely have attitudes that are much more negative toward errors. By contrast, because there is much more tolerance for deviant behavior, errors are more natural in loose cultures, resulting in attitudes that are more positive toward error management. As a result, individuals will be more likely to want to suppress errors and limit that others see that they have occurred in tight versus loose cultures. Emotional processing may be further hindered in tight societies because of higher degrees of social sanctions for norm deviation. While the punishment of deviants occurs in all cultures, it is likely more pronounced in tight cultures. For example, Japanese children who are different from their schoolmates often experience *ijime*, or bullying with the purpose of punishing those who are deviant. To the extent that others also view errors negatively and punish them when they occur, it will collectively detract problem solving for how to manage errors once they have occurred. Accordingly, we predict the following:

Proposition 3: Tightness-looseness will be positively related to error prevention and detection but negatively related to error management. Error prevention and detection will be higher in tight versus loose cultures, yet error management will be lower in tight as compared to loose cultures.

This discussion raises yet another possible cultural paradox: Cultures high on tightness are better able to prevent, detect, and communicate errors but have a lower ability to manage them once they have occurred.

Fatalism

The cultural dimension of fatalism reflects the extent to which people believe that external factors dictate life events. Fatalism is associated with the beliefs that it is not possible to fully control personal outcomes (Aycan et al., 2000), and that whatever happens must happen (Bernstein, 1992). While people in fatalistic cultures such as Russia and India may point to a variety of external sources that exert ultimate control over their lives, including God, fate, or chance, they are united by a common recognition of the role of these external factors in their lives (Caplan & Schooler, 2007). In contrast, people in less-fatalistic cultures, including Germany and the United States, are more likely to endorse the belief that they maintain personal control over their outcomes and lives. At the national level, fatalistic beliefs are negatively correlated with gross domestic product (GDP) per capita, life expectancy at birth, urbanism, percentage of GDP spent on health, voter turnout, and environmental sustainability, human development, and human rights, and women's status in society and positively correlated with heart disease death rate and suicide rate (Leung & Bond, 2004).

Fatalism and Error Processes

Fatalism has been linked to beliefs and behaviors related to health, safety, and coping, which have implications for error prevention, detection, and management. In particular, we expect that fatalism is related to low prevention of errors, given that fatalism has been related to inadequate or hindered preparation for negative events. For example, Hardeman, Pierro, and Mannetti (1997) found that fatalism related to HIV was related to safe sex practices, such that participants who believed that they had control over whether they became infected with the virus were more likely to report intentions to use condoms during sexual encounters. Other studies have shown that fatalistic beliefs regarding earthquakes (i.e., beliefs that the amount of damage incurred during earthquakes was under the control of external forces) were less likely to prepare for earthquakes (Turner, Nigg, & Paz, 1986). Likewise, individuals who exhibit stronger beliefs in fatalism and destiny (Rotter, 1966) tend to use seatbelts less frequently (Colón, 1992; Council, 1969; but see Byrd, Cohn, Gonzalez, Parado, & Cortes, 1999).

People in highly fatalistic cultures may be less likely to create strategies to prevent errors or develop technology for error detection since the occurrence of errors is believed to be under external control and not affected by human interventions or monitoring. In contrast, people in low-fatalism cultures believe that the occurrence of errors is under personal or organizational control and thus may be more likely to establish technological and social mechanisms to prevent and detect errors. In addition to the link between fatalism and preventive measures, fatalism may affect the monitoring process that is important for the detection of errors. For example, Rundmo and Hale (2003) found that managers' fatalism was one of the most important predictors of their monitoring of safety on their jobs. Managers high on fatalism were less likely to engage in actions like safety observations and audits, inspection routines, and supervision of housekeeping as compared to managers low on fatalism. In all, fatalism is expected to result in low error prevention and error detection.

Fatalism may also be related to error management. Fatalistic beliefs are associated with greater external attribution for events. For example, Kouabenan (1998) found that, in a sample of Ivory Coast drivers, those who were more fatalistic were more likely to attribute traffic accidents to external forces, such as headlight care, poorly maintained roads, and other's violation of traffic signals, than to internal forces, such as driver carelessness or impatience. Based on the increased tendency for people in fatalistic cultures to attribute errors to external forces, there may also be less communication about errors because there is little that can be done to prevent similar errors in the future. Thus, although negative or defensive emotional processes are likely to be lower in high-fatalism cultures (in part because they are expected and attributed to the situation), there is less likelihood that they will be managed since they are seen as stemming from external sources and predetermined. In this respect, our very definition of errors—namely, that they could have potentially been avoided—runs counter to fatalistic thinking and reducing error prevention, detection, and management. Thus, although blaming in case of an error will be much lower in cultures high on fatalism, which may make error communication easier since there is little reason to talk about errors because one fatalistically accepts them, there is little motivation for error communication. Accordingly, we predict the following:

> *Proposition 4*: Fatalism is negatively related to error prevention, detection, and management. Error prevention, detection, and management will be lower in cultures high versus low on fatalism. The capacity for catastrophic error consequences is therefore more pronounced in cultures high versus low on fatalism.

This discussion raises yet another possible cultural paradox: Fatalism is the only cultural dimension that is uniformly negative on all aspects of the error process, even though in principle, fatalism should make error communication easier.

Power Distance

Power distance refers to the degree to which members of a society expect and accept inequalities (Hofstede, 1980). In high-power-distance cultures, such as Morocco and Nigeria, people accept large power differentials between levels of the social hierarchy. On the other hand, people in lower-power-distance cultures, including Denmark and the Netherlands, are less accepting of such power differentials. As a core cross-cultural difference, power distance is related to key historical, political, social, and geographic factors and has far-reaching consequences for organizational behavior. For example, Hofstede (1980) noted that cultures with high power distance are likely to have experienced occupation, colonialism, and imperialism, whereas low-power-distance cultures tend to have histories of independence and federalism. In addition, high-power-distance cultures often have traditions of centralized power in the hands of a monarchy or oligarchy, a small middle class, and agrarianism, while low-power-distance cultures are characterized by histories of representative governments, a large middle class, and more modern industry. Hofstede (1980) also linked power distance with geography, finding a negative relationship between geography and latitude; this relationship is likely attributable to the higher levels of social mobility and technological innovation associated with the more extreme latitudes.

Power Distance and Error Processes

Power distance is also associated with unique patterns of organizational behavior, many of which have important implications for the error process. People in high power distance cultures see subordinates and superiors as inherently separate and unequal groups, leading to greater social distance between these groups in the workplace. Superiors are seen as largely unquestionable authorities, and in this respect, we would expect that superiors in high power distance cultures would be particularly concerned with preventing errors "under their watch." At the same time, subordinates in high power distance cultures are less likely to engage in planning for errors. First, subordinates tend to avoid voicing concerns to superiors and uncritically accept their directives and actions (Hofstede, 1991). Human resource and decision-making practices are less participative in high power distance cultures (Aycan, Kanungo, & Sinha, 1999; Brockner et al., 2001; Newman & Nollen, 1996); likewise, subordinates in these cultures do not experience the desire for participative practices as do subordinates in lower power distance cultures (Brockner et al., 2001; Eylon & Au, 1999; Huang & Van de Vliert, 2003; Huang, Van de Vliert, & Van der Vegt, 2005; Robert, Probst, Martocchio, Drasgow, & Lawler, 2000). The general lack of subordinate participation in organizations in high power distance cultures may have a negative impact on error planning. Put simply, since subordinates may not feel comfortable voicing their concerns or ideas to their superiors, who may likewise discourage subordinate participation, error communication may suffer in high power distance cultures (Helmreich et al., 2001). Thus, in high power distance cultures, superiors and subordinates will have differential levels of prevention of errors as compared to those in low power distance cultures.

Power distance may also have an impact on the error detection process. Again, although people of high status will be motivated to detect errors given their enhanced sense of authority and responsibility, in the highly centralized and stratified organizations typical in high power distance cultures, employees may not be motivated to step outside their roles and take on the extra responsibility of monitoring for errors if it is not part of their job (Slater & Narver, 1995). In particular, power distance is negatively associated with national levels of role ambiguity (Peterson & Smith, 1997), suggesting that employees in high power distance cultures have a clear understanding of the responsibilities associated with their jobs. Thus, if those responsibilities do not include error detection, it may be less likely that employees in high power distance cultures will take on that role, especially if this action may be interpreted as encroaching on a superior's role. In contrast, employees in low power distance cultures are likely to have less-clear boundaries between occupation roles (Hofstede, 1980), which may induce them to monitor for errors even when this role is outside the scope of their job. Thus, as with error prevention, there will be discrepancies in detection of errors between superiors and subordinates in high versus low power distance cultures.

Finally, power distance has important implications for error management. In high power distance cultures, superiors and subordinates alike might be reluctant to manage errors once they have occurred. For superiors, there is a great loss of face if errors occur on one's watch. That is, high-status members of high power distance societies may not want to lose face if they have caused an error, particularly because of their greater responsibility in preventing them in the first place given their enhanced authority. In addition, there is likely to be less communication about errors in high power distance cultures. Not only are subordinates in these cultures unlikely to point out their leaders' errors, but also they may avoid probing their supervisors for feedback about their behavior related to errors (Morrison, Chen, & Salgado, 2004). Accordingly, subordinates in high power distance cultures are also unlikely to voice their opinions, feelings, or concerns about their own or others' errors, thus having a negative impact on error communication. In sum, we predict the following:

Proposition 5: The nature of one's status is highly linked to error processes in high versus low power distance cultures. In high power distance cultures, low-status individuals will have lower error prevention, high detection, and low management as compared to high-status individuals. There are fewer differences between high- and low-status individuals in error prevention, detection, and management in low power distance cultures.

As with the other cultural dimensions, this discussion raises yet another possible cultural paradox: High-power parties in high power distance cultures are likely worried about errors and losing face, but they are not likely to learn that they have occurred and thus cannot manage them.

Individualism-Collectivism

The last dimension we consider is individualism and collectivism, which reflects the nature of the relationship between the individual and the group. There has been extensive discussion of these constructs in sociology (e.g., Parsons, 1949), anthropology (Kluckhohn & Strodtbeck, 1961; Mead, 1961) and psychology (Chinese Culture Connection, 1987; Hofstede, 1980; Markus & Kitayama, 1991; Schwartz, 1994; Triandis, 1995). Although sometimes using slightly different parlance, across all of these disciplines there is agreement that cultures can be differentiated on the extent to which people are autonomous versus embedded in groups (Schwartz, 1994). Furthermore, within psychology, there is increasing evidence that the nature of the self, and consequently information processing, varies across individualistic and collectivistic cultures (Heine & Lehman, 1999; Markus & Kitayama, 1991; Triandis, 1989). For example, in individualistic cultures like those of Denmark, New Zealand, and the Netherlands, individuals value their autonomy from groups, want to stand out, and focus on their individual rights and their own needs and interests. On the other hand, people in collectivistic cultures, such as those of Philippines, Georgia, Iran, and India, are tightly embedded in social groups, want to "blend in" and maintain harmony and face, and tend to emphasize duties and obligations to the group over personal needs or desires (Gelfand, Bhawuk, Nishii, & Bechtold, 2004).

Individualism and collectivism are related to several societal-level constructs. Individualism tends to appear in either hunting-and-gathering societies or wealthy, industrialized societies, while collectivistic societies are based more on agriculture and may be classified as developing. Individualism is also associated with capitalism, democracy, and social mobility. Individualism and collectivism are also related to family structure, with the nuclear family being more central in individualistic cultures and the extended kinship network being more central in collectivistic cultures (Hofstede, 1980; see also Gelfand, Bhawuk, et al., 2004). In addition, individualism tends to be associated to a faster pace of life than collectivism (Levine & Norenzayan, 1999). Finally, communication in collectivistic cultures tends to be more indirect, which speaks to the need to preserve group members' face, while people in individualistic cultures tend to communicate their views in more direct ways (Holtgraves, 1997).

Individualism-Collectivism and Error Processes

Important elements of individualism and collectivism that have an influence on the error process are prevention versus promotion focus, attributions and individual blaming and accountability, situational blaming, face, self-criticism, error communication, and motivational processes directed toward individuals and toward the group. Based on extant research, we predict that error prevention will be higher in collectivistic cultures. Collectivists tend to be more concerned with avoiding failure and minimizing potential losses than individualists (Elliot, Chirkov, Kim, & Sheldon, 2001; Hamilton & Biehal, 2005; but see Hsee & Weber, 1999, for an exception), and research has found that collectivists are more prevention focused and individualists are more promotion focused (Lalwani, Shrum, & Chiu, 2009; Lee, Aaker, & Gardner, 2000). Collectivists' tendency to focus on preventing losses and failures likely manifests itself in the development of error prevention mechanisms and strategies, which serves to help avoid the negative experience of failure. In contrast, individualists' drive to attain gains and achieve successful outcomes may distract them from implementing necessary mechanisms to prevent failures or losses, leading to underdeveloped error prevention strategies.

Second, error detection is likely to occur more easily within collectivistic as compared to individualistic cultures due to an enhanced focus on self-criticism versus self-enhancement, respectively. In particular, research has shown that people in individualistic cultures are less attentive to their failures than to their successes and have a pervasive tendency to see themselves as better than others in the domains of academics (e.g., McAllister, 1996; Mizokawa & Ryckman, 1990; Yan & Gaier, 1994), athletics (e.g., De Michele, Gansneder, & Solomon, 1998; Mark, Mutrie, Brooks, & Harris, 1984), driving (Svenson, 1981), health (e.g., Weinstein, 1980), and social relationships (e.g., Ross & Sicoly, 1979; Sedikides, Campbell, Reeder, & Elliot, 1998), among others. By contrast, self-criticism is much more common in collectivistic cultures, in which improving oneself serves as an affirmation of one's interdependence with others and of being a "good" cultural self (Heine & Lehman, 1999). For example, the practice of *hansei*, or critical self-reflection, is pervasive throughout schools in Japan (Heine, Lehman, Markus, & Kitayama,

1999). Rather than helping children to identify their positive characteristics through praise and compliments, Japanese educators encourage students to focus on their shortcomings and weaknesses and to be critical of themselves to adapt and fit in with others. In addition, from a structural point of view, because of enhanced social interdependence and high network density, people in collectivistic cultures may be better able to notice when others make errors than those in individualistic cultures. Accordingly, we expect that errors are more likely to be detected in collectivistic as compared to individualistic cultures.

Predictions regarding individualism and collectivism and error management are more complex. On the one hand, one might expect that collectivists would be less defensive regarding errors due to their increased proclivity to make external versus internal attributions (Choi, Nisbett, & Norenzayan, 1999; Miller, 1984; Morris & Peng, 1994). By contrast, individualists will have particularly negative reactions to their own errors given that errors are inconsistent with the cultural mandate to be the best and because the blaming process is oriented toward the individual. Likewise, dialectical thinking, which is found among collectivists (e.g., that contradictions are possible), also produces a lower degree of surprise when an error happens. This also decreases the idea of hindsight bias—a bias to assume that one would have done it right (in contrast to others) when asked about an action retrospectively (Choi & Nisbett, 2000). Thus, attitudes toward errors may be more positive in collectivistic versus individualistic cultures.

Yet, on the other hand, we expect less communication of errors in collectivistic cultures. First, collectivism is often linked to concerns for face and reputation (Ting-Toomey, 1991), which should make individuals more reluctant to report errors for fear of losing face in the "eyes" of others. Second, individuals in collectivist cultures, who are in a more extended "web of accountability" (Gelfand, Lim, & Raver, 2004), assume that their errors will reflect not only on themselves but also on the *groups to which they belong*. Indeed, research has shown that while Americans tended to indicate that a single individual caused an event, East Asians were more likely to hold many people, particularly groups, accountable for a given action (Chiu & Hong, 1992; Chiu, Morris, Hong, & Menon, 2000; Menon, Morris, Chiu, & Hong, 1999). East Asians have been also shown to use "proxy logic" when assigning blame for an accident or error, in which they first assign blame to the collective responsible for the error rather than the individual and then to the leader/person who represents the collective (Zemba, Young, & Morris, 2006). Accordingly, given concerns for the accountability of the larger group and leaders for one's own actions, we expect that individuals will have more denial of their own errors in collectivistic cultures. Moreover, given that collectivists might self-enhance on collectivistic dimensions (such as their beliefs about their group) (Sedikides, Gaertner, & Vevea, 2007), they might be more defensive about errors committed by their group to avoid harming the reputation of the group. At the extreme, groupthink and collective denials of errors are more likely in collectivistic cultures. Accordingly, we predict the following:

> *Proposition 6*: Collectivism is positively related to error prevention and error detection but negatively related to certain error management processes. Error prevention and error detection will be higher in collectivistic cultures as compared to individualistic cultures. Although errors are less surprising and elicit processing that is less defensive in collectivistic cultures, concerns with face and group accountability will reduce error communication as compared to individualistic cultures.

This discussion raises yet another possible cultural paradox: Collectivistic cultures should be better in error management than individualistic cultures given their strong ties, prevention focus, and penchant for situational attributions. However, communication processes may reverse these processes because of concern for face and concern for the group, which can inhibit error communication and ultimately produce disastrous consequences in the long term.

Implications for Multicultural Teams

Thus far, we discussed cross-cultural variation in the error process, from error prevention, to error detection, to error management. Yet, understanding cultural influences on errors is also critical for contexts in which cultures interact—such as in multicultural and virtual teams that transcend national borders. On the one hand, multicultural team contexts are ripe with potential for more errors with more disastrous consequences than their homogeneous team counterparts, given that there are highly divergent expectations regarding errors in such contexts along with additional processes that make it hard to coordinate in multicultural teams, such as ethnocentrism (Cramton & Hinds 2005), ingroup biases (Salk & Brannon 2000), and high levels of task or emotional conflict (Elron, 1997; Von Glinow, Shapiro, & Brett, 2004), which all lead to reduced (constructive) communication.

However, as with others who argue that multicultural teams can provide strategic advantages for organizations (see Earley & Gibson, 2002; Shapiro, Von Glinow, & Cheng, 2005), we suggest that *cultural diversity can actually help with error processes,* and multicultural teams can even far surpass homogeneous teams due to the complementarities that exist regarding error prevention, detection, and management. Put simply, given that many cultural contexts fall short in general in maximizing all of these processes, it is possible that some multicultural teams, due to their natural complementarities on different error foci, will maximize the prevention, detection, and management of errors. Thus, in the right combination, cross-cultural teams may be able to compensate for the weaknesses/lack of focus of other members. By contrast, homogeneous teams that focus on select parts of the error process will fail to fully prevent and manage errors. Thus, by tapping the different strengths associated with national cultures, the right combination of team members could collectively maximize the effectiveness of each stage of the error process. Next, we discuss some examples of complementarities that are theorized to exist across different error processes to illustrate our argument.

For example, as noted, effective error prevention requires the establishment of individual anticipatory strategies, the development of error detection technology, and clear communication and planning for errors. As discussed, it is expected that different cultural dimensions will be related to preventive measures (or the lack thereof). We suggested that team members from cultures that are high on uncertainty avoidance, low on humane orientation, high on tightness, low on fatalism, and high on collectivism may excel at error prevention. These cultural values are all related to control and its establishment over the environment either to avoid uncertainty or to establish standards for conduct from close or generalized others. In contrast, cultures with high levels of humane orientation, low levels of uncertainty avoidance, high levels of looseness and fatalism, and high individualism are likely to avoid planning or taking preventive action because of a higher tolerance of errors due to a variety of reasons particular to these different cultures. Thus, teams comprised of people from these different cultures may provide a good balance that should result in strong, formal preventive measures that support and protect organization members and their relationships.

Error detection is another part of the error process that may benefit from collaboration between culturally heterogeneous team members. Team members from tight cultures are likely to focus on error detection given the focus on rules, avoidance of rule violation, and high potential for punishment in those cultures. On the other hand, team members from high humane orientation cultures are again less likely to focus on error detection because they care more about people than performance. Similarly, high-fatalism cultures may find it unnecessary to detect errors or to report errors. Moreover, in cases of high power distance, there may be little error communication upward (to people with higher hierarchical positions), which may also decrease error detection. A team comprised of people from tight cultures and high humane orientation cultures, as an example, may compensate for each other such that they develop a focus on error detection that is simultaneously considerate of individuals. In contrast, a combination of people from high humane orientation and loose cultures would likely have a negative impact because there would be a general lack of attention to error detection. As with error prevention, cultural differences in groups can actually help to ensure attention to this aspect of the error process.

Finally, certain multicultural teams may be particularly adept at handling errors. For example, team members from high humane orientation and loose cultures are likely to be more accepting of errors and less likely to engage in defensive attributional processes. In contrast, team members from high UAI and tight cultures are less accepting of errors and more likely to engage in defensive distributional processes. Accordingly, these negative error-handling behaviors could be offset by team members from cultures high on humane orientation or looseness.

Across all error stages, one can deduce important complementarities that arise from cultural differences. For one example, based on the discussion, teams comprised of members from cultures that are on high humane orientation and tightness would have much complementary across all stages of the error process. On the one hand, people from tight cultures are likely to focus on establishing rules and engendering adherence to them, which can help focus the group on error prevention and detection. On the other hand, people from cultures with high humane orientation will temper this focus with an acceptance of the fallibility of people, which should especially attenuate the negative aspects of error handling in tight cultures.

To be sure, we have not presented an exhaustive list of possibilities; rather, we illustrated that cultural differences can create synergies across different stages of the error process. As an important caveat, however, we would add that although these compensatory combinations are theoretically possible, they would likely be accompanied with significant conflict, especially early in the team life cycle. As in other multicultural team contexts, the performance of culturally heterogeneous teams may be enhanced when leaders work to ensure effective communication (Ayoko, Hartel, & Callen, 2002) and help team members uncover hidden knowledge (Baba, Gluesing, Ratner, & Wagner, 2004).

Theoretical and Practical Implications

To err is human; errors are universal. Yet, the error process, as we have argued in this chapter, can be highly influenced by cultural factors. Whether it is preventing errors, detecting them, or managing them, cultures can have different approaches to this human universal and therefore have different strengths and vulnerabilities for errors and their consequences.

As we hope we have illustrated, the field of culture and errors is largely unchartered and has the potential to make wide theoretical and practical contributions. Given the paucity of theory and research in this area, many of our predictions are speculative and are in need of verification. As a first step, constructs and measures on errors are in need of large-scale cross-cultural validation to examine attitudes and beliefs about each stage of the error process and how they are influenced by culture. For example, the Error Orientation Questionnaire (EOQ; Rybowiak, Garst, Frese, & Batinic, 1999), which includes scales for error competence, learning from errors, error risk taking, error strain, error anticipation, covering up errors, error communication, and thinking about errors, has received some cross-cultural validation in the Netherlands and Germany. Whether this construct and measurement is construct relevant, deficient, or contaminated in East Asian, South American, Middle Eastern, and African cultures, to name a few, remains to be tested. Other implicit measures of detection of errors and reactions to errors, such as those using reaction times or nueroscientific methods, could also shed light on cultural differences in error prevention and detection and reaction to errors. For example, the neural basis of human error processing has been linked to activity in the anterior cingulate cortex (Holroyd & Coles, 2002; see also Yeung, Botvinick, & Cohen, 2004), and cultural differences in error processing would presumably reflect such brain-related activity.

Whether studies involve surveys, experiments, implicit or explicit methods it is important that culture and error research also ultimately take a dynamic approach to look at how individual differences (e.g., gender, age) as well as contextual factors (the nature of the task, industry, region) interact with cultural factors to affect errors. For example, the predictions given in this chapter might be amplified in conditions of high time pressures or high cognitive load—factors that can cause individuals to rely on well-learned cultural schemas and norms (see Gelfand, Leslie, & Fehr, 2008). While we do not discuss these distinctions vis-à-vis societal culture and their potential interactions in this chapter, they are fertile ground for future research, as we point out in the general discussion that follows.

The previous discussion of culture and errors also has implications for the design of *error management training* in organizations. Error management training is a training procedure that allows and even encourages error making in the training process to learn optimally from error experiences (Keith & Frese, 2005, 2008). There are a number of conceptual reasons why it seems worthwhile to integrate errors into training. Errors are informative feedback for knowledge and skill acquisition because they show what part of one's mental model is incorrect and needs to be modified. Also, errors prevent trainees from premature automatization of inappropriate action strategies because they interrupt the course of action and make trainees rethink their strategies. Finally, errors occur not only during training but also back at work, where there is less support than during training. Learning how to deal with errors effectively will therefore be a useful skill in itself. This issue is captured in the principle of transfer-appropriate processing, which postulates that those processes required on transfer tasks should be practiced in training (Ivancic & Hesketh, 2000).

Error management training has been shown to lead to higher performance than error-free training in several domains (training of computer skills, social skills, and other skills, such as firefighter skills) (Frese, Muelhausen, Wiegel, & Keith, 2010; Heimbeck, Frese, Sonnentag, & Keith, 2003; Joung, Hesketh, & Neal, 2006). Important aspects of this error management training are the so-called error management instructions (e.g., "The more errors you make, the more you learn! "I have made an error, great! Because now I can learn!") and the fact that the tasks are not graded in difficulty. Rather, those tasks that also appear in reality are given, and the participants of the error management training learn to deal with them by making errors, correcting them, and learning how to deal with difficult situations.

Error management training is effective for a variety of reasons. First, it decreases negative emotions associated with errors and their derivative consequences (Keith & Frese, 2005). Negative emotional reactions to errors may be mentally stressful, leading to secondary tasks of anxiety control, and thus leading to higher cognitive load and reduction of the ability to learn from an error and learning new skills (cf. Hockey, 1983). A second reason for the positive performance effects of error management training lies in the considered approach to the training material, resulting in higher metacognitive activity as a result of error management training (Keith & Frese, 2005).

While error management training might be universally effective, we argue that it will need to be tailored to particular cultural contexts. For example, different instructions may be needed in different cultures that specifically target the source of resistance that goes along with lack of prevention, detection, or management. For example, error management training that tries to help the trainee reduce negative emotions (e.g., "I made an error—great"; Frese et al., 1991), may not counter ineffective emotions in collectivistic cultures, where face and reputational concerns loom large when errors are committed. As such, different training instructions might be needed to help reduce this particular source of negative emotions.

Likewise, in tight cultures, in which punishment is often associated with deviating from normative expectations, such instructions may need to be supplemented with other interventions in the organizational environment to promote psychological safety surrounding errors. In effect, we would expect (cultural) attribute by treatment interactions for error management training. There are a number of studies that have found that error management training was not equally effective across different individuals (e.g., Treatment × Attribute interactions). For example, Heimbeck et al. (2003) found that people high on prove and avoidance goal orientation performed particularly well in *error-avoidant* conditions, for which it is easier to stand out (particularly important for those on prove orientation) and easier to avoid any negative evaluations from others (particularly important for those high on avoidant orientation. Extending this to the culture level, we would expect that error management needs to be tailored to particular cultural contexts to be maximally effective.

Likewise, although studies that have shown better performance as a result of error management training in comparison to error-free training, such training may be ineffective in cultures that already have high tolerance for errors. Put simply, error management training is geared toward reducing negative emotional processes, yet this may be less of a concern (and even counterproductive) in certain cultures. For example, in very loose, humane-oriented, fatalistic, and very low uncertainty avoidance cultures, there are less negative emotional effects of errors in general. In those cultures, people are not afraid of errors and do not experience any negative effects of errors; they are therefore not strained more in an error situation. Moreover, there may be less learning from errors in highly error-tolerant cultures given differences in metacognitive activities during error management training. That is, in very loose, humane-oriented, fatalistic, and very low uncertainty avoidance cultures, there may be less reason to increase metacognitive activities; people just accept errors as a natural part of their activities; therefore, there is little incentive to think again about the errors. Thus, high error tolerance not only may help people be relaxed in error situations but also they may be less motivated to learn in such situations because errors are so much seen as a natural part of one's activities. Thus, in these societies, error management training and error-free training may have similar effects, or in the extreme, error management training may make such individuals even more tolerant of errors and produce even less management of them.

At a more general level, this discussion is consistent with organizational learning, which is often based on negative events. Joung et al. (2006), for example, showed that most learning takes place from negative events and not from positive events. There is also evidence that it is more successful to have been in an error situation oneself than to hear about other people's errors (Ivancic & Hesketh, 2000). All of this may not appear in societies that are highly error tolerant. In such societies, we would argue, error management training might even benefit from training people to have reactions that are more negative (without becoming too extreme) so that they are more attentive to and prepared to manage errors.

These notions not only are important for training per se but also may be important in the function of errors in the learning process at work. If we are right, there needs to be a certain degree of *error intolerance* to learn from errors; on the other hand, if error intolerance becomes *too high*, the strain effects are so high that there is little learning from errors. Thus, error management training must be designed in such a way to target error tolerance or intolerance in different cultures to accomplish this delicate balance between error tolerance and intolerance.

Concluding Remarks

In conclusion, in this chapter we set forth a research agenda on culture and error management processes. While errors are a universal, the specific ways in which people approach errors—in terms of prevention, detection, and management—was theorized to vary widely across the globe. Cultures, because of their differences in values, beliefs, norms, and orientations, bring particular strengths and vulnerabilities to the error process. Given their theoretical and practical importance, understanding cultural influences on errors is a critical frontier of this burgeoning literature.

Acknowledgment

This research is based on work supported by the U.S. Army Research Laboratory and the U.S. Army Research Office under grant number W911NF-08-1-0144.

References

Aiello, J. R., & Kolb, K. J. (1995). Electronic performance monitoring and social context: Impact on productivity and stress. *Journal of Applied Psychology, 80,* 339–353.

Allwood, C. M. (1984). Error detection processes in statistical problem solving. *Cognitive Science, 8*, 413–437.
Aoki, M. (1988). *Information, incentives, and bargaining in the Japanese economy*. New York: Cambridge University Press.
Aycan, Z., Kanungo, R. N., Mendonca, M., Yu, K., Deller, J., Stahl, G., et al. (2000). Impact of culture on human resource management practices: A 10-country comparison. *Applied Psychology: An International Review, 49*, 192–221.
Aycan, Z., Kanungo, R. N., & Sinha, J. B. P. (1999). Organizational culture and human resource management practices: The model of culture fit. *Journal of Cross-Cultural Psychology, 30*, 501–526.
Ayoko, B. O., Hartel, C. E., & Callen, V. J. (2002). Resolving the puzzle of productive and destructive conflict in culturally heterogeneous workgroups: A communication accommodation theory approach. *International Journal of Conflict Management, 13*, 165–195.
Baba, M. L., Gluesing, J., Ratner, H., & Wagner, K. H. (2004). The contexts of knowing: Natural history of a globally distributed team. *Journal of Organizational Behavior, 25*, 546–587.
Beckmann, D., Menkhoff, L., & Suto, M. (2008). Does culture influence asset managers' views and behavior. *Journal of Economic Behavior and Organization, 67*, 624–643.
Bernstein, M. H. (1992). *Fatalism*. Lincoln: University of Nebraska Press.
Brockner, J., Ackerman, G., Greenberg, J., Gelfand, M. J., Francesco, A. M., Chen, Z. X., et al. (2001). Culture and procedural justice: The influence of power distance on reactions to voice. *Journal of Experimental Social Psychology, 37*, 300–315.
Burke, M. J., Chan-Serafin, S., Salvador, R., Smith, A., & Sarpy, S. A. (2008). The role of national culture and organizational climate in safety training effectiveness. *European Journal of Work and Organizational Psychology, 17*, 133–152.
Byrd, T., Cohn, L. D., Gonzalez, E., Parada, M., & Cortes, M. Seatbelt use and belief in destiny among Hispanic and non-Hispanic drivers. *Accident Analysis and Prevention, 31*, 63–65.
Campbell, D. T. (1965). Variation and selective retention in socio-cultural evolution. In H. R. Barringer, G. Blanksten, & R. Mack (Eds.), *Social change in developing areas* (pp. 19–49). Cambridge, MA: Schenkman.
Caplan, L. J., & Schooler, C. (2007). Socioeconomic status and financial coping strategies: The mediating role of perceived control. *Social Psychology Quarterly, 70*, 43–58
Chan, D. K.-S., Gelfand, M. J., Triandis, H. C., & Tzeng, O. (1996). Tightness-looseness revisited: Some preliminary analyses in Japan and the United States. *International Journal of Psychology, 31*, 1–12.
Chen, X. (2000). Social and emotional development in Chinese children and adolescents: A contextual cross-cultural perspective. In F. Columbus (Ed.), *Advances in psychology research* (Vol. 1, pp 229–251).
The Chernobyl accident: What happened. (2000, June 5). BBC News. Retrieved July 31, 2009, from http://www.bbc.co.uk/
The Chernobyl Forum. (2006). *Chernobyl's legacy: Summary report*. Vienna, Austria: International Atomic Energy Agency. Retrieved July 31, 2009, from http://www.iaea.org

Chinese Culture Connection. (1987). Chinese values and the search for culture-free dimensions of culture. *Journal of Cross-Cultural Psychology, 18,* 143–164.

Chiu, C., & Hong, Y. (1992). The effects of intentionality and validation on individual and collective responsibility attribution among Hong Kong Chinese. *Journal of Psychology, 126,* 291–300.

Chiu, C., Morris, M. W., Hong, Y., & Menon, T. (2000). Motivated cultural cognition: The impact of implicit cultural theories on dispositional attributions varies as a function of need for closure. *Journal of Personality and Social Psychology, 87,* 247–259.

Choi, I., & Nisbett, R. E. (2000). Cultural psychology of surprise: Holistic theories and recognition of contradiction. *Journal of Personality and Social Psychology, 79,* 890–905.

Choi, I., Nisbett, R. E., & Norenzayan, A. (1999). Causal attribution across cultures: Variation and universality. *Psychological Bulletin, 125,* 47–63.

Civil Aviation Authority. (1998). *CAP 681 global fatal accident review 1980–1996.* London: Civil Aviation Authority.

Colón, I. (1992). Race, belief in destiny, and seat belt usage: A pilot study. *American Journal of Public Health, 82,* 875–877.

Council, F. M. (1969). *Seat belts: A follow-up study of their use under normal driving conditions.* Chapel Hill, NC: University of North Carolina Highway Safety Research Center.

Cramton, C. D., & Hinds, P. J. (2005). Subgroup dynamics in internationally distributed teams: Ethnocentrism or cross-national learning? *Research in Organizational Behavior, 26,* 231–263.

De Michele, P. E., Gansneder, B., & Solomon, G. B. (1998). Success and failure attributions of wrestlers: Further evidence of the self-serving bias. *Journal of Sport Behavior, 21,* 242–255.

Earley, P. C., & Gibson, C. B. (2002). *Multinational teams: A new perspective.* Mahwah, NJ: Erlbaum.

Elliot, A. J., Chirkov, V. I., Kim, Y., & Sheldon, K. M. (2001). A cross-cultural analysis of avoidance (relative to approach) personal goals. *Psychological Science, 12,* 505–510.

Elron, E. (1997). Top management teams within multinational corporations: Effects of cultural heterogeneity. *Leadership Quarterly, 8,* 393–412.

Erumban, A. A., & de Jong, S. B. (2006). Cross-country differences in ICT adoption: A consequence of culture? *Journal of World Business, 41,* 302–314.

Eylon, D. & Au, K. Y. (1999). Exploring empowerment cross-cultural differences along the power distance dimension. *International Journal of Intercultural Relations, 23,* 373–285.

Frese, M., Brodbeck, F., Heinbokel, T., Mooser, C., Schleiffenbaum, E., & Thiemann, P. (1991). Errors in training computer skills: On the positive function of errors. *Human-Computer Interaction, 6,* 77–93.

Frese, M., Muelhausen, S., Wiegel, J., & Keith, N. (2010). Comparing two training approaches to charismatic communication: Social learning versus error management. Unpublished manuscript, University of Giessen, Germany.

Freud, S. (1954). *Zur Psychopathologie des Alltagslebens.* Frankfurt, Germany: Fischer. (Original work published 1901)

Frink, D. D., & Klimoski, R. J. (1998). Toward a theory of accountability in organizations and human resources management. *Research in Personnel and Human Resources Management, 16*, 1 51.

Gelfand, M. J., Bhawuk, D. P. S., Nishii, L. H., & Bechtold, D. J. (2004). Individualism and collectivism. In R. J. House, P. J. Hanges, M. Javidan, P. W. Dorfman, & V. Gupta (Eds.), *Culture, leadership, and organizations: The GLOBE study of 62 societies* (pp. 437–512). Thousand Oaks, CA: Sage.

Gelfand, M. J., Erez, M., & Aycan, Z. (2007). Cross-cultural organizational behavior. *Annual Review of Psychology, 58*, 1–35.

Gelfand, M. J., Leslie, L. M., & Fehr, R. (2008). To prosper, organizational psychology should adopt a global perspective. *Journal of Organizational Behavior, 29*, 493–517.

Gelfand, M. J., Lim, B.-C., Raver, J. L. (2004). Culture and accountability in organizations: Variations in forms of social control across cultures. *Human Resource Management Review, 14* (Special issue), 135–160.

Gelfand, M. J., Nishii, L. H., & Raver, J. L. (2006). On the nature and important of cultural tightness-looseness. *Journal of Applied Psychology, 91*, 1225–1244.

Gelfand, M. J., Raver, J. L., Nishii, L., Leslie, L. M., Lun, J., Lim, B. L., et al. (2011). Differences between tight and loose cultures: A 33-nation study. Unpublished manuscript.

Gigerenzer, G., & Todd, P. M. (1999). *The ABC Research Group. Simple heuristics that make us smart*. Berlin: Oxford University Press.

Gray, S. J. (1988). Towards a theory of cultural influence on the development of accounting systems internationally. *Abacus, 24*, 1–15.

Hamilton, R. W., & Biehal, G. J. (2005). Achieving your goals or protecting their future? The effects of self-view on goals and choices. *Journal of Consumer Research, 32*, 277–283.

Hanges, P. J., & Dickson, M. W. (2004). The development and validation of the GLOBE culture and leadership scales. In R. J. House, P. J. Hanges, M. Javidan, P. W. Dorfman, & V. Gupta (Eds.), *Cultures, leadership and organizations: A 62 nation GLOBE study* (pp. 122–151). Thousand Oaks, CA: Sage.

Hardeman, W., Pierro, A., & Mannetti, L. (1997). Determinants of intentions to practice safe sex among 16–25 year-olds. *Journal of Community and Applied Social Psychology, 7*, 345–360.

Heimbeck, D., Frese, M., Sonnentag, S., & Keith, N. (2003). Integrating errors into the training process: The function of error management instructions and the role of goal orientation. *Personnel Psychology, 56*, 333–362.

Heine, S. J., & Lehman, D. R. (1999). Culture, self-discrepancies, and self-satisfaction. *Personality and Social Psychology Bulletin, 25*, 915–925.

Heine, S. J., Lehman, D. R., Markus, H. R., & Kitayama, S. (1999). Is there a universal need for positive self-regard? *Psychological Review, 106*, 766–794.

Helmreich, R. L. (2000). Culture and error in space: Implications from analog environments. *Aviation, Space, and Environmental Medicine, 71*, 133–139.

Helmreich, R. L., & Merritt, A. C. (1998). *Culture at work in aviation and medicine: National, organizational, and professional influences*. Aldershot, UK: Ashgate.

Helmreich, R. L., Merritt, A. C., & Wilhelm, J. A. (1999). The evolution of crew resource management training in commercial aviation. *International Journal of Aviation Psychology, 9*(1), 19–32.

Helmreich, R. L., Wilhelm, J. A., Klinect, J. R., & Merritt, A. C. (2001). Culture, error and crew resource management. In E. Salas, C. A. Bowers, & E. Edens (Eds.), *Improving teamwork in organizations: Applications of resource management training* (pp. 305–331). Hillsdale, NJ: Erlbaum.

Hockey, R. (1983). *Stress and fatigue in human performance*. Chichester, UK: Wiley.

Hofstede, G. (1980). *Culture's consequences*. London: Sage.

Hofstede, G. (1984). *Culture's consequences: International differences in work-related values*. Beverly Hills, CA: Sage.

Hofstede, G. (1991). *Culture and organizations: Software of the mind*. Berkshire, UK: McGraw Hill.

Hofstede, G. (2001). *Culture's consequences, second edition: Comparing values, behaviors, institutions, and organizations across nations*. Thousand Oaks, CA: Sage.

Holroyd, C. B., & Coles, M. G. H. (2002). The neural basis of human error processing: Reinforcing learning, dopamine, and the error-related negativity. *Psychological Review, 109*, 679–709.

Holtgraves, T. (1997). Styles of language use: Individual and cultural variability in conversation indirectness. *Journal of Personality and Social Psychology, 73*, 624–637.

Hopwood, D. (1988). *Syria 1945–1986: Politics and society*. London: Unwin Hyman.

House, R. J., Hanges, P. J., Javidan, M., Dorfman, P. W., & Gupta, V. (Eds.). (2004). *Cultures, leadership and organizations: A 62 nation GLOBE study*. Thousand Oaks, CA: Sage.

House, R. J., & Javidan, M. (2004). Overview of GLOBE. In R. J. House, P. J. Hanges, M. Javidan, P. W. Dorfman, & V. Gupta (Eds.), *Cultures, leadership and organizations: A 62 nation GLOBE study* (pp. 9–28). Thousand Oaks CA: Sage.

Hsee, C., & Weber, E. U. (1999). Cross-national differences in risk preferences and lay predictions. *Journal of Behavioral Decision Making, 12*, 165–179.

Huang, X., & Van de Vliert, E. (2003). Where intrinsic motivation fails to work: National moderators of intrinsic motivation. *Journal of Organizational Behavior, 24*, 159–179.

Huang, X., Van de Vliert, E., & Van der Vegt, G. (2005). Breaking the silence culture: Stimulation of participation and employee opinion withholding cross-nationally. *Management and Organization Review, 1*, 459–482.

Human error blamed in Tanzania crash. (2002, September 11). BBC News. Retrieved July 31, 2009, from http://www.bbc.co.uk/

Ivancic, K., & Hesketh, B. (2000). Learning from errors in a driving simulation: Effects on driving skill and self-confidence. *Ergonomics, 43*, 1966–1984.

Jennings, P. D., Cyr, D., & Moore, L. F. (1995). Human research management in the Pacific Rim: An integration. In L. F. Moore & P. D. Jennings (Eds.), *Human resource management on the Pacific Rim: Institution, practices, and attitudes* (pp. 351–379). New York: de Gruyter.

Jing, H.-S., Lu, C. J., & Peng, S.-J. (2001). Culture, authoritarianism, and commercial aircraft accidents. *Human Factors and Aerospace Safety, 1*, 341–359.

Joung, W., Hesketh, B., & Neal, A. (2006). Using "war stories" to train for adaptive performance: It is better to learn from error or success? *Applied Psychology: An International Review, 55*, 282–302.

Kahneman, D., & Klein, G. (2009). Conditions for intuitive expertise: A failure to disagree. *American Psychologist, 64*, 515–526.

Keith, N., & Frese, M. (2005). Self-regulation in error management training. Emotion control and metacognition as mediators of performance effects. *Journal of Applied Psychology, 90*, 677–691.

Keith, N., & Frese, M. (2008). Performance effects of error management training: A meta-analysis. *Journal of Applied Psychology, 93*, 59–69.

Kluckhohn, F., & Strodtbeck, F. (1961). *Variations in value orientation*. Evanston, IL: Row, Peterson.

Kouabenan, D. R. (1998). Beliefs and the perception of risks and accidents. *Risk Analysis, 18*, 243–252.

Kroeber, A. L., & Kluckhohn, C. (1952). *Culture: A critical review of concepts and definitions*. Cambridge, MA: Harvard University Peabody Museum.

Lalwani, A. K., Shrum, L. J., & Chiu, C. (2009). Motivated response styles: The role of cultural values, regulatory focus, and self-consciousness in socially desirable responding. *Journal of Personality and Social Psychology, 96*, 870–882.

Lee, A. Y., Aaker, J. L, & Gardner, W. L. (2000). The pleasures and pains of distinct self-construals: The role of interdependences in regulatory focus. *Journal of Personality and Social Psychology, 78*, 1122–1134.

Leung, K., & Bond, M. H. (2004). Social axioms: A model for social beliefs in multicultural perspective. *Advances in Experimental Social Psychology, 36*, 119–197.

Levine, R. V., & Norenzayan, A. (1999). The pace of life in 31 countries. *Journal of Cross-Cultural Psychology, 30*, 178–192.

Li, W.-C., Harris, D., Li, L.-W., & Wang, T. (2009). The differences of aviation human factors between individualism and collectivism culture. In J. A. Jacko (Ed.), *Human-computer interaction* (Part 4, pp. 723–730). Berlin: Springer.

Lynn, R., & Hampson, S. L. (1975). National differences in extraversion and neuroticism. *British Journal of Social and Clinical Psychology, 14*, 131–137.

Maddux, W. W., & Yuki, M. (2006). The "ripple effect": Cultural differences in perceptions of the consequences of events. *Personality and Social Psychology Bulletin, 32*, 669–683.

Mark, M. M., Mutrie, N., Brooks, D. R., & Harris, D. V. (1984). Causal attributions of winners and losers in individual competitive sports: Toward a reformulation of the self-serving bias. *Journal of Sport Psychology, 6*, 184–196.

Markus, H., & Kitayama, S. (1991). Culture and self: implications for cognition, emotion, and motivation. *Psychological Review, 98*, 224–253.

Martin, K., & Freeman, R. E. (2003). Some problems with employee monitoring. *Journal of Business Ethics, 43*, 353–361.

McAllister, H. A. (1996). A self-serving bias in the classroom: Who shows it? Who knows it? *Journal of Educational Psychology, 88*, 121–131.

McIntire, M. (2003, November 1). History of human error found in ferry accidents. *The New York Times*. Retrieved July 31, 2009, from http://www.nytimes.com

Mead, M. (1961). *Cooperation and competition among primitive people*. Boston: Beacon Press.

Menon, T., Morris, M. W., Chiu, C., & Hong, Y. (1999). Culture and the construal of agency: Attribution to individual versus group dispositions. *Journal of Personality and Social Psychology, 76*, 701–717.

Merkin, R. S. (2006). Uncertainty avoidance and facework: A test of the Hofstede model. *International Journal of Intercultural Relations, 30,* 213–228.
Millendorfer, J. (1976). *Mechanisms of socio-psychological development.* Vienna, Austria: Studia.
Miller, J. G. (1984). Culture and the development of everyday social explanation. *Journal of Personality and Social Psychology, 46,* 961–978.
Mizokawa, D. T., & Ryckman, D. B. (1990). Attributions of academic success and failure: A comparison of six Asian-American ethnic groups. *Journal of Cross-Cultural Psychology, 21,* 434–451.
Money, R. B., & Crotts, J. C. (2003). The effect of uncertainty avoidance on information search, planning, and purchases of international travel vacations. *Tourism Management, 24,* 191–202.
Morishima, M. (1995). The Japanese human resource management system: A learning bureaucracy. In L. F. Moore & P. D. Jennings (Eds.), *Human resource management on the Pacific Rim: Institution, practices, and attitudes* (pp. 351–379). New York: de Gruyter.
Morris, M., & Peng, K. (1994). Culture and cause: American and Chinese attributions for social and physical events. *Journal of Personality and Social Psychology, 67,* 949–971.
Morrison, E., Chen, Y., & Salgado, S. (2004). Cultural differences in newcomer feedback seeking: A comparison of the United States and Hong Kong. *Applied Psychology: An International Review, 53*(1), 1–22.
Newman, K. L., & Nollen, S. D. (1996). Culture and congruence: The fit between management practices and national culture. *Journal of International Business Studies, 27,* 753–779.
Parsons, T. (1949). *Essays in sociological theory: Pure and applied.* New York: Free Press.
Pelto, P. J. (1968, April). The differences between "tight" and "loose" societies. *Transaction,* pp. 37–40.
Peterson, M. F., & Smith, P. B. (1997). Does national culture or ambient temperature explain cross-national differences in role stress? No sweat! *Academy of Management Journal, 40,* 930–946.
Reason, J. (1990). *Human error.* New York: Cambridge University Press.
Reason, J. (1995). A systems approach to organizational error. *Ergonomics, 38,* 1708–1721.
Robert, C., Probst, T. M., Martocchio, J. J., Drasgow, F., & Lawler, J. J. (2000). Empowerment and continuous improvement in the United States, Mexico, Poland, and India: Predicting fit on the basis of the dimensions of power distance and individualism. *Journal of Applied Psychology, 85,* 643–658.
Ross, M., & Sicoly, F. (1979). Egocentric biases in availability and attribution. *Journal of Personality and Social Psychology, 37,* 322–336.
Rotter, J. B. (1966). Generalized expectations for internal versus external control of reinforcement. *Psychological Monographs, 80*(609).
Rundmo, T., & Hale, A. R. (2003). Managers' attitudes towards safety and accident prevention. *Safety Science, 41,* 557–574.
Ryan, A. M., McFarland, L., Baron, H., & Page, R. (1999). An international look at selection practices: Nation and culture as explanations for variability in practice. *Personnel Psychology, 52,* 359–391.

Rybowiak, V., Garst, H., Frese, M., & Batinic, B. (1999). Error Orientation Questionnaire (EOQ): Reliability, validity, and different language equivalence. *Journal of Organizational Behavior, 20*, 527–547,

Salk, J. E., & Brannen, M. Y. (2000). National culture, networks, and individual influence in a multi-national management team. *Academy of Management Journal, 43*, 191–202.

Salter, S. B., & Niswander, F. (1995, Second Quarter). Cultural influence on the development of accounting systems internationally: A test of Gray's [1988] theory. *Journal of International Business Studies*, pp. 379–397.

Schlösser, O., Frese, M., Al Najjar, M., Arciszewski, T., Besevegis, E., Bishop, G. D., et al. (2007). *Humane orientation as a new cultural dimension—A validation study on the GLOBE Scale in 25 Countries.* Unpublished manuscript, University of Giessen, Germany.

Schwartz, S. H. (1994). Beyond individualism and collectivism: New cultural dimensions and values. In U. Kim, H. C. Triandis, C. Kagitçibasi, S.–C. Choi, & G. Yoon (Eds.), *Individualism and collectivism: Theory, method, and applications* (pp. 85–122). Newbury Park, CA: Sage.

Sedikides, C., Campbell, W. K., Reeder, G. D., & Elliot, A. J. (1998). The self-serving bias in relational context. *Journal of Personality and Social Psychology, 74*, 378–386.

Sedikides, C., Gaertner, L., & Vevea, J. L. (2007). Evaluating the evidence for pan-cultural self-enhancement. *Asian Journal of Social Psychology, 10*, 201–203.

Sellen, A. J. (1994). Detection of everyday errors. *Applied Psychology: An International Review, 43*, 475–498.

Shapiro, D. L., Von Glinow, M., & Cheng, J. L. C. (2005). *Managing multinational teams: Global perspective.* Oxford, UK: Elsevier Science.

Slater, S., & Narver, J. C. (1995). Market orientation and the learning organization. *Journal of Marketing, 59*, 63–74.

Stone, R. (2006, April). Inside Chernobyl. *National Geographic.* Retrieved July 31, 2009, from http://www.nationalgeographic.com/

Sully de Luque, M. F., & Sommer, S. M. (2000). The impact of culture on feedback-seeking behavior: An integrated model and proposition. *The Academy of Management Review, 25*, 829–849.

Sutton, R. I., & Galunic, D. C. (1996). Consequences of public scrutiny for leaders and their organizations. In B. M. Staw & L. L. Cummings (Eds.), *Research in organizational behavior: An annual series of analytical essays and critical reviews* (Vol. 18, pp. 201–250). New York: Elsevier Science/JAI Press.

Svenson, O. (1981). Are we all less risky and more skillful than our fellow drivers? *Acta Psychologica, 47*, 143–148.

Tetlock, P. E. (1985). Accountability: The neglected social context of judgment and choice. In L. L. Cummings & B. M. Staw (Eds.), *Research in organizational behavior* (Vol. 7, pp. 297–332). Greenwich, CT: JAI Press.

Ting-Toomey, S. (1991). Intimacy expressions in three cultures: France, Japan, and the United States. *International Journal of Intercultural Relations, 15*, 29–46.

Triandis, H. C. (1972). *The analysis of subjective culture.* New York: Wiley.

Triandis, H. C. (1977). *Interpersonal behavior.* Monterey, CA: Brooks/Cole.

Triandis, H. C. (1989). The self and social behavior in differing cultural contexts. *Psychological Review, 96*, 506–520.

Triandis, H. C. (1995). *Individualism and collectivism*. Boulder, CO: Westview.
Triandis, H. C., Kurowski, L. L., & Gelfand, M. J. (1994). Workplace diversity. In M. D. Dunnette & L. M. Hough (Eds.), *Handbook of industrial and organizational psychology* (Vol. 4, pp. 769–827). Palo Alto, CA: Consulting Psychologists Press.
Triandis, H. C., & Suh, E. M. (2002). Cultural influences on personality. *Annual Review of Psychology, 53,* 133–60.
Turner, R. H., Nigg, J. M., & Paz, D. (1986). *Waiting for disaster: Earthquake watch in California.* Berkeley: University of California Press.
United Nations Scientific Committee on the Effects of Atomic Radiation. (2000). *UNSCEAR 2000 report to the General Assembly: Sources and effects of ionizing radiation* (Vol. 2, Annex J). Retrieved July 31, 2009, from http://www.unscear.org
Von Glinow, M., Shapiro, D. L., & Brett, J. M. (2004). Can we talk, and should we? Managing emotional conflict in multicultural teams. *Academy of Management Review, 29,* 578–592.
Weinstein, N. D. (1980). Unrealistic optimism about future life events. *Journal of Personality and Social Psychology, 39,* 806–820.
Yan, W., & Gaier, E. L. (1994). Casual attributions for college success and failure: An Asian–American comparison. *Journal of Cross-Cultural Psychology, 25,* 146–158.
Yeung, N., Botvinick, M. M., & Cohen, J. D. (2004). The neural basis of error detection: Conflict monitoring and error-related negativity. *Psychological Review, 111,* 931–959.
Zapf, D., Brodbeck, F. C., Frese, M., Peters, H., & Prumper, J. (1992). Errors in working with office computers: A first validation of a taxonomy for observed errors in a field setting. *International Journal of Human-Computer Interaction, 4,* 311–339.
Zemba, Y., Young, M. J., & Morris, M. W. (2006). Blaming leaders for organizational accidents: Proxy logic in collective- versus individual-agency cultures. *Organizational Behavior and Human Decision Processes, 101,* 36–51.

12

A New Look at Errors: On Errors, Error Prevention, and Error Management in Organizations

Michael Frese and David A. Hofmann

Where We've Been

Whenever humans are involved in producing outcomes, errors will be inevitable (Reason, 1990). The chapters in this volume illustrate that there are many different ways that errors can influence and be influenced by organizations. Our goal for this volume was for each of the chapters to offer a new perspective on errors given the domain about which the authors were writing. Specifically, we wanted each of the authors to move away from the traditional view of errors as nuisances that need to be eliminated to a more balanced view recognizing that errors can have both negative and positive effects, and that errors are both more important and more complex to deal with than suggested in the past.

This is the basic theme of Chapter 1 of this volume, in which we argue that errors are ubiquitous; therefore, it is unlikely to eliminate them completely. Given this, we need to consider in much more depth how to manage effectively the errors that continue to occur despite efforts to prevent them. One key distinction that we make is between errors and violations for which the differentiating factor is the intentionality of the actor. Errors are unintentional, whereas violations are intentional acts. It is important to recognize that, although violating behaviors are intended, any significant negative consequences that result are not intended or necessarily foreseen and, thus, often the result of an error. Violating the highway speed limit to ensure that one arrives at the airport in time to catch a flight is an intended act. The major accident that occurs due to excessive speed is both unintended and, in most cases, viewed as unlikely when the behavior (speeding) is undertaken. Finally, we describe a taxonomy of errors that has a number of different potential uses in future error research.

In Chapter 2, Keith argues cogently that an error management approach to training allows participants to better respond to the emotional impact of errors, and it also allows participants to maximize learning. Within the training context, an error management approach adopts the view that errors cannot be completely avoided in the training process, and that a pure error prevention approach leads to less learning, at least under the more realistic situation of having to transfer the training to new tasks on the job.

Another area in which errors can play a central—and sometimes positive serendipitous—role is the realm of creativity and innovation. Frequently, great inventions come about because of accidental discoveries resulting from errors. In Chapter 3, Hammond and Farr describe this process as well as how errors can happen during the innovation process.

Another alternative view of errors is provided by Mousavi and Gigerenzer in Chapter 4. Specifically, they describe how the traditional, experimental research documenting the various "cognitive biases" may itself be biased by not taking into account practical goals of human judgment (ecological validity). They forward the idea of ecological rationality as the matching of decision-making strategies to the structure of information in the environment. This perspective is then applied to efforts to "debias" decision-making efforts with specific application to the interpretation of medical statistics.

Bell and Kozlowski (Chapter 5) describe how errors in teams develop, the consequences they may have, and how they should be managed. One important distinction that Bell and Kozlowski make is between individual errors occurring within the team and team-level errors. Again, the presupposition is that errors—emerging from either individuals within the team or at the team level—are a foregone conclusion. In light of this perspective, Bell and Kozlowski spend significant time discussing the error management process within teams as well as future research needs.

In Chapter 6, Weaver, Bedwell, and Salas describe how team training can increase reliability and improve error management. They review a number of different team training techniques and describe how they link to improving team reliability. They conclude with a series of key theoretical and practical things to consider when designing a team training initiative.

MacPhail and Edmondson (Chapter 7) suggest that the work context is a key determinant in identifying how best to learn from errors. Given that errors are inevitable, it is critically important that individuals, units, and organizations effectively learn from these errors. This not only feeds into future error prevention efforts but also can lead to the development of improved error management systems. Another key component to learning from errors is the development of a culture signaling psychological safety.

Chapter 8 shifts our focus to the top floor of the organization. In this chapter, Shimizu and Hitt discuss errors made at the strategic level. These errors are particularly difficult to identify because of the inherent uncertainty in the broader context of the decision. This context often provides unclear signals, long time lags, and no clear benchmarks against which to judge the outcome of the decision. Shimizu and Hitt also consider a number of contributing factors to top management team errors, such as cognitive limitations and biases and coordination problems in the implementation process.

Hollnagel (Chapter 9) describes three ages of safety—the age of technology, human factors, and safety management—and how each relates to the attribution of failure and, relatedly, the way in which each age dealt with these failures. The last age, the age of safety management, is characterized by the concept of the resilient organization, which becomes increasingly necessary as systems become more complex and less tractable.

Ramanujam and Goodman (Chapter 10) take up the issue of the relationship between organizational errors and adverse consequences to the organization. One primary focus is on the notion of latent errors or organizational deviations from rules and operating procedures that can potentially generate adverse outcomes when they interact with trigger events. They further describe, from a systems viewpoint, the antecedents of organizational feedback processes for error correction and error amplification. Two cases—one from a bank, the other from a hospital—are used to illustrate how latent errors and triggering events can result in significant negative consequences.

Finally, in Chapter 11 Gelfand, Frese, and Salmon examine the cross-cultural implications for both error prevention and error management. A number of cross-cultural dimensions (e.g., uncertainty avoidance, humane orientation, individualism, tightness-looseness) are discussed and how they might have an impact on error prevention, error detection, and error management. A number of cultural paradoxes that result from such a conceptualization are discussed.

Moving Forward

Although our focus has been on taking a new perspective on errors, it is important to point out that the traditional preventive focus on errors has resulted in tremendous improvements across a wide variety of industries. For example, some of the early work on error prevention from both a system design and training perspective by Fitts and colleagues (Fitts, 1962) has had long-term impacts in the airline, aircraft, and nuclear power industries. Since the early discussions of the widespread nature of errors in medicine (Bogner, 1994; Leape, 1994), there have been a number of areas in which simple models of error prevention, such as making prescriptions readable, reducing germs by hand washing, and using different plugs and sockets for different functions (e.g., in machines used for anesthesia), can result in significant improvements (e.g., Gawande, 2009; Sprunt, Redman, & Leidy, 1973).

Thus, even though throughout this volume we have attempted to encourage a new perspective on errors, the traditional approach to errors should not be overlooked. As the volume draws to a close, perhaps it is informative to consider how the traditional preventive focus can be integrated into the new perspective on errors discussed herein. It is to this task that we now turn, first by discussing potential areas of future research and then presenting an integrative model of managing human risks within organizations.

The issue of organizational culture and climate has been mentioned throughout this volume (Chapters 1–8, 11; see also Keith & Frese, in press; van Dyck, Frese, Baer, & Sonnentag, 2005; Zohar & Hofmann, in press). Across these various treatments, a number of new research questions emerge. One question is how to precisely measure all of the constructs (i.e., safety climate, high-reliability organization, error management, and latent errors) and identify the appropriate level of measurement and aggregation (Chan, 1998). Other related questions are how these climates and cultures develop, how they are interrelated, and how they can be changed by management efforts (Chapter 6, this volume; see also Helmreich & Merritt, 2000; Zohar & Hofmann, in press).

Another area of future research entails applying the concepts here beyond the industries in which they have been typically discussed. With rare exception, much of the research on human error (and safety issues) has been done in industries in which there is a high degree of either physical or environmental risk (e.g., nuclear, manufacturing, health care, oil and gas). But, we believe that many of the concepts discussed here might easily translate to industries for which the risk is financial as opposed to physical or environmental. The migration and application of these concepts to the financial industry seems all the more called for in light of the financial meltdown in the first decade of the 21st century, which, by all accounts, resulted from a mélange of errors and violations as well as insufficient organizational approaches to risk issues. Even though there has been some limited discussion of error analysis in the trading and financial sector (Fenton-O'Creevy, Nicholson, Soane, & Willman, 2005; Soane, Nicholson, & Audia, 1998), we believe that there is a great deal of remaining opportunity to explore how some of the discussions in this volume might apply to this industry.

Although a number of new research questions have been described in each chapter, additional research questions could come about by integrating concepts across chapters. For example, it is interesting to consider how the individual-level chapters might provide implications for the organizational chapters and vice versa. For example, the notion of learning from errors is likely to involve some individual-level components (Chapter 2), some team-level components (Chapters 5–7), and potentially organizational-level amplifying processes. Similarly, one can imagine how the cultural influences on error prevention and management (Chapter 11) might play out both within and across teams and within and across organizations (or units of a global organization).

The relationship between failure and success is another area that can be discussed across chapters by considering how errors may be necessary for high performance and how adverse consequences need to exist on a societal level for a successful society—this goes from churning of entrepreneurial failures in a society to the question of how important crises and its ensuing processes of development are necessary, up to the point that Festinger (1983) has emphasized that all of human development is due to failures. The issue of top management team errors (Chapter 7, this volume) and how they are related to errors and learning occurring at other levels in the organization is also interesting to consider. One could consider, for example, how individual errors occurring within the senior management ranks can have an impact on errors occurring at that level and how these individual errors establish the context for units and teams within the organization and can result in not only that error cascading throughout the organization but also new errors and violations occurring at lower levels within the organization.

In addition to these areas for future research, we believe there is an opportunity to think more broadly about integrating the more traditional prevention-focused approach to errors with some of the ideas discussed in this volume. Looking across the distinctions between errors and violations (Chapter 1, this volume) and how organizations attempt to cope with these types of failures, two underlying dimensions seem to emerge. The first dimension focuses on the intentionality of the actor's behavior enacted to achieve the desired outcome. Specifically, was the actor trying to accomplish some expected and desired outcome (e.g., such as increasing productivity or lifting a patient out of bed) through a generally accepted process or through a deviation from this accepted protocol (i.e., correct vs. incorrect intentions regarding the behavior)? While this first dimension involves an assessment of the actor's behaviors used to accomplish some desired or expected outcome, the second dimension focuses on the different orientations to failures that organizations can adopt. Specifically, this dimension describes orientations that are either prevention or management focused. The distinction here is that systems focused on prevention attempt either to stop or to encourage behavior prior to action, whereas systems focused on management attempt to facilitate the recovery and learning from behavior that has already been performed. Aligning these

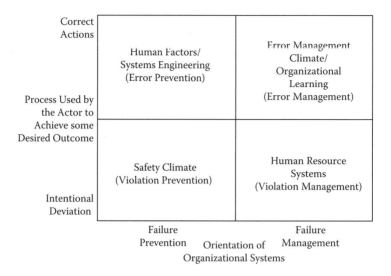

FIGURE 12.1
Integrated view of actor's behavior and organizational context.

two dimensions in a 2 × 2 matrix results in the four cells depicted in Figure 12.1.

Figure 12.1 provides a general framework to think about different types of failure and the ways in which organizations attempt to cope with them. Safety climate is more oriented toward the prevention of intentional violations or, alternatively, encouraging adherence to accepted safety protocol. An error management climate, on the other hand, is more focused on coping with unintentional outcomes after they occur; that is, the error management. Now, turning to the diagonal elements, we can draw a distinction between the focus of human factors/systems engineering and human resource management systems. The science of human factors and systems engineering is focused on the study and improvement of human–system interface and effectiveness. In the context of failure, the goal of human factors and system engineering is the design of systems such that correct actions are encouraged, and deviant actions are prevented. Take, for example, one of the lead quotations from *To Err Is Human* (Kohn et al., 2000): "Human beings, in all lines of work, make errors. Errors can be prevented by designing systems that make it hard for people to do the wrong thing and easy for people to do the right thing" (p. ix). From a perspective of coping with failure, this is clearly a failure prevention orientation designed to eliminate the possibility of incorrect actions. Thus, human factors and system engineering primarily focus on the prevention of errors.

The other diagonal element, human resource management systems, consists of initiatives designed to manage the aftermath of intentional deviations from accepted protocol. Reason (1997), in his discussion of organizational accidents, emphasized how the context can influence individual violations. This orientation—that there are larger systemic or contextual factors at play when violations occur—is consistent with other discussions regarding organizational climate as it relates to safety (e.g., Hofmann & Stetzer, 1996; Zohar, 2003). But, one should recognize that a just, or fair, culture is not a permissive culture. In other words, one must realize that "it would be naïve not to recognize that, on some relatively rare occasions, accidents can happen as the result of the unreasonably reckless, negligent, or even malevolent behaviour of particular individuals" (Reason, 1997, p. 205). In this case, appropriate disciplinary actions are necessary. Other research, for example, has noted that appropriately applied discipline is associated with increased perceptions of fairness and justice (e.g., Ball, Trevino, & Sims, 1994; Trevino & Ball, 1992**)**. Within our context, human resource management systems that enact targeted disciplinary activities in relevant situations are designed to manage these violations after they occur. Policies that either terminate the individual or apply other disciplinary actions (e.g., leave without pay, written reprimands) are reactive, post hoc actions taken to manage the implications of behavior after it occurs. Thus, this represents the management of intentional violations.

Taken together, we believe that the various types of failure and ways of coping with these failures illustrated in Figure 12.1 have both practical implications for organizations as well as theoretical implications for researchers interested in safety and errors in organizations. From a practical standpoint, the integration of these various ways to cope with failure suggests that each approach makes a needed contribution to the management of failures in organizations. Human factors and system engineers focus on the human technology interface and attempt to implement structural fixes designed to encourage correct behavior. Safety climate, and the informal norms and values constituting this climate, help to ensure that employees are actively mindful of safety and are encouraged—implicitly and explicitly—to adhere to safety standards and protocol. Error management climate serves to facilitate the learning from failures that results from unexpected and unintended outcomes of employees trying to do the right thing. Finally, human resource management systems help organizations deal with the, albeit rare, acts of extreme recklessness or rampant disregard for safety policies and procedures. Each of these different approaches to failure is necessary and plays a critical role.

From a theoretical perspective, Figure 12.1 suggests that our measures of safety and error management climate could be reconsidered in light of both the contributing type of individual behavior and the larger system within which they are embedded. Specifically, it suggests that our measures of these two climates should be more closely linked to their respective contributing behaviors. In particular, perhaps future assessments of safety and error climate should more clearly link safety climate with the adherence to safety protocol as well as the improvement of these processes, whereas future measures of error management climate should more clearly define the notion of error as an unintentional and unexpected outcome with a correct underlying motivation and intention. Clearly specifying these underlying distinctions, and perhaps revising our measures to better highlight this distinction, might enable researchers to better differentiate safety from error climates. Researchers should also consider the broader integrative system within which these climates operate. For example, adding items to a safety and error climate assessment asking employees' perceptions regarding system and engineering design efforts as well as safety-related human resource management systems might provide a much richer and broader assessment of how the organization is actively seeking to prevent, manage, and cope with failures that occur.

Concluding Remarks

Our goal for this volume was to provide a new look at errors occurring in organizations from a number of different perspectives and to identify potential areas of future research. We believe that we have accomplished this goal, and we hope you, the reader, agree. In this chapter, we have tied together some of the general themes cutting across the various chapters and identified additional points for future research and integration. Our hope and belief is that the various chapters in this volume will result in new and significant research into errors and violations as well as their prevention and management. Practically, we hope that the research sparked by this volume will result in safer, more reliable, and more innovative organizations with a high learning orientation in the future.

References

Ball, G., Trevino, L., & Sims, H. (1994). Just and unjust punishment: Influences on subordinate performance and citizenship. *Academy of Management Journal, 37*, 299–322.

Bogner, M. S. (Ed.). (1994). *Human error in medicine.* Hillsdale, NJ: Erlbaum.
Chan, D. (1998). Functional relations among constructs in the same content domain at different levels of analysis: A typology of composition models. *Journal of Applied Psychology, 83*, 234–246.
Fenton-O'Creevy, M., Nicholson, N., Soane, E., & Willman, P. (2005). *Traders: Risks, decisions, and management in financial markets.* Oxford; UK: Oxford University Press.
Festinger, L. (1983). *The human legacy.* New York: Columbia University Press.
Fitts, P. M. (1962). Functions of man in complex systems. *Aerospace Engineering, 21*, 34–39.
Gawande, A. (2009). *The checklist manifesto: How to get things right.* New York: Metropolitan Book.
Helmreich, R. L., & Merritt, A. C. (2000). Safety and error management: The role of crew resource management. In B. J. Hayward & A. R. Lowe (Eds.), *Aviation resource management* (pp. 107–119). Aldershot, UK: Ashgate.
Hofmann, D. A., & Stetzer, A. (1996). A cross-level investigation of factors influencing unsafe behaviors and accidents. *Personnel Psychology, 49*, 307–339.
Hofmann, D. A., & Stetzer, A. (1998). The role of safety climate and communication in accident interpretation: Implications for learning from negative events. *Academy of Management Journal, 41*, 644–657.
Keith, N., & Frese, M. (in press). Enhancing firm performance and innovativeness through error management culture. In N. M. Ashkanasy, C. P. M. Wilderom, & M. F. Peterson (Eds.), *Handbook of organizational culture and climate* (2nd ed.). Thousand Oaks, CA: Sage.
Kohn, L. T., Corrigan, J. M., & Donaldson, M. S. (2000). *To err is human: Building a safer health system.* Institute of Medicine, Committee on Quality of Health Care in America. Washington, DC: National Academy Press.
Leape, L.L. (1994). The preventability of medical injury. In M. S. Bogner (Ed.), *Human error in medicine* (pp. 13–25). Hillsdale, NJ: Erlbaum.
Maurino, D. E., Reason, J., Johnston, N., & Lee, R. B. (1995). *Beyond aviation human factors.* Hants, UK: Ashgate.
Reason, J. (1990). *Human error.* New York:Cambridge University press.
Reason, J. (1997). *Managing the risks of organizational accidents.* Aldershot, UK: Ashgate.
Soane, E., Nicholson, N., & Audia, P. G. (1998). *Baring's collapse—A case study.* London: London Business School.
Sprunt, K., Redman, W., & Leidy, G. (1973). Antibacterial effectiveness of routine hand washing. *Pediatrics, 52*, 264–271.
Trevino, L. K., & Ball, G. A. (1992). The social implications of punishing unethical behavior: Observers' cognitive and affective reactions. *Journal of Management, 18*, 751–768.
van Dyck, C., Frese, M., Baer, M., & Sonnentag, S. (2005). Organizational error management culture and its impact on performance: A two-study replication. *Journal of Applied Psychology, 90*, 1228–1240.
Weick, K. E., & Sutcliffe, K. M. (2001). *Managing the unexpected—Assuring high performance in an age of complexity.* San Francisco: Jossey-Bass.

Zohar, D. (2003). The influence of leadership and climate on occupational health and safety. In D. A. Hofmann & L. E. Tetrick (Eds.), *Health and safety in organizations: A multilevel perspective* (pp. 201–230). San Francisco, CA: Jossey-Bass.

Zohar, D., & Hofmann, D.A. (in press). Organizational climate and culture. In S. W. J. Kozlowski (Ed.), *The Oxford handbook of industrial and organizational psychology*. New York: Oxford University Press.

Author Index

A

Aaker, J.L., 299
Abrahamson, E., 216
Ackerman, P.L., 7, 53, 54
Acomb, D.B., 116
Acton, B., 159
Adner, R., 199, 203, 205, 208, 211, 218
Aguinis, H., 152
Aiello, J.R., 293
Al Najjar, M., 289
Alexander, R.C., 203
Allwood, C.M., 19, 276
Alonso, A., 152
Altman, A., 161
Alvesson, M., 29
Amabile, T.M., 70, 78
Amason, A.C., 216
Anderson, J.R., 7, 8, 180
Anderson, N., 68, 69
Anderson, P.A., 206, 216
Andrews, D., 155
Aoki, M., 292
Apostolakis, G., 230
Arazoe, M., 236
Arciszewski, T., 289
Argyris, C., 177
Asch, S.E., 87
Atomic Energy Commission, 229
Au, K.Y., 296
Audia, P.G., 320
Augenstein, J.S., 124
Axtell, C.M., 68, 71, 72, 73, 77, 78
Aycan, Z., 279, 294, 296
Ayoko, B.O., 303
Ayton, P., 100

B

Baas, M., 71
Baba, M.L., 303
Baer, M., 31, 46, 73, 79, 80, 177, 320
Baetge, M.M., 116

Baker, D.P., 116, 146, 147, 152
Baldwin, T.T., 60, 163, 164
Ball, G., 323
Bandura, A., 45, 46, 49, 56, 60, 77
Barnett, J., 120
Barnett, S.M., 50
Baron, H., 285
Barrett, L.F., 87
Bartlett, C.A., 207
Bartunek, J.M., 87
Basadur, M.S., 72
Basel Committee on Banking
    Supervision, 245
Bates, D.W., 180
Batinic, B., 35, 304
Batra, B., 199
Batt, R., 146
Bauer, J., 45, 61, 115, 123, 127
Baum, J.A.C., 200, 204
Baumaard, P., 213
Baxter, J.S., 4
Bea, R.G., 146
Bechtold, D.J., 298
Beckman, J., 10
Beckmann, D., 284, 287
Bedwell, W.L., 318
Behson, S.J., 161
Beier, M.F., 56
Beinhocker, E.D., 202, 217
Bell, B.S., 49, 52, 53, 54, 55, 56, 114, 119, 121, 123, 125, 126, 130, 136, 149, 318
Bellows, J., 187
Benjamin, A.S., 59
Berenholtz, S.M., 152
Bernstein, M.H., 294
Berwick, D., 166
Besevegis, E., 289
Bettis, R., 209
Bhawuk, D.P.S., 298, 299
Biehal, G.J., 299
Bierly, P., 144

327

Bigley, G.A., 246–247, 254
Birnbach, D.J., 124
Bishop, G.D., 289
Bjork, R.A., 49, 50, 56, 57, 59
Blair, C., 70, 86
Blatt, R., 246
Blickensderfer, E.L., 147, 158, 159, 161, 167
Block, R.A., 102
Bobrow, D.G., 32
Boeker, W., 199, 207, 215, 216
Bogner, M.S., 143, 319
Bohmer, R.M., 131, 188
Bond, M.H., 294
Bossidy, L., 208
Botero, I., 123
Botvinick, M.M., 304
Bourgeois, L.J., III, 207, 210, 211, 214, 218
Bowen, H., 191
Bowers, C.A., 120, 150, 151, 159
Brannick, M.T., 149, 167
Brannon, M.Y., 301
Bransford, J.D., 51, 54, 159, 160
Brett, J.M., 301
Broad, M., 164
Brockner, J., 296
Brodbeck, F.C., 1, 3, 45, 46, 275, 305
Brooks, D.R., 299
Brown, A.L., 9, 54
Brown, S.L., 199, 201, 205, 208, 217, 218
Brown,, S.L., 199
Bruce, R.A., 79
Bruner, J.S., 47
Brunsson, N., 216, 219
Budescu, D.V., 104
Bunderson, J.S., 130, 133
Burgelman, R.A., 218
Burke, C.S., 118, 122, 129, 131, 132, 146, 150, 151, 158, 159, 162
Burke, M.J., 286, 287
Byrd, T., 294
Byrne, J.A., 201

C

Caldwell, D.F., 35
Callahan, P., 203, 215
Callen, V.J., 303

Campbell, D.T., 278
Campbell, K.A., 4
Campbell, M., 56
Campbell, W.K., 299
Campione, J.C., 54
Cannon, M.D., 123, 126, 132, 133, 177, 196, 200, 201
Cannon-Bowers, J.A., 123, 146, 147, 148, 149, 151, 152, 153, 157, 158, 159, 161, 167
Cao, C.G., 134
Caplan, L.J., 294
Cardinal, L.B., 201, 202, 211, 217
Carlson, R.A., 159
Carmeli, A., 77, 80
Carnegie, D., 115, 149
Carrithers, C., 46
Carroll, J.M., 46
Carroll, J.S., 200, 202, 204
Ceci, S.J., 50
Chan, D. K.-S., 291, 320
Chan, D.W.L., 60
Chan-Serafin, S., 286
Chang, A., 180
Chang, C.J., 199, 204, 217
Chapple, E.D., 28
Charan, R., 208
Chatman, J.A., 35
Chen, X., 292
Chen, Y., 297
Cheng, J.L.C., 301
Chernobyl accident: What happened, 273
Chernobyl Forum, 273
Chillarege, K.A., 49, 53, 59
Chinese Culture Connection, 298
Chinn, D.J., 180
Chirkov, V.I., 299
Chiu, C., 299, 300
Choi, I., 300
Christianson, M.K., 246
Cianciolo, A.T., 7
Civil Aviation Authority, 274
Clark, R.E., 55
Clegg, C.W., 37
Cohen, D.J., 164
Cohen, J.D., 304
Cohen, M.D., 213

Author Index

Cohn, L.D., 294
Coles, M.G.H., 304
Colón, I., 294
Conn, A.B., 72
Connolly, T., 83
Converse, S.A., 149
Cooke, N.J., 115, 131, 133, 135, 153
Corrigan, J.M., 322
Cortes, M., 294
Corver, S., 230
Cosmides, L., 99
Council, F.M., 294
Cowan, D.A., 70
Cramton, C.D., 301
Crandall, B., 242
Crandall, S., 164
Croft, G.S., 51
Cronin, M.A., 122
Crossan, M.M., 215
Croteau, R., 180
Crotts, J.C., 285
Cummings, A., 68, 78
Cyert, R.M., 25, 206, 207, 209, 210, 212, 214, 217
Cyr, D., 292

D

D'Aveni, R.A., 209
Dahlin, K.B., 200, 204
Dailey, L., 70, 71, 86
Daily, C.M., 216
Davidi, I., 74
Davies, D.R., 22
Davies, J.M., 143
Davies, N., 2
Davoudian, K., 230
Dawes, R.M., 105
Day, R., 116, 146, 152, 162
De Dreu, C.K.W., 68, 71, 88
de Haan, E.H., 46
De Jong, S.B., 287
De Michele, P.E., 299
Debowski, S., 49, 53, 57, 58, 60
Delbecq, A.L., 118
Deller, J., 294
DeMiguel, V., 106
DeNeysem J,O, 137

DeNisi, A., 53
DeNnis, A.R., 83
Denrell, J., 200, 207, 208, 209, 217
DeShon, R.P., 119
DiazGranados, D., 151, 167
Dickinson, T.L., 147, 149, 174
Dickson, M.W., 290
Diehl, M., 85
Donaldson, M.S., 322
Dorfman, P.W., 280
Dormann, T., 48, 52
Dörner, D., 6, 12, 32, 36
Dougherty, E.M., Jr., 229
Drasgow, F., 296
Drazin, R., 69, 74, 75, 76
Driskell, J.E., 151, 157
Druskat, V.U., 118
Dunbar, M., 134
Dunnette, M.D., 11
Duran, J.L., 131
Dutton, J.E., 26, 205, 210, 216
Dweck, C.S., 55, 78
Dwyer, D.J., 159, 161

E

Earley, P.C., 301
Edmondson, A.C., 35, 115, 116, 123, 125, 126, 131, 132, 133, 177, 180, 181, 184, 185, 187, 188, 196, 200, 201, 202, 204, 207, 208, 215, 318
Ehrenstein, A., 58
Eisenhardt, K.M., 199, 201, 205, 206, 207, 208, 210, 214, 217, 218
Elliot, A.J., 299
Ellis, A.P.J., 121, 123
Ellis, S., 74
Ellstrand, A.E., 216
Elron, E., 301
Endsley, M.R., 147, 157
Entin, E.E., 159, 160, 163
Erev, I., 104
Erez, M., 279
Ericsson, K.A., 61
Erumban, A.A., 287
Eulberg, J.R., 164
Eylon, D., 296

F

Falshaw, J.R., 202
Farber, J.M., 1
Farjoun, M., 1
Farmer, S.M., 71, 80
Farr, J.L., 68, 72, 74, 78, 79, 80, 81, 88, 318
Fay, D., 77
Fehr, R., 304
Feldman, L., 188
Fenton-O'Creevy, M., 320
Ferlins, E.M., 188
Ferrara, R.A., 54
Festinger, L., 321
Finkelstein, M.O., 101
Finkelstein, S., 206
Fiore, S., 150
Fischer, I., 100
Fischhoff, B., 101, 102
Fisher, C.D., 58
Fisher, C.W., 117
Fiske, S.T., 203, 205, 217
Fitts, P.M., 319
Five case studies on successful teams, 150
Flavell, J.H., 9
Fletcher, T.D., 131
Floyd, D.B., 1
Ford, C.M., 69, 72, 74
Ford, J.K., 45, 54, 55, 59, 152, 157, 162, 163, 167
Ford, N., 75
Foster, A., 75
Foushee, H.C., 116, 122
Fowlkes, J.E., 131, 152, 161
Franks, J.J., 51, 160
Freeman, J., 143, 214
Freeman, R.E., 293
Frese, M., 1, 3, 6, 7, 9, 10, 11, 12, 31, 32, 35, 37, 45, 46, 47, 48, 49, 50, 51, 52, 53, 54, 55, 58, 59, 61, 73, 77, 79, 80, 116, 117, 119, 121, 123, 125, 126, 127, 128, 129, 135, 136, 143, 161, 177, 201, 202, 203, 205, 208, 209, 212, 216, 217, 245, 247, 249, 275, 289, 292, 293, 304, 305, 319, 320
Freud, S., 274

Frink, D.D., 292, 293
Futtrell, D., 137

G

Gaertner, L., 301
Gaier, E.L., 299
Gaissmaier, W., 107
Galanter, E., 6
Galbraith, J.R., 25
Galinsky, A.D., 77
Galunic, D.C., 293
Gansneder, B., 299
Gardner, W.L., 299
Garlappi, L., 106
Garst, H., 35, 304
Gawande, A., 320
Gelfand, M.J., 279, 291, 292, 298, 299, 300, 304, 319
George, A., 180
George, J.M., 70, 81, 83
Gersick, C.J.G., 25, 28, 120, 126, 181, 209, 217
Ghandhi, T.K., 180
Ghemawat, P., 209
Ghodsian, D., 59
Ghoshal, S., 207, 217
Gibson, C.B., 301
Gielnik, M.M., 61
Gigenrenzer, G., 97, 99, 100, 103, 106, 107, 108, 275, 318
Gittell, J.H., 80
Glaister, K.W., 202
Glassop, L.I., 150
Gluck, P., 166
Gluesing, J., 303
Glynn, M.A., 69, 74
Goeschel, C.A., 152
Goldstein, I.L., 152, 157, 162, 163, 167
Gonzalez, E., 294
Gooding, R., 207
Goodman, J., 49
Goodman, P.S., 33, 122, 245, 247, 252, 258, 319
Goodwin, G.F., 150, 151
Gorman, J.C., 131
Gould, S.J., 100
Grant, R.M., 201, 202

Gray, S.J., 288
Greenberg, P., 166
Greve, A., 27, 207, 211
Greve, H.R., 200
Griffin, D.W., 103, 104
Griffin, R.W., 4
Groeneweg, J., 230
Gross, S.E., 150
Guilford, J.P., 75
Gully, S.M., 49, 54, 55, 117, 123
Gupta, V., 280

H

Hacker, W., 6, 7, 52
Hackman, J.R., 25, 28, 114, 120, 126, 181
Hale, A.R., 227, 241, 295
Haleblian, J., 206
Hall, R.I., 30
Hambrick, D.C., 199, 207, 209, 210, 217
Hamilton, R.W., 299
Hammond, M.M., 68, 70, 78, 318
Hampson, S.L., 285
Hanges, P.J., 280, 290
Hannan, M.T., 143, 214
Hannover, B., 9
Hanson, S.J., 1
Hansson, P., 102
Hardemann, W., 294
Hargadon, A., 75
Harkins, S., 121
Harper, D.R., 102
Harrington, E., 68
Harris, D., 274
Harris, D.V., 299
Hartel, C.E., 303
Haunschild, P.R., 200, 201, 202, 208, 212
Hayward, M.L.A., 204, 210, 217, 218
Heckhausen, H., 10
Hedberg, B.L.T., 27
Heggestad, E.D., 54
Heimbeck, D., 45, 46, 47, 48, 49, 52, 55, 59, 247, 305, 306
Heinbokel, T., 45, 305
Heine, S.J., 298, 299
Heinen, B., 162
Heinrich, H.W., 228, 232

Helmreich, R.L., 31, 128, 152, 167, 273, 274, 277, 279, 287, 297, 320
Hensel, D., 180
Hertwig, R., 108
Hesketh, B., 45, 46, 48, 49, 50, 51, 52, 53, 54, 56, 57, 59, 305, 306
Higgins, E.T., 12
Hill, R.K., 152
Hilling, C., 65
Hinds, P.J., 301
Hitt, M. A., 199, 204, 208, 209, 210, 211, 212, 216, 217, 318
Hobbs, A., 150
Hockey, R., 305
Hodgson, T.L., 51
Hoffrage, U., 103, 108
Hofmann, D.A., 5, 26, 35, 47, 78, 80, 116, 117, 119, 120, 121, 126, 127, 128, 129, 135, 136, 143, 179, 201, 202, 203, 205, 208, 209, 212, 216, 217, 245, 247, 249, 275, 292, 293, 320, 323
Hofstede, G., 278–279, 280, 284, 285, 286, 287, 296, 297, 298, 299
Holding, D.H., 163
Hollenbeck, J.R., 116
Hollnagel, E., 229, 231, 233, 234, 235, 242, 319
Holroyd, C.B., 304
Holtgraves, T., 299
Holtzman, A., 152
Hong, Y., 300
Hopwood, D., 292
Hornsby, J.S., 218
Hotz, R.L., 114
Hough, L.M., 11
House, R.J., 279, 280
Hovden, J., 227
Howse, W. R., 129
Hsee, C., 299
Hsieh, T.C., 180
Huang, X., 296
Huff, A.S., 210
Huff, J.O., 210, 211
Huffcutt, A.I., 89
Hug, K., 99, 100
Huguenin, R.D., 5
Hui, C., 115

Human error blamed in Tanzania crash, 273
Hutchins, E., 20, 21, 129

I

Iijima, Y., 236
Ikin, C., 2
Ilgen, D.R., 58, 116, 136
Institute of Medicine, 245, 246
International Civil Aviation Organization, 156
Ireland, R.D., 218
Irmer, C., 19
Ito, H., 236
Ivancic, K., 45, 46, 48, 49, 50, 51, 52, 53, 54, 58, 59, 305, 306

J

Jackson, S.E., 205, 210, 216
Jacobs, R., 143
Janis, I.L., 35, 87, 206, 216
Janssen, O., 78, 89
Jarrell, G.A., 202
Javidan, M., 279, 280
Jennings, P.D., 292
Jensen, J.M., 123
Jensen, M.C., 202
Jentsch, F., 120
Jing, H.-S., 274, 279
Johansson, G., 101
Johnson, J.L., 216, 217
Johnson, R.A., 199, 204
Johnston, J.H., 146, 161
Johnston, N., 33
Joint Commission on Accreditation of Healthcare Organizations, 119, 156
Jones, C.F., 74
Jones, G.R., 81, 83
Joung, W., 48, 58, 60, 305, 306
Juslin, P., 102, 104, 105, 106

K

Kaber, D.B., 134
Kaempf, G.L., 152
Kahneman, D., 8, 12, 22, 34, 100, 210, 234, 275
Kakebeeke, B.M., 53
Kanfer, R., 20, 53, 54
Kanter, R.M., 73
Kanungo, R.N., 294, 296
Kaplan, S., 202, 217
Kark, R., 80
Karl, K.A., 162
Karoly, P., 53
Kavanagh, M.J., 167
Kazanjian, R.K., 69, 74
Kazemi, R., 155
Kecklund, L.J., 232
Keith, N., 45, 47, 48, 49, 50, 51, 52, 53, 54, 59, 61, 161, 177, 247, 305, 317, 320
Kelloway, E.K., 162
Kemeny, J.G., 229
Kerr, N.L., 83, 87
Kessels, R.P., 46
Keyser, D.J., 180
Kiesler, S., 29, 213, 214, 215
Kim, B., 123
Kim, Y., 299
King, H., 116, 124, 150, 152
Kingma, B.R., 117
Kirschner, P.A., 55
Kirwan, B., 229, 241
Kitayama, S., 298, 299
Kiyokawa, K., 236
Kjellen, U., 241
Klein, C., 124, 132, 151, 167
Klein, G.A., 234, 242, 275
Klein, J.T., 79
Klein, K.J., 72, 73, 120
Kleinbölting, H., 103
Kleine, B.M., 61
Klimoski, R.J., 292, 293
Klinect, J.R., 128, 274
Kluckhohn, C., 278, 298
Kluger, A.N., 53
Knight, K.E., 25
Koenig, R., 118
Kohn, L.T., 322
Kolb, K.J., 293
Koles, K.L.K., 49
Kontogiannis, T., 4, 20, 21, 127, 129, 136

Kosarzycki, M.P., 115, 149
Kossiavelou, Z., 4
Kouabenan, D.R., 295
Kozlowski, S.W.J., 45, 49, 52, 53, 54, 55, 56, 114, 117, 119, 120, 121, 123, 125, 126, 130, 131, 136, 149, 163, 318
Kraiger, K., 152
Krampe, R.Th., 61
Krause, D.E., 78
Kraut, R.E., 1
Kreuzig, H.W., 12
Kroeber, A.L., 278
Kubota, R., 236, 237, 238
Kung-Mcintyre, K., 71
Kuratko, D.F., 218
Kurowski, L.L., 279
Kurz-Milke, E., 107

L

La Mettrie, J.O., 230
Lalwani, A.K., 299
Landy, F., 143
Lane, H.W., 215
Lant, T.K., 199, 211, 213
Lantané, B., 121
Latham, G.P., 6, 60, 164, 165
Lauber, J.K., 116
Lavric, A., 51
Lawler, J.J., 296
Lawrence, R.J., 2
Lawton, R., 4
Lazzara, E.H., 161
Leape, L.L., 166, 168, 319
Lee, A.Y., 299
Lee, F., 185
Lee, J., 2
Lee, R.B., 33
Leggett, E.L., 55
Lehman, D.R., 298, 299
Leidy, G., 320
Leplat, J., 231
Leritz, L.E., 70
Lerner, J.S., 214
Leslie, L.M., 291, 304
Leung, K., 294
Leveson, N., 233

Levin, B., 101
Levine, R.V., 299
Levinthal, D.L., 199, 203, 205, 208, 211, 212, 213, 218
Levitt, B., 25, 27, 181, 210, 212, 213, 219
Levy, P.E., 9
Li, L.-W., 274
Li, W.-C., 273, 279
Lichtenstein, S., 102
Licuanan, B.F., 86
Lim, B.-C., 300
Lim, B.L., 291
Lindblom, C.E., 234
Locke, E.A., 6
Loeb, J., 180
Lonergan, D.C., 71, 72
Lord, R.G., 9
Lorenzet, S.J., 48, 162
Louis, M.R., 25, 205, 206, 209, 212, 217
Lounamaa, P., 213
Lu, C.J., 274
Lun, J., 291
Lundy, D.H., 159
Lynn, R., 285
Lynn, S., 209

M

MacCallum, G.A., 160
MacDonald, J., 48
MacGregor, D., 101
MacPhail, L., 178, 187, 318
Maddux, W.W., 77, 292
Magjuka, R.J., 164
Maier, G.W., 19
Major, D.A., 131
Mann, S., 166
Mannetti, L., 294
Manstead, A.S.R., 4
March, J.G., 12, 25, 27, 32, 106, 181, 200, 201, 206, 207, 208, 209, 210, 212, 213, 214, 216, 217, 218, 219, 234, 248
Mark, B.A., 35, 247
Mark, M.M., 299
Marks, M.A., 158
Markus, H., 298, 299
Marmaras, N., 4

Mars Climate Orbiter Mishap
 Investigation Board, 114
Martin, K., 293
Martocchio, J.J., 162, 296
Marx, D.A., 155
Mason, P.A., 199
Mathieu, J.E., 149
Matsushima, H., 236
Maurino, D.E., 33
McAllister, H.A., 299
McCrae, R.R., 76
McDaniel, R., 25
McDonnell, L., 134
McFarland, L., 285
McGrath, R.G., 200, 205, 214
McHugh, P.P., 123
McIntire, M., 273
McIntyre, R.M., 147, 149, 156
McKechnie, P.I., 206
McPherson, J.A., 159
Mead, M., 298
Meckling, W.H., 202
Mednick, S.A., 75
Meijman, T.F., 34
Mendonca, J., 294
Menkhoff, L., 284
Menon, T., 300
Merkin, R.S., 287
Merriam-Webster Online Dictionary, 3
Merritt, A.C., 31, 128, 152, 167, 273, 274,
 277, 279, 287, 320
Milanovich, D.M., 152, 159
Millendorfer, J., 285
Millenson, M., 188
Miller, C.C., 201, 202, 211, 217
Miller, D., 30
Miller, G.A., 6
Miller, J.G., 300
Miller, S., 200
Milliken, F.J., 1, 29, 199, 201, 206
Milner, K.A., 119
Mintzberg, H., 25, 202, 217
Mitroff, I.L., 29
Mizokawa, D.T., 299
Mohaghegh, Z., 155
Money, R.B., 285
Montgomery, D.B., 213
Moore, L.F., 292

Mooser, C., 45, 305
Moran, P., 207
Moray, N., 143
Morgan, B.B., Jr., 159, 160
Morgan, R.L., 49
Morgeson, F.P., 26, 120
Morimoto, T., 180
Morishima, M., 292
Morone, J.G., 209
Morris, C.D., 51, 54
Morris, M., 300
Morris, M.W., 300
Morrison, E., 297
Moscovici, S., 87
Mosier, K.L., 245
Mosleh, A., 155
Moss, S.A., 78
Mousavi, S., 318
Muelhausen, S., 305
Mulder, G., 34, 127
Mulder, R.H., 45, 61, 115, 123
Mulford, M., 105
Mullen, J.E., 162
Mumford, M.D., 70, 71, 72, 81, 84, 85,
 86, 88, 90
Murphy, C.E., 116, 150
Musson, D., 152
Mutrie, N., 299

N

Nadkarni, S., 199, 208, 209
Narayanan, V.K., 199, 208, 209
Narver, J.C., 297
Nason, E.R., 117
National Transportation Safety Board,
 113, 120, 151
Neal, A., 48, 305
Needham, D.M., 152
Neff, N.L., 68
Nelson, R.R., 25, 27, 28, 210, 213
Nembhard, I.M., 125
Neubert, M.J., 58
Neuwirth, E., 187
Newman, K.L., 296
Newstrom, J.W., 164
Nichols, D.R., 151
Nicholson, D., 129

Author Index

Nicholson, N., 320
Nielsen, P., 166
Nietschze, F., 226
Nigg, J.M., 294
Nijstad, B.A., 68, 71
Nisbett, R.E., 300
Nishii, L.H., 291, 298
Niswander, F., 288
No survivors found, 257
Nollen, S.D., 296
Nordstrom, C.R., 49, 53, 59
Norenzayan, A., 299, 300
Norman, D.A., 8, 10, 12, 18, 32, 35

O

O'Brien, J., 74
O'Connor, E.J., 164
O'Leary, D., 180
O'Leary-Kelly, A., 4, 162
O'Reilly, C.A., 35
Obstfeld, D., 144, 186, 255
Ocasio, W., 209, 210
Odean, T., 100
Okhuysen, G.A., 199
Oldham, G.R., 68, 71, 78, 83
Olson, H., 102
Osburn, H.K., 70
Oser, R.L., 151, 159, 160, 161

P

Page, R., 285
Panko, R.R., 2
Parada, M., 294
Parasuraman, R., 22
Paries, J., 233
Parsons, T., 298
Paulson, A.S., 209
Payne, S.C., 49, 161
Paz, D., 294
Pearsall, M., 121
Pelto, P.J., 291
Peng, K., 300
Peng, S.-J., 274
Pentland, B.T., 25
Perlow, L.A., 199, 209, 211
Perrow, C., 1, 188, 189, 194, 246, 249, 256

Perry-Smith, J.E., 79
Pescosolido, A., 118
Peters, B.J., 187
Peters, G.A., 187
Peters, H., 3, 275
Peters, L.H., 164
Peters, T., 33
Peterson, M.F., 297
Pfafferott, I., 5
Pfeffer, J., 206, 207, 219
Pfeifer, P.E., 104
Philips, L.D., 102
Phillips, J.J., 158
Phillips, P.P., 158
Pidgeon, N., 230
Pierro, A., 294
Pisano, G.P., 131
Pliske, R., 242
Porter, M.E., 216
Posner, M.I., 76
Poucet, A., 229
Powell, S.M., 152
Prahalad, C.K., 209
Pratt, S., 166
Pribram, K.H., 6
Priem, R.L., 206
Priest, H., 122, 146, 155, 159
Prince, C., 149, 151, 152, 161
Pritchard, R.D., 4
Probst, T.M., 296
Pronovost, P.J., 152, 166
Prümper, J., 1, 3, 46, 275
Puffer, M., 206
Putka, D.J., 78
Pyun, J., 1

Q

Quiñones, M.A., 58, 167

R

Ramanujam, R., 33, 180, 245, 246, 247, 252, 258, 319
Rappensperger, G., 19
Rasmussen, J., 7, 8, 10, 18, 116, 249
Ratner, H., 303
Raver, J.L., 291, 300

Reason, J., 1, 3, 4, 5, 6, 8, 9, 10, 13, 18, 19, 22, 31, 32, 33, 34, 116, 117, 120, 127, 143, 150, 194, 230, 232, 245, 248, 254, 275, 276, 317, 323
Redman, W., 320
Reeder, G.D., 299
Regehr, G., 48
Reither, F., 12
Repenning, N.P., 199, 255, 256, 257
Rhee, M., 200
Rhodenizer, L., 151
Richardson, C., 180
Ringenbach, K.L., 78
Risser, D., 166
Ritossa, D.A., 78
Robert, C., 296
Roberto, M.R., 185, 188
Roberts, K.H., 146, 186, 189, 230, 246, 247, 249, 254
Rochlin, G.L., 144
Roe, E., 247
Rogers, D.A., 48, 49
Roll, R., 210, 217
Rooney, J.J., 156
Rosen, M.A., 115, 129, 130, 150, 159, 161
Rosenthal, M.M., 246
Ross, J., 209, 210
Ross, L., 203, 214, 215
Ross, M., 299
Rothenberg, A., 75
Rotter, J.B., 294
Rouiller, J.Z., 167
Rousseau, D.M., 146
Rubin, G., 180
Rudolph, J.W., 255, 256, 257
Rueter, H.H., 25
Rumar, K., 101
Runco, M.A., 71, 72
Rundmo, T., 295
Russ-Eft, D.F., 60
Russo, E.J., 219
Russo, J.E., 102
Ryan, A.M., 285
Rybowiak, V., 35, 304
Ryckman, D.B., 299

S

Saari, L.M., 60
Sabella, M.J., 158
Sagan, S.D., 249
Salas, E., 48, 54, 115, 116, 118, 119, 120, 122, 123, 124, 129, 130, 131, 132, 133, 135, 146, 147, 148, 149, 150, 151, 152, 153, 155, 156, 157, 158, 159, 160, 161, 162, 163, 166, 167, 318
Salbador, R., 286
Salgado, S., 297
Salinger, R.D., 165
Salisbury, M., 116, 124, 150
Salk, J.E., 301
Salmon, E., 319
Salter, S.B., 288
Sandelands, L.E., 26
Sarpy, S.A., 286
Sayles, L.R., 28
Schaub, H., 6
Schaubroeck, J., 77
Schleiffenbaum, E., 45, 305
Schlösser, O., 289, 290
Schmidt, A.M., 55, 119
Schmidt, R.A., 49, 50, 56, 57
Schneider, W., 7, 159
Schooler, C., 294
Schulman, P., 247, 255
Schulz, M., 212
Schwall, A.R., 68
Schwartz, D.L., 159
Schwartz, L., 107
Schwartz, S.H., 298
Schyve, P., 180
Scott, G.M., 71, 72
Scott, R.W., 247
Scott, S.G., 79
Seamster, T.L., 152
Sedikides, C., 299, 301
Seger, A.C., 180
Sego, D.J., 116, 167
Sellen, A., 19, 21, 276
Seo, M.-G., 87
Serfaty, D., 159, 160, 163
Seward, J.K., 202
Sexton, J.B., 152

Shalley, C.E., 70, 79, 81, 83, 96
Shapira, Z., 201
Shapiro, D.L., 301
Shappell, S.A., 180
Shea, J.B., 49
Sheldon, K.M., 299
Shepperd, J., 85
Sherwood, R.D., 160
Shiffrin, R.M., 7
Shimizu, K., 199, 204, 208, 209, 210, 211, 212, 216, 217, 318
Shin, S.J., 71
Shneiderman, B., 1
Shoemaker, P.J., 102, 219
Shoemaker, P.J.H., 219
Shrivastava, P., 29
Shrum, L.J., 299
Shuffler, M., 162
Sicoly, F., 299
Silvestri, S., 161
Simison, R.L., 200
Simon, H.A., 12, 32, 205, 206, 234
Simonson, R.L., 217
Simonton, D.K., 75, 77
Sims, D.E., 118, 119
Sims, H., 323
Sin, H.P., 68
Sinha, J.B.P., 296
Sirio, C., 180
Sitkin, S.B., 1, 32, 74, 201, 208, 212, 215, 218
Skinner, B.F., 46, 56
Skitka, L.J., 134
Slater, S., 297
Slonim, A.D., 155
Slovic, P., 12, 100
Smelcer, J.B., 2
Smith, A., 286
Smith, D.K., 203
Smith, E.M., 45, 48, 51, 54, 59, 117, 167
Smith, P.B., 297
Smith-Jentsch, K., 147, 159, 161, 167
Snook, S.A., 256
Snow, R.E., 55
Soane, E., 320
Solomon, G.B., 299
Sommer, S.M., 286

Sonnentag, S., 31, 45, 46, 55, 61, 79, 177, 247, 305, 320
Sorra, J.S., 72, 167
Spear, S., 191
Spector, P.E., 159
Spender, J.C., 144
Speziali, J., 231
Sproull, L., 29, 213, 214, 215
Sprunt, K., 320
Stagl, K.C., 129, 130, 151
Stahl, G., 294
Starbuck, W.H., 1, 27, 29, 213
Stasser, G., 79, 87
Stäudel, T., 12
Staw, B.M., 26, 68, 88, 203, 206, 209, 210, 213, 217
Sterman, J.D., 256
Stetzer, A., 5, 323
Stewart, J., 9
Stigler, S.M., 105
Stone, R., 273
Stout, R.J., 148, 151, 152, 153, 159, 161
Stradling, S.G., 4
Strodtbeck, F., 298
Stroebe, W., 85
Suh, E.M., 278
Sullivan, B.N., 200, 201, 202, 208, 212
Sully de Luque, M.F., 286
Sundstrom, E., 137
Sutcliffe, K.M., 133, 144, 145, 186, 189, 246, 247, 255
Suto, M., 284
Sutton, R.I., 25, 205, 206, 209, 212, 217, 219, 293
Svenson, O., 101, 232, 299
Sweller, J., 55

T

Tamuz, M., 214
Tannenbaum, S.I., 48, 61, 115, 147, 148, 149, 157, 161, 162, 167
Tatoglu, E., 202
Taylor, M.S., 58, 131
Taylor, P.J., 60
Taylor, S.E., 203, 205, 217
Teng, E., 77
Tesch-Römer, C., 61

Tesluk, P.E., 68
Tetlock, P.E., 214, 292
Thiemann, P., 45, 305
Thomas, H., 210
Thomas, M.J.W., 116, 128, 136
Thomke, S., 185
Thompson, C.M., 78
Thompson, D.A., 152, 180
Tierney, P., 71, 80
Tindale, R.S., 84, 87
Ting-Toomey, S., 300
Tischner, C., 78
Titus, W., 79, 87
Tjosvold, D., 115, 126, 135
Todd, P.M., 275
Toomey, L., 152
Tracey, B.J., 167
Trevino, L., 323
Triandis, H.C., 11, 278, 279, 291, 298, H.C.291
Trivers, R., 100
Tucker, A.C., 196
Tucker, A.L., 125, 185
Turner, R.H., 294
Tuttle, D.B., 116
Tversky, A., 12, 100, 103, 104, 210, 234
Tzeng, O., 291

U

Unger, J.M., 61
Ungson, G.R., 25, 26
United Nations Scientific Committee on the Effects of Atomic Radiation, 273
Unsworth, K., 68, 69, 70
Uppal, R., 106

V

Valacich, J.S., 83
Van de Ven, A.H., 118
Van de Vliert, E., 296
Van der Linden, D., 55
Van der Vegt, G.S., 130, 296
Van Dyck, C., 31, 33, 35, 46, 55, 79, 89, 177, 320

Van Schaardenburgh-Verhoeve, K.N.R., 230
Van Yperen, N.W., 78
Vancouver, J.B., 78
Vanden Heuvel, L.N., 156
VandeWalle, D., 55
Vaughan, D., 1, 200, 204, 246, 251, 256, 258, 267, 269
Vega, L.A., 72
Vevea, J.L., 301
Volpe, C.E., 147, 148, 157, 159
Von Glinow, M., 301
Vye, N.J., 160

W

Wachter, R.M., 114
Wagner, J., 207
Wagner, K.H., 303
Wall, T.D., 68
Wallstein, T.S., 104
Walsh, J.P., 25, 26, 202, 205
Wang, T., 274
Wason, P.C., 98
Waterson, P.E., 68
Watola, D., 123
Weaver, S.J., 149, 318
Weber, E.U., 299
Webster, J.L., 134
Wegner, D.M., 121
Weick, K.E., 29, 144, 145, 186, 189, 201, 206, 212, 213, 214, 246, 247, 255
Weinstein, N.D., 299
Weisbach, M.S., 199, 203, 217
Weissbein, D.A., 54
Wendland, D., 53
Wennerholm, P., 102
West, M.A., 79, 80, 88
Westphal, J.E., 155
White, R.E., 215
Whiteman, J.A.K., 49
Wiechmann, D., 119
Wiegel, J., 305
Wiegmann, D.A., 180
Wiersema, M., 201, 203, 214, 215, 216
Wijnen, C.J.D., 77
Wilde, G.J.S., 5, 125
Wildman, J.L., 149

Wilhelm, J.A., 274, 277
Wilhem, J.A., 128, 152
Williams, K.B., 49, 53, 121
Willman, P., 320
Wills, A.J., 51, 52
Wilson, J.M., 122, 123, 124, 125, 126, 129, 136, 151, 153, 155, 156, 157, 158, 159, 160, 161, 163
Wilson, K.A., 116, 122, 131, 134, 146, 150
Wilson-Donnelly, K., 122
Winman, A., 102, 105
Winter, S.G., 25, 27, 28, 210, 213
Woehr, D.J., 89
Woloshin, S., 107
Wong, S., 130
Wood, R.E., 49, 53
Woodman, R.W., 73
Woodruff, W., 161
Woods, D.D., 233, 242
Worline, M., 185
Wreathall, J., 233
Wright, M.C., 134
Wu, J.-S., 230
Wu, T.S., 161

Y

Yan, W., 299
Yeung, N., 304
Young, M.J., 300
Yu, K., 294
Yu, Z., 115
Yuki, M., 292
Yukl, G., 61

Z

Zaccaro, S.J., 158, 162
Zapf, D., 1, 3, 6, 7, 9, 10, 11, 13, 18, 19, 20, 21, 24, 32, 46, 52, 58, 117, 125, 275
Zavalloni, M., 87
Zbaracki, M.J., 199, 206, 217
Zeisig, R.L., 159
Zemba, Y., 300
Zhao, X., 68
Zhou, J., 68, 70, 71, 72, 73, 83
Zohar, D., 320, 323

Subject Index

A

Absolute risk, 107
Access, 20
Accident investigation methods, 231, 232
　step-by-step, 231
Accidental discoveries, 318
Accidents, 33, 204, 249
　in absence of malfunctions, 232
　discovery through, xii
　identifying causes, 133
　prevention of industrial, 228
　railway, 273
Accountability
　across group boundaries, 192
　and tight societies, 292
Acquisition failures, 199, 204, 216, 217
Acquisitions
　Renaullt-Nissan, 207
　Sony-Columbia Pictures, 207
Action cycles, 25–26, 36, 119
　cognitive regulation levels, 26
　collective, 26–28
　error management and error prevention in, 31–37
　feedback, 22
　individual *vs.* collective, 26
　organizational, 27
　trapping errors during, 128
Action-oriented mental models, 52
Action processes, and individual behavior in organizations, 6–10
Action regulation, 120
　bottom-up aspect, 9
　levels, 7
　as top-down process, 9
Action stages, in creativity/innovation, 68–69
Action teams, overrepresentation of, 137
Action theory, 10, 25, 52, 117

application to collectives, 30
Actor goals, 20, 21
Actor interdependence, 193
　and learning domain, 182
Actor's behavior, integrated view, 322
Adaptability, 148
　CRM targets, 152
　TCA questions, 154
Adaptation training, 160
Adaptive goals, and rationality, 100
Adaptive transfer, 50, 57
　error encouragement and, 50–51
　error management training for, 61
　metacognition and posttraining tasks, 54
Adverse consequences
　Barings Bank case, 259–263
　delineating latent errors, 247–252
　linking latent errors to, 252–259, 259–268
　Mid-Western Hospital case, 263–265
　organizationally significant, 248
　research opportunities, 268–270
　synergistic effects of antecedents, 259
Adverse events, 165
　causes, 233
　links to organizational errors, 245–247
　medical, 178
Affiliation, fulfillment of needs related to, 289
Agriculture, in collectivist societies, 298
Airline accidents, 28, 200, 202
Ambiguity, 286
　causal, 219
　of causality, 212, 213
　cultural tolerance for, 280, 285
　in feedback, 22
　normative argument issues, 101
Analogical transfer, 50

341

behavior-modeling training and, 60
    error-avoidant training for, 61
Analogies, 9, 10
    organizational level, 27
Anchoring heuristics, 12
AND/OR gates, 156
Antecedents, 258–259, 265
Assertiveness, 147
    CRM targets, 152
Assessment
    positive skewing of, 211
    TMT barriers to, 210–211
Associative network, diversity in, 77
Asymmetric distributions, and
        overconfidence bias, 101
Atomic Energy Commission Reactor
        Safety Study, 228–229
Attention, 20
    attracting with errors, 52
    low degree of, 212
    triggering with error management
        training, 51
Attentional checks, 7, 8, 9, 29
    omitted, 28
Attentional resources, 29
    demanded by change, 258
Attribution error, 193, 203
Authoritarianism
    and humane orientation, 290
    and power distance, 296
Automatic behavior patterns
    as inhibitors of innovation, 81
    as skill level of organization, 25
    switching to conscious processing
        from, 52
Automatic movements, 117
Automatic scanning, 213
Automation, role in error avoidance,
        134
Automation bias, 134
Automobile recalls, 200
Autonomy, 90, 280, 298
    and creativity/innovation, 78–79
Availability biases, 213
Aviation
    CRM and, 146, 151–152
    latent errors, 245

professional culture of
    invulnerability, 132
teams in, 115
Avoidability, 201, 218
    of bad errors, 109
    of errors, 2, 5, 116
    of medical errors, 108
Avoidance goal orientation, 55, 215

B

Backup behavior, 122, 147, 150, 158
    TCA questions, 154
Barings Bank case, 246, 259–263
    contrasting with Mid-Western
        Hospital case, 266
    trigger event, 262
Barriers, 240
Behavior, in organizations, 3
Behavior-modeling training, 60
Behavioral modeling, 162
Benchmarking, 27
Better-than-average bias, 101–102
Bhopal disaster, 252
Bias, 100
    self-serving interpretation, 210–211
Biased norms, 106
Biased task selection, 103–104
Bilateral kinship systems, 291
Blame, cultural assignment of, 300
Blame-shifting, 192–193
Boards of directors, unfreezing rigid
        cognitive maps by, 216
Bottomry, 227
Bounded rationality, 12, 31–32, 206
Brainstorming
    inhibitory effects on creativity,
        70–71
    social errors in, 84
British Rail accidents, 4
Bullying, 293

C

Calculative errors, 98
Calibration curve, 104
Catastrophes, 6
    in high UAI cultures, 289

Subject Index

and slowness of error detection, 288
Causal maps, 30
Causality
   ambiguous, 213
     oversimplified assumptions about, 226
Cause and effect, error of confusing, 226
Cause-effect chats, 156
Causes, finding/inventing for errors, 226
Certainty illusion, 106
Challenger disaster, 200, 204, 230, 252
Change
   attentional resources demanded by, 258
   error and need for, 74
   group resistance to, 89
   inevitability of errors with, 45
Change conditions, 163, 165
Cheating detection task, 99
Checking
   and efficiency-thoroughness tradeoff, 240
   optimum intervals, 242
Checklists, 37
Chernobyl disaster, 4, 230, 273
   risk-error interactions and, 6
Chief executive officer (CEO), 199. *See also* Top management
China Airlines Flight 140, 117–118
Closed-loop communications, 147, 150, 156, 158, 165
   TCA questions, 154
Code of Hammurabi, 226–227
Cognition style, error examples, 14
Cognitive biases, 29, 97, 209, 217, 219, 318
   as barriers to learning, 213
   habit errors perpetuated by, 120
   by top management, 205–206
Cognitive error studies, 110
   revisiting errors in, 97–98
Cognitive errors, 86
Cognitive limitations, in top management, 205–206
Cognitive load, 304
Cognitive maps, 205
   reinforcement, 211
   rigidity of, 216
   self-reinforcing nature of, 209
   sharing and inertia, 213
   unfreezing mechanisms, 216
Cognitive norms, 97
Cognitive persistence, with prevention focus, 71
Cognitive processes, in error management training, 52
Cognitive regulation, 10
   success of error detection and, 24
Cognitive reliability and error analysis method (CREAM), 237
Cognitive schemas, 205
Collaboration, 194
Collective action cycles, 26–28
   error management and prevention in, 31–37
Collective efficacy, 150
   TCA questions, 155
Collective errors, xii, 2, 25–26
   conclusions and generalizations, 30–31
   habit/thought types, 31
   in idea evaluation, 87–88
   in idea generation, 84–85
   in implementation, 88–89
   in problem identification, 83–84
Collective failure, 113–115
   collective error and, 119–122
   error effects and team learning, 122–126
   error emergence, 135–137
   future research directions, 134–135
   individual errors in teams, 116–119
   within-team error management, 127–134
Collective failure modes, 28–30
Collective-level processes, 73
Collective mapping errors, 121
Collective organization, 225
Collective orientation, 147
Collective team identification, 130
Collective work, 183
Collectivism, 280, 283–284, 298–299
   and error processes, 299–301

Commitment, escalation of, 203, 210, 261
Commitment to resilience, 144, 145, 147, 158, 160
Communication errors
  in collectivist cultures, 300
  in medicine, 156
  in teams, 122
Communications
  breakdowns in team-level, 155
  closed-loop, 147
  in collectivist cultures, 299
  complexity in, 228
  CRM targets, 152
  formalized, 288
  in team-based organizations, 150
Compensation, outcome-based, 202
Complacency, 212
  as result of success, 74
Complex decisions, errors based on, 13, 18
Complex interactions, 194
Complex linear models, 232
Complex systems, error-free performance in, 189
Complex tasks, thought and memory errors, 18
Complex technological systems, 228
Complex work environments, 143
Complexity
  of error detection, 21–24
  growth in communication and transportation, 228
  tasks with low, 183
Comprehensibility, of tractable/intractable systems, 232
Comprehensive unit-based safety program (CUSP) model, 166
Computer skill acquisition, 47
  error-avoidant training for, 46
  posttraining performance in, 49
  using error management training, 58
Computer tasks, error recovery time, 1
Conceptual framework, 252–253
Conditional probability, 99
  vs. natural frequency, 107–109
Conditional variance, 104

appearance of systematic bias in, 105
Confidence, balancing with realism, 219
Confidence intervals, 103
  overconfidence due to narrow, 102–103
Confidence judgments, 104
  noisy data in, 104–105
Confirmatory bias, 86
Confirming evidence, 34
Conflict, as result of innovation, 89
Conformity, 291
Conscientiousness, 72
Conscious attention, 8
  switching from automatic processing to, 52
Conscious regulation, 7, 8, 10, 117
Consensual assessment technique, 70
Consequences, barriers between errors and, 46–47
Consistency, and error management, 36
Constraints, safety by, 240–242
Constructs, function *vs.* structure, 26
Content-blind norms, 99, 109
Content-sensitive norms, 98, 100
Context, importance in organizational learning from error, 177–179
Contextual scanning, 70
Contingent learning processes, 194
Contradictory goals, 4
Contradictory ideas, integrating creatively, 75
Contributory creativity, 69
Control
  and fatalism, 294
  and uncertainty avoidance, 285
Control theories of motivation, 20
Cooperation, TCA questions, 155
Coordination
  across boundaries, 186
  defined, 186
  TCA questions, 154
  in team-based organizations, 150
Coordination of actions, 30
Corning, optical fiber business launch, 209

Corrective action, 156
Creative endeavors, errors in, 80–81
Creative self-efficacy, 77
Creativity
   and autonomy, 78–79
   climate for, 79–80
   collective-level predictors, 78–80
   and error management culture, 79
   future research directions, 89–90
   idea evaluation phase, 71–72
   idea generation phase, 70–71
   implementation phase, 72–73
   individual-level predictors, 76–78
   and job complexity, 78
   and learning goal orientation, 78
   openness to experience and, 76–77
   and personal initiative, 77
   problem/opportunity identification phase, 69–70
   and psychological safety, 80
   psychology roots, 68
   role of errors, 67
   and self-efficacy, 77–78
   and social networks, 79
   workplace, 68–73
Crew resource management (CRM), 146, 151, 160, 276
   and aviation, 151–152
Crisis coping, innovation through, 74
CRM. *See* Crew resource management (CRM)
Cross-functional teams. *See* Interdisciplinary review
Cross-training, 158, 159, 160, 167
   three levels, 158
Cultural diversity, xiii
   and help with error processes, 301–302
Cultural influences, 319
   on error processes, 278–280
   on errors, 273–274
   fatalism, 283, 294
   humane orientation, 281–282, 289–290
   implications for multicultural teams, 301–303
   individualism-collectivism, 283–284, 298–299
   power distance, 283, 296
   theoretical and practical implications, 304–307
   theory summary, 281–284
   tightness-looseness, 282, 291
   uncertainty avoidance, 280, 284, 285–289
Cultural stability, and tight societies, 291
Culture
   as collective programming, 279
   defined, 278
Currency conversions, errors in, 2
Cursor movements, unnecessary, 1
Customer satisfaction, in team-based organizations, 150

D

Dana-Farber Cancer Institute, 188
Debriefings, 124, 132
Decision automation, 134
Decision making, 217
   debiasing, 318
   improvised, 179
   influence of power structure on, 207
   trade-offs in, 234
Deepwater Horizon, 38
Deference to expertise, 144, 145, 147, 157
Deliberate practice, 61
Dependencies, 238
Derailment, 84, 85
Destiny. *See* Fatalism
Deviance
   cultural tolerance for, 293
   normalization of, 251, 256, 266
Deviations, 116, 248
   cultural norms regarding, 291
   detection by measurement systems, 253
   errors as, 3
   intentional, 3, 4
   by multiple parties, 261, 264
   spiral of, 262
   unintended, 3, 245
Devil's advocate, 216

Directory-updating errors, 124
Disasters, xi, 204
Disconfirming evidence, 34
Discovery learning, xii, 47
Dissension, and decision-making accuracy, 87
Distraction, 183
Distributed work, 258, 266
Divergent thinking
and openness to experience, 76
and social networks, 79
Diversification failures, 199, 217
Diversity. *See* Team diversity
Division of labor, 260
Domain knowledge, role in innovation, 70
Domino model, 232
Driving simulator training, 49
Dyadic level errors, 120
Dynamic environments, 37, 146, 213, 218
Dynamic team leadership, 123

E

Ecological rationality, xii, 98, 106, 109–110
Economic development, and humane orientation, 289
Efficacy, collective, 155
Efficiency
encouragement by organizations, 242
maintaining by constraining performance variability, 241
Efficiency-Thoroughness Trade-Off (ETTO) principle, 234–236
ETTO rules, 239
naked machinery example, 236–240
organizational setting, 240
Electric maze, 185
Electronic training delivery, 164
Emergent outcomes, 232
Emotion control, 54, 56
Emotional processes
in error management training, 52–53
in tight societies, 293

Employee turnover, reductions in team-based organizations, 150
Equivocality, in feedback, 22
Error-amplifying feedback processes, 245–247, 252, 255–257, 256, 258, 261, 265, 319
interaction with error-correcting processes, 257–258
organizational antecedents, 258–259
Error aversion, 35
and uncertainty avoidance, 286
Error avoidance, during training, 46–47
Error-avoidant conditions, 306
Error-avoidant training, 46, 48
for analogical tasks, 61
comparison with error management training, 59–60
for high-prove and high-avoid goal orientations, 55
processing requirements, 52
Error-based training, 48
Error cascades, reducing through error management, 34
Error classification, 226
Error climates, 324
Error consequences, 19, 31
disconnecting errors from, 127
in dynamic environments, 149
long-term, 33
negative, 1, 2
reducing with error management and prevention, 32
system-level, 32
Error-constructing mechanisms, 258
Error-correcting feedback processes, 245–247, 252, 253–255, 259, 264
absence at MWS, 267
effectiveness determinants, 254
failure of, 261
interaction with error-amplifying processes, 257–258
organizational antecedents, 258–259
Error correction, 128, 265, 319
challenges with limiting function, 21
error detection and, 19

Subject Index

external support, 24
for intellectual regulation errors, 24
sensorimotor errors, 24
Error cultures, 35
Error detection, 19, 127, 208, 276
challenges, 2
collective level, 127
in collectivist cultures, 299, 301
complexity, 21–24
by external sources, 20, 21
and fatalism, 294–295
and humane orientation, 290
limiting function type, 21
modes, 19–21
in multicultural teams, 302
and power distance, 297, 298
related to feedback process, 37
slowness in high UAI cultures, 287
as stimulus of team learning, 135
in tight/loose societies, 292
timing, 128
at TMT level, 209–211
and uncertainty avoidance, 288
Error encouragement, 48, 49, 50, 55, 56, 58
Error examples, heuristic, 14–15
Error explanation, 128, 276
Error-filled training, 48
Error-innovation pathways, 74
problem/opportunity identification, 74–75
serendipity, 75–76
Error intolerance, 307
Error literature, 2
Error management, xi, 1–2, 32, 90, 276, 317–318, 321
in collectivist cultures, 301
and creativity/innovation, 79
factors influencing, 34–37, 136–137
and fatalism, 294–295
during goal development, 36
goals, 33–34
hindrances from formalized communication, 288
and humane orientation, 290
improving at TMT level, 216–217
in individual and collective action cycles, 31–37

interpersonal fear as barrier, 181
limitations in high UAI cultures, 287
as local phenomenon, 181
in multicultural teams, 303
and power distance, 297, 298
quick damage control, 33
reducing error cascades through, 34
secondary error prevention through, 34
stress and frustration reduction through, 128
team training and, 143–144
as team value, 148
within teams, 127
through team training, 150–151, 152–165
in tight/loose cultures, 293
at TMT level, 208
during training, 46–47
transparency and, 36
and uncertainty avoidance, 288
vs. error prevention, 38
Error management instructions, positive error training in, 48
Error management process, 127–129
Error management training, 45, 305
cognitive processes in, 52
comparison with error-avoidant training, 59–60
and cultural diversity, 304
effectiveness, 48–51, 53, 57
emotional and motivational processes in, 52–53
implications for training research/ practice, 56–58
individual differences and, 55–56
posttraining performance with, 49
processes promoting transfer, 51–54
promotion of adaptive transfer through, 57
psychological mechanisms and performance, 46
resemblance to transfer situation, 51
self-regulation in, 53–54
task-generated feedback in, 58–59

transfer-appropriate processing in, 53–54
vicarious method, 58
Error metrics, 166
Error mitigation strategies, 116, 144
Error of false causality, 226
Error of free will, 226
Error of imaginary causes, 226
Error orientation, 275
Error Orientation Questionnaire, 304
Error prevention, xi, 1–2, 32, 38, 276, 318, 321, 322
  in collectivist cultures, 299, 301
  consequences for long-range plans, 37
  and fatalism, 294–295
  in high UAI cultures, 287
  and humane orientation, 290
  impact on industries, 319
  improvements due to, 320
  in individual and collective action cycles, 31–37
  in multicultural teams, 302
  and power distance, 298
  secondary, 128
  in teams, 127
  through error management, 34
  in tight/loose societies, 293
  and uncertainty avoidance, 285, 288
Error process, 275–276, 275–277
  and cultural diversity, 301–302
  cultural influences, 278–301
  error detection/management, 276
  error prevention, 276
  fatalism and, 294–295
  humane orientation and, 290–291
  individualism-collectivism and, 299–301
  latent errors, 277
  neural basis, 304
  power distance and, 296–298
  tightness-looseness and, 292–293
  time and errors, 278
Error recovery time, in computer tasks, 1
Error reporting, 146
  underreporting problems, 166
Error response, at TMT level, 209–211

Error review, 180
  detrimental political influences on, 192
  interdisciplinary *vs.* expert-based, 178–189
Error self-efficacy, 275
Error signals, 19, 22
  feedback distributions, 23
  nesting in teams, 126
  unclear, 131
Error taxonomy, 1–2, 10–13, 116
  reliability and validity, 13, 18–19
  schematic, 11
Error tolerance, 306, 307
Error-tolerant cultures, 306
Error training, 48
Error typologies, 180
Error underreporting, 166, 168
Errors, 1–2, 38, 317. *See also* Collective errors; Individual errors
  attracting learners' attention through, 52
  avoidability of, 3
  bad and good, 97
  categorization, 204
  as cause of failure, 225–226
  challenges of differentiating from violations, 4
  changes in frequency, 266
  changes in magnitude, 266
  classification approaches, 180
  in cognitive error studies, 97–98
  collective denials of, 301
  collective facilitation effects, 76
  collective level, 25–26
  committed by researchers, 100
  conceptual distinctions, 275
  in creative and innovative endeavors, 80–81
  cultural influences, 273–274
  defined, 3, 179, 275
  defining at TMT level, 218
  defining in terms of action *vs.* consequences, 179
  deliberate practice based on, 61
  determining by outcome, 202
  as deviations, 3
  differentiating from risks, 5

difficulty of predicting, 199
distinguishing good from bad, 218
due to bounded rationality, 12
due to inattention, 4
ecological view, 109–110
effects on team learning, 135–136
emotional attitudes toward, 275
equating with punishment, 45
evident, 20–21
ex ante *vs.* ex post studies, 252
expectation of, 34
facilitation effect on innovation, 74–80
fear of revealing, 33
in funds transfers, 2
good, bad, nonerrors, 109
and group dynamic processes, 206–207
at handoff points, 187
health-related outcomes, 35
as hindrance to innovation, 80
impact on retrieval, 126
inconsistent currency conversions, 2
incorporating into training, 47–48, 61
individual facilitation effects, 76
and inefficiencies, 3–6
inevitability with change, 45, 143
interaction with violations, 4
knowledge based, 11
learning power of, 144, 212, 276
learning through, 45–46
mechanisms for quantifying and diagnosing, 165
need to discuss openly, 33
negative consequences, 1–2
negative emotions accompanying, 34, 52
negative feedback from, 45
organizational, 242
organizational learning from, 177–179, 181–182
in organizations, 1–2, 179–182
outcomes *vs.* consequences, 32
overconfidence bias, 100–106
perception of impossibility, 35
positive consequences, 1, 46, 74
positive framing of, 47, 48
prevention, detection, management, 273–274
punishment of, 33
quick communication of, 37
reality of learning from, 215–216
and risk, 3–6
role in creative/innovative processes, 67
size of consequences, 204
sources at TMT level, 205
sources in learning domains, 190
in spreadsheets, 2
strategies, 46–47
summary in innovation process, 89
during team learning, 124
time and, 278
timely detection, 209
at TMT level, 200–208
tolerance for, 290
by top management, 199–200, 201–205
ubiquity of, xi, 46, 128, 304, 317
universality of, 273
and violations, 3–6
vs. risk, 249, 275
Errors of leniency, 86, 87
Escalation of commitment, 203, 210, 261
ETTO fallacy, 240
ETTO rules, 235, 239. *See also* Efficiency-Thoroughness Trade-Off (ETTO) principle
collective counterparts, 235–236
Evaluation apprehension, 84, 85
Event probability, 156
Evident errors, 20–21
Ex ante studies, 252, 265
Ex post studies, 252
Execution errors, 251, 260
Execution stage, 127, 251
Expected creativity, 69
Experimentation, 180, 185, 204, 218
Expert-based review, 178
Expert performance, 61
Expertise
deference to, 144, 145, 147
and innovation limits, 84

openness to, and creativity, 76–77
Experts, habit errors by, 18
Explicit knowledge, 124
Exploration
    as anxiety-provoking situation, 55
    as beneficial detour, 57
    and posttraining performance, 52
Exploratory learning, 47, 48
External sources, correction with aid of, 20

F

Face-saving
    in collectivist cultures, 298, 299
    and communication of errors, 300
    in high UAI cultures, 288
    and power distance, 298
Face-to-face training, 164
Face-to-face work, 258
Factor trees, 156
Failure
    caused by organizational error, 238
    and development of resilient organizations, 231–234
    as flip side of success, 225
    impossibility of proving, 211
    need for error as cause, 225–226
    organizational coping, 322
    preoccupation with, 144, 148
    relationship to success, 321
    as source of future success, 212
    and three ages of safety, 226–231
    triggering organizational learning through, 196
    and variability of normal performance, 234–240
Failure mode and effects analysis (FMEA), 228
Failure modes, collective, 28–30
Failure to commit, 88
Failures
    in acquisitions, 199
    in diversification, 199
    in international expansion, 199
    learning from, 80
    root causes, 115
    in subfunctioning teams, 115
    systematic, 114
    team shared beliefs about, 132, 133
    willingness to acknowledge, 132
False causality, error of, 226
False-positive rate, 107
Fatalism, 279, 283, 294
    and error processes, 294–295
Fault tree analysis, 228
Fault tree models, 156
Federal Aviation Administration (FAA), 151
Feedback, 11, 19, 161, 246, 319. *See also* Error-amplifying feedback processes; Error-correcting feedback processes
    ambiguity, 22
    from errors, 45
    errors as team, 126
    insensitivity to negative, 209–210
    negative, 52
    organizational antecedents, 258–259
    suppression of negative, 37
    task-generated, 58, 59
    between team training evaluation and reliability, 167
    through simulation-based training, 133
Feedback distributions, and error signaling, 23
Feedback loops, role in accidents, 257
Feedback processing, 12, 27, 36
    error examples, 15
Feedback signaling, 22
Feedforward, 233
Financial industry, 320
Flat management structure, 150
Flexibility, 148
    barriers in high UAI cultures, 287
    CRM targets, 152
    disallowing in strategic planning, 202
Flexible action patterns, 7, 8, `10, 11, 18, 24, 25
    error examples, 17
    errors at level of, 12
Flynn effect, 102
Framing heuristics, 12
Free will, error of, 226
Frequency gambling, 13

Frustration, effect of error management training on, 53
Fukushima Nuclear Power Plant, 38
Functional overconfidence, 105–106
Fundamental attribution error, 203, 214. *See also* Attribution error

G

Generic error-modeling system (GEMS), 117
Goal conflicts, 36
Goal development, 6, 11, 27, 36
    hierarchically ordered actions based on, 6
Goal directedness, 117
Goal errors, 14, 15
Goal orientation, 78. *See also* Learning goal orientation
    and benefits of error management training, 55
    error examples, 14
Goal-oriented actions, 3, 6
Goal prioritization, errors through, 4
Goal realization, practical limitations to, 205
Goal setting errors, 12
Goals
    contradictory, 4
    multiple, 4
Governance mechanisms, 202, 216
Grammar of action, 6, 25
Group conflicts, due to innovation, 89
Group discussion, 125
Group dynamic processes, and TMT errors, 206–207
Group invulnerability, illusion of, 206
Group norms, 83
Groupthink, 35, 206
    in collectivist cultures, 301
    and failure of mutual learning, 130
Guidance
    in error management training, 53
    minimizing in training, 48
Guided error training, 159, 161–162
Guided training, 46
    alternative approaches, 60
Gut feelings, 12

H

Habit errors, 11, 13, 17, 24, 28, 31
    by experts, 18
    through lack of attention, 29
Habitual routines, 120
Handoffs, vulnerability to error, 187
Hansei, 299
Harm, 180
Hazard and operability analysis (HAZOP), 228
Health statistics, need for transparency in, 107
Heedful interrelating, 118
Helping behaviors, 289
Heparin administration, 263
Heuristics, 7, 9, 11, 12, 100, 117
    error examples, 14–16
    for feedback processing, 29
    in organizations, 234
High reliability, 144
    path from team training to, 149
    principles, 143, 144, 146
High-reliability culture, 255
High-reliability organizations (HROs), 144, 189
    cultural values, 145
    five value dimensions, 144
    latent errors in, 246
    redundancy, flexible structures, and culture in, 254
High-reliability teams (HRTs), 132, 133, 146
    values, 147–148
High-risk environments, 144, 188, 195
Hindsight, limitations in safety management, 233
Hindsight biases, 213
Hobbes, Thomas, 225
Homogeneity
    and team error management, 130
    and tight societies, 291
Hospital errors, 200, 202, 247
    latent error examples, 245
HRO cultural values, 145
Human cognition
    advantages, 31
    limitations, 31–32

Human error, 237, 273
  application to financial industry, 320
  vs. ETTO rules, 239
Human factors engineering, 116, 227, 231, 319, 322, 323
  age of, 229–230
Human-machine interaction design, 230
Human resource management systems, 322, 323
Humane orientation, 279, 281–282, 289–290
  and error processes, 290–291
  and racism, 290
Hunter-fisher-gatherer cultures, 291
  individualism in, 298

I

"I know best" cognitive style, 14, 15
Idea evaluation, 71–72
  collective errors in, 87–88
  common errors in, 82
  estimation errors in, 86
  individual errors in, 84–87
  three stages of, 72
Idea generation, 70–71
  collective errors in, 84–85
  common errors in, 82
  individual errors in, 84
  individual factors in, 71
Ideation stage, 68
If-then statements, 8, 98
Imaginary causes, error of, 226
Immersion level, 163, 164
Imperfect correlations, 105
Implementation climate, 73
Implementation effectiveness, 72
Implementation phase, 68, 72–73
  collective errors in, 88–89
  collective nature of, 73
  common errors in, 82
  individual errors in, 88
Implementation processes, coordination problems, 207–208
Improvisation, 184, 188

Inappropriate responses, 242
Inattention, 4, 13
Inclusive review process, 178
Individual behavior
  foundations of, 6–10
  goal-directed view, 2
  in task execution domain, 183
Individual differences, and error management training effectiveness, 55–56
Individual errors
  in idea evaluation, 85–87
  in idea generation, 84
  in implementation, 88
  in problem identification, 81, 83
  in routine work contexts, 184
  in teams, 116–119
Individualism, 280, 283–284, 298–299
  and error processes, 299–301
Industrial psychology, 229
Industrial Revolution, 227
Inefficiencies, 3–6, 38
  differentiation from errors, 3
Inequality. *See* Power distance
Inertia
  as barrier to learning, 213–214
  organizational, 210
Information access, facilitating, 108
Information exchange, 147, 153, 156, 158
  restriction in high UAI cultures, 288
  TCA questions, 154
Information integration, 11
Information-processing errors, 117
Information search, 6, 27, 36
  cheating defined according to, 99
  error management during, 36
  and individualism-collectivism, 298
  and innovation, 70
  serendipity through, 75
  and uncertainty avoidance, 285
Infrastructure errors, 251, 260
Innovation, 318
  and autonomy, 78–79
  climate for, 79–80
  collective-level errors, 82
  collective-level predictors, 78–80

Subject Index 353

common errors by stage, 82
economics/sociological roots, 68
and error management culture, 79
errors during, xii
facilitation effect of errors on, 74–80
future research directions, 89–90
idea evaluation phase, 71–72
idea generation phase, 70–71
implementation phase, 72–73
individual-level errors, 82
individual-level predictors, 76–78
and job complexity, 78
and learning goal orientation, 78
and openness to experience, 76–77
organizational support for, 90
and personal initiative, 77
problem/opportunity identification phase, 69–70
process, 69
and psychological safety, 80
as result of playfulness, 97
role of errors, 67
and self-efficacy, 77–78
serendipity pathway, 75–76
and social networks, 79
through error, 46
workplace, 68–73
Innovation effectiveness, 72
Innovative endeavors, errors in, 80–81
Instructional strategies
effects of transfer climate on, 167
mapping to learning objectives, 158
required for error reduction, 159
Intellectual regulation, 11
error examples, 15–16
Intellectual regulation errors, 24
Intention, 321
in innovation, 80–81
Intentional deviations, 3, 4
and goal prioritization, 4
Intentional violations, and risk assessments, 5
Interdepartmental politics, 178
Interdependence
in collectivist cultures, 299
cultural, 280

Interdependent workflow arrangements, 118, 119, 149, 181, 186
and level of individual errors, 118
Interdisciplinary review, 178, 179, 182, 187, 192
Interest rate predictions, 102–103
Intergroup communication protocols, 192
Internal goal comparison, 19, 20
in tight societies, 292
International expansion, retreat from, 204
Interpersonal coordination domain, 186–187
characteristics, 190
differentiation from system interaction domain, 188
learning strategies, 192–193
Interpersonal error, 120
Interpersonal fear, as barrier to error management, 181
Interruptions, and stress/performance, 255, 256
Intractable systems, 232
Intrapersonal error, 120
Intrinsic motivation, 53, 72
role in innovation, 70
Intuition, 12
Inventions, 67
Investment allocation decisions, 105–106
IQ measurements, year effect, 102
Isolation, and tight societies, 291

J

Janusian thinking, 75
Job complexity, 90
and creativity/innovation, 78
Judgment domain, 184–185, 194
characteristics, 190
learning strategies, 191–192
Judgment errors, 12, 16, 194
absolute risk *vs.* relative risk issues, 107
avoiding, 107

conditional probability *vs.* natural frequency, 107–109
logical truth violations as, 98–100
research problems, 97
vs. perceptual errors, 97

K

KLM 747 crash, 29
Knowledge, skills, and attitudes (KSAs), 146, 157
  instructional strategies targeting, 159
  requirements for error reduction, 159
Knowledge based action regulation, 7, 11, 18, 194
Knowledge errors, 11, 12, 13, 18, 205
  handling time requirements, 18
Knowledge-intensive service delivery, 185
Knowledge requirements, TCA questions, 154
Knowledge sharing, 123, 136

L

Labels
  misapplying to disparate phenomena, 100
  vs. models for confidence judgments, 103
Lapses, 18
Latent errors, xiii, 33, 265, 277
  in Barings Bank case, 260
  conceptual framework, 252–253
  distinguishing from adverse outcomes, 247–252
  execution-related, 251
  frequency, 251, 252, 259, 262, 268
  and individual actions, 250
  linking to adverse consequences, 252–259, 259–268
  magnitude, 251, 262
  in monitoring stage, 251
  organizational characteristics, 250
  positive outcomes, 270
  severity, 250, 252, 268
  ubiquity of, 254
  variety, 250–251, 252, 262, 268
  vs. accidents, 249
  vs. risk, 259
  vs. safety, 249
  vs. violations, 249
Leadership training, 162
Learn-how activities, 124
Learn-what activities, 124
Learning. *See also* Organizational learning
  and ambiguity of causality, 213
  attribution error as barrier to, 214
  benefits from incorporating errors, 56, 74
  cognitive biases as barriers to, 213
  differences in politics as barriers to, 214
  dysfunctional, 269
  with error encouragement instructions, 55
  from errors, 215–216
  event-level barriers, 213
  fundamental attribution error as barrier to, 214
  group-level barriers, 213–214
  individual and team components, 321
  individual/group-level barriers, 213
  inertia as barrier to, 213–214
  organizational, 177–179, 181–182
  organizational-level barriers, 213–214
  and posttraining performance, 56
  social-level barriers, 214
  structuring issues, 195
  success bias as barrier to, 214
  through errors in training, 45–46
  TMT barriers, 213
  and within-training performance, 56
Learning by avoidance, 215
Learning domains, 177–179, 182–183
  characteristics, 190
  common sources of error, 190
  contributions, 194–195
  corrective solutions, 190
  crossing, 189

defined, 179
discussion, 193–194
and errors in organizations, 179–182
framework, 178
interpersonal coordination, 186–187
judgment, 184–185
learning strategies, 189–193
limitations, 195–196
schematic, 183
system interactions, 187–189
task execution, 183–184
Learning goal orientation, 72, 90
and creativity/innovation, 78
Learning objectives, 157–158
mapping instructional strategies to, 158
Learning strategies, 182, 189–190
interpersonal coordination domain, 192–193
judgment domain, 191–192
system interactions domain, 193
task execution domain, 190–191
Leeson, Nick, 259–263, 261
Leniency, errors of, 86, 87
Limiting functions, 21
Living abroad, and creativity, 77
Logic
confusing rationality with, 98
limitations as universal yardstick, 109
Logical errors, 98
Logical truth
as irrelevant measure of erroneous choice, 99
violations as judgment errors, 98–100

M

Man-technology-organization (MTO) model, 232
Management intervention, 25
Management support, 163, 167
role in innovation, 72–73
of training, 164
Mapping errors, 12, 15, 29, 30
collective, 121

Mars Climate Orbiter (MCO), 113–114
Mars Surveyor program, 113–114
Mean confidence, and percentage correct, 103–104
Measurement systems
absence at Mid-Western Hospital, 264
detection of deviations by, 253
Medical errors, 187, 246, 263, 319
Dana-Farber Cancer Institute, 188
and need for verification, 269
preventing through nurse verification, 264
Medical statistics, 318
Medicine
Dana-Farber Cancer Institute case, 188
error example, 114
error prevention, 162
errors in, 200
functional overconfidence in, 106
inconsistent dosing instructions, 187
morphine dose miscalculation, 185
root cause analysis in, 156
task execution errors in, 184
team communications example, 157
team diversity consequences, 130
teams in, 115
view of error as personal failure in, 119
Megacognitive strategies, 117
Memory errors, 12, 16, 18
Mental models, 52
sharing in teams, 129, 147, 158
Metacognitive regulation, 7, 9, 11, 12, 56
through error management training, 54
Mid-Western Hospital case, 263–265
contrasting with Barings Bank case, 266
Military, teams in, 115
Mindfulness, 123, 247, 255
Minuteman Launch Control System, 228
Miscalibration, and overconfidence bias, 104–105

Misclassification, 13
Misconduct, 180
Missile defense systems, 228
Modernization, and humane orientation, 289
Monitoring, 11, 158, 260, 264
  four types of, 89
  and latent errors, 251
  limitations of distributed work settings for, 258
  in loose societies, 293
  redundancy in, 257
  shared, 160
  in tight societies, 292
Monitoring errors, 251
Monitoring styles, error examples, 15
Motivation, control theories of, 20
Motivational processes, in error management training, 52–53
Movement errors, 11, 13, 17
Muddling through, 234
Multicultural teams, 301–304
Multiple goals, 4
Mutual performance monitoring, 118, 129, 131, 150
  TCA questions, 154
Mutual trust, TCA questions, 155

N

Naive sampling model, 102–103
NASA, 113–114
National Transportation Safety Board, 113, 151
Natural frequency, *vs.* conditional probability, 107–109
Near-miss reporting, 146
Near misses, 124, 180
  learning power of, 144, 168
  reporting, 168
Negative emotions, 305
  accompanying errors, 34, 52
  as inhibitors of innovation, 83
  roles in innovation/creativity, 70
Negative feedback
  suppression of, 37
  TMT insensitivity to, 209–210
Negative models, 60

Negative performance, as antecedent of organizational change, 199
Negative reporting, 235
Negligence, 113, 273
Nietzsche, Friedrich, 226
Noisy data, 104, 105
Noisy variables, 213
Nonconscious regulation, 7, 9, 13
Nonrepresentatitive tasks, logical biases, 103–104
Normal performance
  characterizing, 234
  ETTO principles in, 234
  variability of, 234–236
Normal response, 238
Normalization of deviance, 251, 256, 266
  and facilitation of error-amplifying processes, 258
Norms
  abstract, incomplete, and irrelevant, 110
  and ambiguity, 101
  cognitive, 97
  content-blind, 99
  content-sensitive, 98
  increased cultural accessibility of, 292
  need for reasonable models, 102
  of researchers, 102
  sensitivity to pragmatics and semantics, 99
  theory of, 100
Noticing, 160
Novel interactions, 179
Novel processes, 179, 185
  demand for creative problems solving in, 190
Novel situations, 184
Novel solutions, 45
Novel tasks, error management training for, 61
Novices, errors made by, 18
Nuclear power plants, 202, 229, 273
  impact of error prevention, 319
  latent errors, 245
  operational violations, 200

Subject Index

organizational factors and safety, 230
PSA application to, 228–229
safety analysis, 229

O

Omission errors, 13, 17, 24
On-the-job performance, 59
Openness to experience, 90
  and creativity, 76–77
Operational error, 204
Operations, sensitivity to, 144, 147
Operator errors, 118
Opportunity identification, 69–70, 89
  as pathway from error to innovation, 74–75
Optimism
  excessive, 100–101
  illusive, 87
Optimism bias, 100–101
Orbiter spacecraft
  engineer errors, 33
  errors resulting in loss of, 2
Organizational accidents, 4, 237, 270
Organizational antecedents, 265
Organizational change
  antecedents, 199
  facilitating through system interaction errors, 193
Organizational climate, 79–80, 90
Organizational culture, 35, 320
  embedding reliability and error management in, 157
Organizational errors, 242, 319
  and adverse consequences, 245–247
  failure caused by, 238
Organizational factors, impact on nuclear power plant safety, 230
Organizational failures, errors resulting in, 1
Organizational frames of reference, 29
Organizational goals, as signaling devices, 258
Organizational learning, 218, 306, 322
  from error, 177–179, 181–182
  failure as trigger of, 196

in judgment domain, 191–192
overcoming social and psychological barriers to, 177
simplification as driver of, 215
in task execution domain, 190–191
through sharing, 215
Organizational memory, 26
Organizational politics, 206, 211, 217
Organizational safety, team contributions, 150
Organizational training, error management approach, 46
Organizations
  action processes in, 6–10
  discussing error in, 181
  error management in, 317
  error prevention in, 317
  errors in, 1–2, 179–182, 317
  failure to analyze and correct errors, 180
  functional clusters, 26–27
  goal-oriented actions in, 3
  as information-processing entities, 25
  latent errors in, 33
  and outcomes, 225
  oversimplicity in time, 30
  and planning, 225
Outcome assessment, 217
Outcome-based assessment, challenges, 203
Outcome-based compensation, 202
Outcome stage, 127
Outcomes
  and change in power balance, 210
  determining errors by, 202
  difficulty in predicting, 211
  emergent, 232
  optimizing performance, 150
  overestimating, 86
  and planning, 225
  undesired, 180, 184
  unintended, 226
Overanalysis, 86, 88
Overattention, 13
Overconfidence, 87, 98, 217
  functional, 105–106
  mislabeling as error, 109

and self-efficacy, 78
Overconfidence bias
   and asymmetric distributions, 101
   better than average bias, 101–102
   due to narrow confidence intervals, 102–103
   as error, 100–101
   functional overconfidence, 105–106
   mean confidence and percentage correct, 103–104
   miscalibration as, 104–105
Overoptimism, 88

P

Pace of life, and individualism-collectivism, 299
Packing instructions, 236
Pan Am 747 crash, 29
Participative practices, 296
Penicillin, invention through error, 74
Perceived safety culture, 259
Percentage correct, mean confidence and, 103–104
Perceptual contrast training, 159, 160, 167
Perceptual errors, *vs.* judgment errors, 97
Performance, 162. *See also* Posttraining performance
   characterizing normal, 234
   expert, 61
   immediate during training, 49
   improving after training, 49
   interruptions/stress and, 255
   with negative-model training, 60
   on-the-job *vs.* posttraining, 59
   with positive-only models, 60
   within-training *vs.* posttraining, 56
Performance appraisal, reducing error effects in, 89
Performance conditions, underspecification of, 233
Performance effects, self-regulatory processes and, 54
Performance goal orientation, 78
Performance improvement strategy, 157
Performance metrics, 163
Performance monitoring, 122, 134, 147
Performance variability, 241
   constraining, 241
   in intractable systems, 233
Perfusionist error, 184
Persistent behaviors, and creativity, 77
Personal initiative, 90
   and creativity, 77
Perspective, 20
Phraseology, TCA questions, 154
Placebo effects, and belief in treatment efficacy, 106
Plan development, 11
   error management during, 36
Plan execution, 6, 27, 36
Plan generation, 6, 27, 36
Plan monitoring, 27
Plan orientation, error examples, 14
Planning, 148
   for correct action, 286
   impacts of power distance on, 297
   and outcomes, 225
   in tight *vs.* loose societies, 292
   and uncertainty avoidance, 285
Planning barriers, 21
Planning stage, error detection during, 128
Politics, 212
   as barriers to TMT learning, 214
   failures and, 207
   and individualism-collectivism, 298
   organizational, 206
Poor reasoning, 185
Positional clarification, 158
Positional modeling, 158
Positional rotation, 158
Positive error management, 90
Positive error training, 48, 53
Positive-feedback loops, 256
Positive framing, of errors, 47, 48
Positive only models, 60
Post-it notes, 97
   invention through error, 74
Postoperative surgery errors, 184
Posttraining performance, 49, 50, 59
   and exploration during training, 52
Power distance, 280, 283, 296

Subject Index

and error processes, 296–298
Pragmatics, and rationality, 99
Prebrief-performance-debrief cycle, 150
Precipitating events, 118
Prediction errors, 51
Preoccupation with failure, 144, 145, 148, 168
Preparatory phases, in creativity/innovation, 69
Prepared mind, 217
Prevalence data, 107
Prevention focus, 12, 299
    and innovation, 71
Prioritizing dilemma, 235
Proactive creativity, 69, 77
Probabilistic mental models theory, 103, 104
Probabilistic risk assessment (PRA), 228, 229
Probabilistic safety assessment (PSA), 228
Probabilistic statements, 21
    and overconfidence bias, 102
Problem identification, 69–70, 82, 89
    collective errors in, 83–84
    individual errors in, 81, 83
    as pathway from error to innovation, 74–75
Procedures, 37, 124
Process knowledge
    familiar, 183
    in groups *vs.* between groups, 187
Process novelty, 185
Process uncertainty, 184, 193, 194
    and learning domains, 182
    sources, 185
Product quality, 150
Production blocking, 84, 85
Productivity
    and efficiency-thoroughness tradeoff, 235
    safety as prerequisite for, 233
    of team-based organizations, 150
Prognosis, 11
Prognosis errors, 12, 16, 29, 30
Promotion focus, 12, 299
    and innovation, 71

Prosocial values, 289
Prospecting, 234
Prove goal orientation, 55
Proxy logic, 300
Psychological safety, 90, 162, 275
    and creativity/innovation, 80
    importance for interpersonal coordination domain, 193
    as requirement for organizational error management, 181
    team-level, 132
    variances across groups, 181

Q

Quick damage control, 33
Quick response, 144

R

Racism, and humane orientation, 290
Railroad Safety Appliance Act, 227
Rare events, focus on, 252
Rater error training, 89
Rational decision making, theories of, 234
Rationality, 97
    and adaptive goals, 100
    confusing logic with, 98
    ecological view, 98, 109–110
    social, 98
Realism, balancing with confidence, 219
Reciprocal altruism, 100
Recognition errors, 13, 17, 24
Recognition-primed decision making, 234
Redundant systems, 116, 122
    and high reliability, 254
    in monitoring, 257
Regression toward the mean, 105
Regulation levels, 11, 117
Relative risk, 107
Reliability. *See also* High reliability
    enhancing through team training, 143–144
    pursuit of, 146–152
    team opportunities to improve, 143

Reliability analysis, 228
Reliability engineering, 228
Reluctance to simplify, 144, 145, 148
Remote Associates Test, 75
Renault, Nissan acquisition, 207
Report and be good rule, 236
Research, and uncertainty avoidance, 285
Researcher errors, 100
    based on biased norms, 106
    overconfidence, 106
Resilience, commitment to, 144, 147
Resilience engineering, 233–234, 241, 242
Resilient organizations, 231–234
Resistance
    change creating, 211
    at TMT level, 211–212
Resource availability, 163, 164
Resource estimation, errors in, 86, 88
Responses, inappropriate, 242
Responsive creativity, 69
Restructuring activities, 199
Retrieval, 136
    and team learning, 125–126
Retrieval cues, 126
Reward systems, 146
Rich insights, loss of, 215
Rigidity
    of cognitive maps, 216
    in organizational responses, 30
    vicious cycle of, 212
Ripple effects, 33
Risk, 3–6, 4–5, 210
    absolute *vs.* relative, 107
    applicability, 109
    change over time, 108
    differentiating from errors, 5, 275
    insurance against, 227
    negative outcomes resulting from, 5
    rejection of, 86
    specifying precisely, 108
    teams and mitigation of, 114–115
    teamwork as key point of, 156–157
Risk assessment, 230, 231, 275
    step-by-step, 231
    systematic, 227
Risk homeostasis theory, 5

Risk size, 108
Ritualistic behaviors, in high UAI cultures, 288
Rogue traders, 260
Roles and responsibilities, unclear differentiation among intersecting groups, 186
Rome-New York latitutde, 103–104
Root cause analysis (RCA), 156, 180, 191, 194
    as preemptive methodology, 168
Routines, 13, 25
    and inertia, 213
    new combinations of existing, 27–28
    predetermined states, 203
Rule based action regulation, 7, 8, 18, 25, 124, 194, 248
Rules and regulations, 240–241
Rules of thumb, 37

S

Safety, 249, 319
    age of technology, 227–229
    by constraints, 240–242
    definitions, 233
    by design, 240–242
    human factors age, 229–230
    as prerequisite for productivity, 233
    safety management age, 230–231
    scientific study of, 227
    three ages, 226–227
Safety culture, 264, 266, 322, 324
    dysfunctional effects of strong, 267
    at Mid-Western Hospital, 263–265
    perceived, 259
Safety management, 227, 230–231, 240–242, 319, 324
    basis in feedforward, 233
Safety training, in high UAI cultures, 286
Scenario-based training, 159, 161
Schema-based information processing, 81
    as inhibitors of innovation, 83
Scotchgard, 74
    serendipitous invention, 67

Subject Index

Search strategy, and intrinsic motivation, 53
Securities trading, latent errors in, 245
Self-attribution bias, 213
Self-correcting performance units, 114, 148, 149
Self-correction training, 134
Self-criticism, in collectivist cultures, 299, 300
Self-efficacy, 90
   creative, 77
   and creativity, 77–78
   and implementation phase, 72
   increased, 162
Self-justification, 212
Self-reflection, 12
Self-regulatory processes
   in deliberate practice, 61
   development through error encouragement, 56
   and performance effects, 54
   in tight societies, 292
Self-report systems, 168
Self-serving attribution, 206, 210
   by top management, 210–211
Senior management, cognitive processes, 29
Sensitivity data, 107
Sensitivity to operations, 144, 145, 147, 158, 160
Sensorimotor errors, 25, 117, 118
   correction time, 24
Sensorimotor regulation, 11, 24
   error examples, 17
Serendipitous juxtapositions, 75
Serendipity, 67, 75–76, 78, 89
Service quality, 150
Shared cognition, 153
Shared information
   group focus on, 87
   and loss of creativity, 79
   and team learning, 122–124
Shared mental models, 83, 147, 158, 160
Shared situation awareness, 147, 160
Signal detection theory, 38
Simplification processes, 84
   as driver of organizational learning, 215
   loss of rich insights through, 215
Simplify, reluctance to, 144, 148
Simulation-based training, 133, 185
Sink rate warnings, 113
Situational awareness, 120, 134
   CRM targets, 152
   shared, 147, 156, 158, 160
Situational constraints, management removal of, 164
Skewed distributions, 101
Skill-based regulation, 18, 27, 194
Skill level, 7
Skinnerean approach, 45
Slips, 18
Social-cognitive theory, 45, 60
Social context, and high UAI cultures, 287–288
Social contract theory, 99, 100
Social differentiation, 291
Social distractions, 116
Social loafing, 85, 116
Social networks
   and creativity/innovation, 79
   diversity of, 90
Social rationality, 98
Sociotechnical probabilistic risk assessment (ST-PRA), 155, 156
Sociotechnical systems, 231–234
   intractability of, 233
Sony, acquisition of Columbia Pictures, 207, 209
Southwest Airlines Flight 1455, 113, 120, 121
Spelling errors, 20–21
Spreadsheet errors, 2
Stability
   employee preferences for, 69
   of tractable/intractable systems, 232
Standard operating procedures, 25
   deviations from, 257
   and high UAI cultures, 287
Standardization, 240
Status, and error processes, 298
Status quo, 89
Stock-picking, functional overconfidence and, 106
Storage, 136
   and team learning, 124–125

Stored knowledge, 125
Strategic planning
  poor implementation, 207
  as primitive tribal ritual, 202
  as sacred cow, 201
  time requirements, 208
Stress, 269
  and interruptions, 255, 256
Structure
  absence in error management training, 53
  and innovation, 70
  in training methods, 47
Structured Query Language (SQL), command code errors, 2
Subactions, 6, 7, 117
Subgoals, 27
Subject matter experts (SMEs), error metrics by, 166
Subplans, 27
Success bias, 214, 215
Superordinate goals, 36
Supervisory support, 163
Surgery, task execution errors during, 184
Surgical skills training, 49
Swiss cheese model, 232
Symmetric distributions, 101
System errors, 186
System interactions domain, 187–189, 194
  characteristics, 190
  learning strategies, 193
Systematic errors, 103
Systematic exploration, 55
Systematic failure, 114
Systems design, xi
Systems engineering, 322

T

Tacit knowledge, 124
Target audience, 163, 164
Task difficulty, and functional overconfidence, 105–106
Task execution domain, 183–184, 193
  characteristics, 190
  learning in, 190–191

Task-generated feedback, 58–59
Task interdependence, 146, 149
  moderating effects, 119
Teaching hospitals
  learning from errors, 178
  medication errors at, 187
Team-based work systems, 114, 146–152
  advantages, 149–150
Team behaviors, 31
Team-blind error, 188
Team characteristics, 129, 130–131
Team climate, 129, 132–133
Team cohesion
  facilitation through leadership training, 162
  TCA questions, 155
Team compilation model, 125
Team coordination, 119
Team coordination audit (TCA), 152–156, 153
  key questions, 154–155
  as preemptive methodology, 168
Team coordination training, 159, 160, 166, 167
Team diversity, 130
  and unclear error signals, 131
Team error management, xii, 127, 318
  error emergence, 135–137
  factors influencing, 129–134, 136–137
  process, 127–129
  team characteristics and, 130–131
  team climate and, 132–133
  team interventions and, 133–134
Team errors, 113–115. *See also* Collective failure
  collective error, 119–122
  effects on team learning, 122–126
  emergence, consequences, and management, 113–115
  individual errors, 116–119
  origin and emergence, 116–122
Team interventions, 129, 133–134
Team learning
  error effects on, 122
  mapping effects of errors on, 135–136

Subject Index 363

retrieval and, 125–126
sharing behaviors and, 122–124
storage and, 124–125
Team life cycle, effect on error management, 131
Team orientation, TCA questions, 155
Team performance, monitoring, 124
Team self-correction, 148
Team self-correction training, 159, 161
Team training, 318
   cross-training, 158, 160
   enhancing reliability with, 143–144
   error management through, 143–144
   as error reduction/management technique, 150–151, 152–165
   framework, 153
   future research directions, 165–168
   guided error training, 161–162
   and high reliability, 144, 146
   and high-reliability teams, 146
   impact on outcomes, 151
   leadership training, 162
   mapping instructional strategies in, 158
   miscellaneous issues, 163–165
   path to high reliabillity from, 149
   perceptual contrast training, 160
   performance improvement strategy, 157
   and pursuit of reliability, 146–152
   root cause analysis and, 156
   scenario-based training, 161
   team-based work, 146–152
   team coordination and adaptation training, 160
   team self-correction training, 161
   and teamwork as key point of risk, 156–157
   training evaluation, 162–163
   training/learning objectives, 157–158
   training requirements/constraints, 164
   training transfer, 163
   transfer climate and, 164–165
Team training strategies, 166
Team typology, 130

Teams
   defined, 149
   opportunities supplied by, 143
   as workload sponges, 150
Teamwork, 149
Teamwork competencies, as error barriers, 165
Technical execution errors, 183
Technology, 319
   age of, 227–229
   role in error avoidance, 134
   UAI barriers to, 287
Thought errors, 12, 16, 18, 31, 86
Threat-rigidity responses, 26
Three Mile Island (TMI), 229
Tight coupling, 194
   loosening, 255
Tightness-looseness, 279, 282, 291
   and error management training, 306
   and error processes, 292–293
Time-based assessment, 203, 217
Time frame
   assessments based on, 203
   errors and, 278
Top management, xiii, 318–319
   barriers to action, 211–212
   barriers to assessment, 210–211
   barriers to attention, 209–210
   barriers to learning, 213
   cognitive limitations and biases, 205–206
   coordination problems, 207–208
   differences in views/politics, 214
   error detection and response, 209–211
   error management, 208–217
   errors by, 199–200, 201–205
   event-level barriers to learning, 213
   fundamental attribution error, 214
   future research issues, 217–219
   group- /organizational-level barriers to learning, 213–214
   group dynamic processes and errors, 206–207
   implementation problems, 207–208
   inertial barriers to learning, 213–214

insensitivity to negative feedback, 209–210
learning from errors, 212
negative consequences of errors, 203
research on errors, 200–201
self-serving interpretation, 210–211
social-level barriers to learning, 214
sources of errors, 205
success bias, 214
uncertainty and resistance, 211–212
visibility and salience, 203
Top management team (TMT), 199. *See also* Top management
Toyota production system, 191
Tractable systems, 232
Trade-offs, in decision making, 234
Trainer intervention, 48
Trainers, avoiding subjective trainee reactions, 57
Training
  error avoidance approach, 46–47
  error management approach, 46–47
  error mitigation through, 116
  error strategies, 46–47
  incorporating errors into, 47–48, 304–305
  learning through errors in, 45–46
  permission to attend *vs.* support of, 164
  self-correction, 134
  simulation-based, 133
Training constraints, 164
Training effectiveness, measuring, 163
Training evaluation, 162–163
Training methods, incorporating errors into, 47–48
Training needs analysis, 152
Training objectives, 157–158
Training outcomes, motivational and cognitive variables, 55–56
Training research, 56–58
  open questions, 58–61
Training transfer, 163
Transactive memory behaviors, 121, 124
Transfer, 45, 162
  adaptive, 50
  analogical, 50
  boosting with error management approach, 46
  effectiveness with error management training, 48–51
  posttraining tasks, 56–57
  processes promoting, 51–52
  of training, 163
  types affected by error management training, 50
Transfer-appropriate processing, 51, 53–54
Transfer climate, 164–165, 167
Transparency, and error management, 35, 36
Transportation
  accidents, 273
  complexity in, 228
Treatment efficacy, functional overconfidence and, 106
Trial and error exploration, 55
Trigger events, 161, 264, 269
  exogenous, 262
Trust, 162
  and efficiency-thoroughness tradeoff, 239–240
  mutual, 155
Two-step questions, 103
Typographical errors, 21

U

Unavoidable adverse events, 180
Uncertainty, 183, 208, 279
  in judgment domain, 19
  and outcomes assessment, 217
  strategic decisions under, 201
  time-based assessment with, 217
  at TMT level, 211–212
Uncertainty avoidance, 280, 281, 284
  and error management, 288
  and error prevention, 288
  and error processes, 285–289
Unconscious habits, and goal realization, 205
Undo function, 31
Unhedged positions, 261
Unilineal kinship, 291

Subject Index   365

Unintended deviations, 3, 5, 116, 179, 184, 245, 247, 274
Unintended outcomes, 226
Unique information
   focus on, and creativity, 79
   group focus on, 87
Unwanted effects, 241
USS Vincennes, 117, 118

V

Validity, error taxonomy, 13, 18–19
Values, as limitation to goal realization, 205
Variability
   naked machinery example, 236–240
   of normal performance, 234–236
Variance, 103
   as biased estimator, 103
Vicarious error management training, 58
Vicious downward cycle, 256
Vigilance, 129, 266
   decline with emphasis on safety culture, 269
   individual and collective, 189
Violation management, 322
Violations, 3–6, 38, 249, 275, 317
   as intentional deviations, 3
   interaction with errors, 4
   in tight societies, 292
Virtual teams, 301
Visibility, 35
Visibility-effectiveness problem, 235–236
Visual errors, 97
Vulnerabilities, in existing work processes, 191

W

WASH-1400 study, 229
Wason selection task, 98, 106
"We know best" cognitive style, 16
Weighted adverse outcome score, 166
Welfare societies, 289
West Virginia coal mining disaster, 257
Within-culture variation, 280
Work context, xii, 193, 318. *See also* Context
Work design, 116
Work-group climates, 83
Work structure, as antecedent of feedback processes, 258
Workflow interdependence, 118–119, 131
Workload sponges, 150
Workplace, creativity and innovation in, 68–69

HD58.82 .E77 2011
Errors in organizations